Plato's Philosophy of Science

Plato's Philosophy of Science

Andrew Gregory

Bloomsbury Academic
An imprint of Bloomsbury Publishing Plc

B L O O M S B U R Y
LONDON • NEW DELHI • NEW YORK • SYDNEY

Bloomsbury Academic
An imprint of Bloomsbury Publishing Plc

50 Bedford Square
London
WC1B 3DP
UK

1385 Broadway
New York
NY 10018
USA

www.bloomsbury.com

BLOOMSBURY and the Diana logo are trademarks of Bloomsbury Publishing Plc

First published in 2001 by
Gerald Duckworth & Co. Ltd.

British Library Cataloguing-in-Publication Data
A catalogue record for this book is available from the British Library.

ISBN: HB: 978-0-7156-2987-1
ePDF: 978-1-4725-0237-7
ePub: 978-1-4725-0238-4

Library of Congress Cataloging-in-Publication Data
A catalog record for this book is available from the Library of Congress.

Typeset by Ray Davies

Contents

Acknowledgements

A book always owes a great many debts of gratitude, so let me begin with a blanket thank you to all those with whom I have discussed Plato and ancient science over the past few years. Special thanks are due to Bob Sharples, Verity Harte, M.M. McCabe and Hasok Chang who have all read, commented on, and discussed parts of this book. I would also like to thank the ancient philosophy reading group at KCL, and audiences to whom I have given papers, especially at the Institute of Classical Studies and at UCL.

On the technical side, Amanda dos Santos has done sterling work in preparing many of the diagrams. I would also like to thank Deborah Blake and Ray Davies at Duckworth for their patience and efficiency in preparing the book for production.

Last, but by no means least, I would like to thank Sheelagh Doherty for her support and patience during the long process of writing this book. Without her, and those mentioned above, this would have been a poorer effort.

Department of Science and Technology Studies
University College London
December 2000

Andrew Gregory

For Sheelagh, with love

Introduction

Plato's conception of the natural world, and how we ought to investigate and explain it have provoked some sharply divergent evaluations. He has received much criticism, the main accusations being that he was antipathetic to the investigation of nature, feeling that it could produce no worthwhile results, that his views were overly teleological, that his views were retrogressive relative to the atomists and the other physiologoi, that his methodology was anti-empirical, that he discouraged careful observation and that he effectively substituted mathematics for physics. Others have praised him, largely for what they see as his role in the mathematisation of nature and of science, while some have sought to strike a balance. The aim of this book will be to produce a reappraisal of Plato's philosophy of science, and to attempt to clarify the basis, nature and justification of these evaluations. The main arguments of the book, which I outline in more detail below, are as follows:

1. Plato's views on the nature of science, and his views on astronomy and cosmology in particular develop, and do so in interesting ways.

2. Plato can best be seen as someone who is struggling with the foundations of scientific realism.

3. Plato has interesting epistemological, cosmological and nomological reasons for teleology, in addition to any other reasons he may have.

4. The anti-empirical charges against Plato's conception of science are ill founded.

5. Plato believes the investigation of nature will produce worthwhile results.

6. Plato has a series of important and cogent criticisms of the atomists and the other physiologoi.

7. Plato's views on science, and his geometrical atomism in particular, are undervalued relative to the atomists and physiologoi.

Some comments on my use of 'science' are perhaps in order. In my view it is quite proper to speak of Greek, and indeed Presocratic science. Clearly there were significant weaknesses, but I believe there is enough in common between the activities of the Greeks and the moderns to call both science. I would argue that the key factor is the Greek conception of the cosmos and their insistence on natural (as opposed to mythological or supernatural) explanation. Whether it is proper to speak of Plato and science will be a matter for this book. Whether he invokes supernatural and mythological explanations will be an important issue for his cosmology. While Plato was clearly not an empirical scientist himself, that is not a bar to his being a theoretical scientist or philosopher of science. I argue that he made considerable contributions on both counts. Where Plato fits into the general development of Greek science and philosophy of science, especially in relation to astronomy and cosmology, will be another important issue for this book. Of course there are weaknesses in his views too, which we will need to consider carefully in relation to contemporary Greek science. I shall be interested in two means of assessing Plato's views on science:

A. Is this position or strategy relevant to modern debates?
B. Was this view viable and plausible in the context of fourth-century Greek science?

Critical to A is the attempt to see Plato as the progenitor of scientific realism. Critical to B is Plato's relation to the atomists and other physiologoi. On some issues, these strategies may be combined. So Plato's position may contain a kernel which is still significant in modern debates, but relative to modern views Plato overplays his position (though the overplaying may be defensible contextually). So, for instance, I shall argue that Plato's strategy for the solution of certain perennial cosmological problems is virtually identical to some modern strategies, though his actual answers are distinctly more teleological. That teleology might be defended as a historical first instance (no hindsight to judge it with) or as being as plausible in context as other approaches, or as being constrained by other areas of Greek thought.

The nature and role of the *Timaeus* will of course be a major issue. One sub-theme will be an argument that the *Timaeus*, in terms of content, style and concerns is somewhat better integrated into the mainstream of Plato's later work than is generally allowed. In particular I believe there to be some important parallels with the *Theaetetus* and other late works which can illuminate some of the key issues that the *Timaeus* raises. I shall also argue that the *Timaeus* is more of a discussion of how to do science, and cosmology in particular, than the statement of a dogmatic position.

As to the dating of Plato's works, I argue that the *Timaeus* is late and that it contributes to an ongoing revision of Plato's thought as part of a coherent programme with other late works. I would shy away from attempting to establish any precise sequence. Partly this is because the evidence is inconclusive, and partly because I suspect there is no such precise sequence. I have yet to see any convincing argument that Plato wrote his works sequentially and discretely, that he began one work, worked on it alone, finished that work and only then began the next one. Indeed I find this rather implausible,[1] and more implausible for the later works, which are arguably the result of long meditation on the early and middle period works, and which in places deliberately evoke and reflect upon them. Further difficulties here are the nature of 'publication' in the ancient world (did Plato revise works once he released them, if so which version do we have?), and the fact that we know little of how Plato's works may have been used in the academy (were they revised in the light of debate/teaching?). There is also the problem of Platonic anonymity. Plato wrote dialogues rather than treatises, and it is not clear to what extent the views he gives to his characters are expressions of his own views. I do not believe that we ought give up discussing the development of Plato's thought, but we do need to realise limits of our conjectures. I do have an argument that the development of Plato's cosmology would indicate that the *Timaeus*, *Philebus*, *Laws* and *Epinomis* form the later stage of his thought in this respect, although that needs to be treated with due caution.

I make use of the *Epinomis*, so a word on authenticity. I accept that the *Laws* was 'on the wax' when Plato died and was subsequently 'published' (and edited?) by Philip of Opus, who also 'published' (and edited/wrote?) the *Epinomis*.[2] Quite possibly he had a greater hand in the creation of the *Epinomis*. With due caution, I take it to be at least indicative of what a close associate of Plato thought ought to be an appendix to the *Laws*.

Unless otherwise stated, all references to Socrates are to the character of Plato's dialogues and not the historical figure. The relation between the two is unclear, and will not be a major concern of this book, though I am probably more sceptical than most about identifying the opinions and life history of the Socrates of the dialogues with the historical figure.[3]

I. The development of Plato's views

I argue that there is significant development in Plato's cosmological ideas, and it is certainly worth looking in detail at the models offered in various works. The major issue here is whether there is a critical difference between the views offered by one group of works, the *Phaedo*, the *Republic* (both in the discussion of astronomy in Book VII and in the Myth of Er), and the *Politicus*, and another group comprising of the *Timaeus*, *Philebus*, *Laws* and *Epinomis*. In the first group the visible heavens are not amenable to precise mathematical description and there is a gradual degeneration of the cosmos. In the second, we can have precise mathematical treatment of the visible heavens and the cosmos is stable. Along with this cosmological stability, there also appears to be a change in Plato's attitude to inevitable political, social and moral decay. Exactly how widespread the idea that the world, humans and society were caught up in some slow but inevitable decay was among authors (e.g. Homer, Hesiod), philosophers (e.g. Empedocles) and the Greek public in general is difficult to judge, but we may fairly say that it was a reasonably common notion, and it is worth noting Plato's allusions to Hesiod in his description of political decay at *Republic* 546e ff. Later Plato though, I argue, is unequivocal. The cosmos is entirely stable, and there is no one and nothing to blame but ourselves for political, social and moral decay. Behind this move there may be some important metaphysical and cosmological realignments, and there are considerable implications for the *Timaeus* if at least some physical entities can behave in an entirely regular manner. I shall also be concerned with ways in which the *technai* (crafts, arts and skills ranging from cobbling and building to astronomy and medicine) and Plato's conception of the good become more mathematical in their orientation.[4] The more Plato's conception of the good rests on ideas of unity, mathematical simplicity and elegance and less on ethical and aesthetical ideas, the closer he will be to some modern realist positions in the philosophy of science. Both would agree that we need to employ criteria beyond that of empirical adequacy in order to determine which theory to adopt. Modifications to Plato's views on these issues have considerable implications for his philosophy of science, and later period Plato may well be immune to many of the criticisms aimed at a middle period work such as the *Republic*.

Plato's position in the development of Greek cosmology is slightly complex, as there is significant development within his own thought. Plato gives us one of the first accounts of a spherical, unsupported earth in the *Phaedo*, and so, to use Furley's term, is important in the development of the centrifocal cosmos. Plato then considers the same sort of questions applied to the cosmos. He moves from a model in the *Republic* which has external bonds, a pivot on which it turns and where celestial motion degenerates to one which is entirely self-sufficient, stable and requires no external bonds or pivots in the *Timaeus*. This, arguably, is significant, parallels Presocratic debates about the nature of the earth, and happens very quickly within Plato. Plato's contribution to Greek cosmology in this respect may be substantially underrated. Another matter that will require careful consideration is the fact that in the *Timaeus* Plato introduces the idea that the celestial bodies are animate, ensouled, immortal deities, an idea which differs significantly from the *Republic*, and from the *Politicus* where the cosmos is specifically animate but mortal when god (Cronus) relinquishes control. Why does Plato wish to do this, and what sort of step is this in the history of Greek cosmology? One more general point that I will emphasise is that Greek cosmology in general was considerably affected by, and indeed perhaps shaped by the absence of a conception of gravity and indeed of force in general. Plato introduces a like-to-like principle operating at the level of the elements (earth, water, air, fire) in the *Timaeus*. It is highly important to understand what this principle can and cannot do. It will account for some terrestrial gravitational effects, but will not help with the motions of the celestial bodies or with cosmology more generally. It is also vital for understanding Plato's cosmogony to analyse very carefully what this principle is and is not capable of in the chaos of the pre-cosmos. Plato's conception of the pre-cosmos and its transition to a cosmos will be an important issue in Plato's attack on the atomists and in understanding his reasons for advocating teleology.

The key development in astronomy within Plato is the introduction of regular circular motions for the sun, moon and planets offset from those of the fixed stars in such a way as to begin to represent their motion along the ecliptic and through the zodiac. This we find in the *Timaeus* and subsequent works but not in previous works. Thus the planets are not really 'wanderers' (or as the Babylonians referred to them, *bibbu*, 'sheep') but move in a regular manner. If so, then it is Plato who provides the theoretical basis for concentric (or homocentric) sphere astronomy and as such facilitates an enormous leap forward for Greek astronomy. We will need to discuss the respective contributions of Plato and Eudoxus carefully, as this is a matter of some controversy, as is the nature of the challenge which Plato is alleged to have issued to those engaged in these studies as reported by Simplicius:

> Which hypotheses of regular, circular and ordered motion are able to save the phenomena of the planets?[5]

This I will refer to as the Platonic challenge. An important issue here is whether Plato considered the model of the *Timaeus* to be a prototype embodying the important principle of regular and ordered circular motion (or combinations of such motions) for each of the celestial bodies, or whether it is supposed to be a dogmatic and finished model. That will form part of a larger debate on exactly how we should treat the *Timaeus*.

II. Why teleology?

A main area of criticism of Plato has been his use of teleology. Indeed, at times the *Timaeus* has been described as a 'teleologist's manifesto'. A standard view from the history of science would have it that only with the revival of the views of the ancient atomists by such seventeenth-century thinkers as Descartes and Gassendi did science free itself from the restrictive tentacles of the teleology of Plato and Aristotle. Since then science has progressed rapidly, relying on a mechanistic conception of the world in which there is no place for teleology. Even Vlastos, who tries to see good in the *Timaeus* where he can,[6] criticises Plato severely on this matter, and speaks of:

> The retrograde turn which Plato gives to cosmological inquiry when he converts so blatantly preconceptions of value into allegations of fact.[7]

Vlastos is very positive about the nature of the Platonic challenge, but comments:

> What emboldened Plato to start that tradition was a metaphysical fairy tale.[8]

Why though does Plato feel that he requires this teleology? There are several answers here that are uncomplimentary to Plato, not least that this is some form of overspill from Plato's programme in ethics. Graham comments:

> Plato's demand for teleological explanation in natural philosophy has roots in Socratic ethics.[9]

Perhaps though Plato has some independent philosophical reasons about the nature of enquiry, cosmology and regular behaviour. These are difficulties which face modern philosophy of science too, and Plato's strategy, and perhaps his answers too may well have some modern relevance. Zeyl comments:

> The metaphysics of Plato's middle period, for example, is now generally appreciated as a serious and largely coherent attempt (whether mistaken or not) to answer genuine and important questions about the nature of knowledge and reality.[10]

One theme of this book is to attempt to show that Plato's teleological approach to the investigation of nature is a serious and coherent attempt to answer genuine and interesting problems in the philosophy of science.

In modern philosophy of science there is a problem known as underdetermination. In short, in order to determine which theory to adopt the data (however good) are insufficient and non-empirical criteria must be employed. I argue that Plato recognised some of the difficulties here, albeit in a rather more general form, in Meno's paradox. Plato then develops a solution involving teleology, but one which nevertheless contains certain important affinities with modern realist attempts to solve the underdetermination problem by applying (at various stages of Plato's development) criteria such as beauty, simplicity and unity. This is important in itself, and also for understanding how Plato goes about solving certain problems in astronomy and cosmology in the *Timaeus*, especially those problems where he argues for a single solution from an indefinite field of possibilities.

I also argue that Plato uses teleology for a second epistemological problem. In the Socrates' dream passage of the *Theaetetus*, he argues that composite entities can be analysed into their parts, and so we can have accounts of them. However, that which is incomposite cannot be given the same sort of account, and on this conception of knowledge is unknowable. Employing one of his favourite analogies, Plato argues that we might have an account of syllables (in terms of the letters which constitute them) but not of letters. Like Meno's paradox, this is presented quite generally, but one problem for science here might well be what sort of account can we have of the ultimate parts of the physical world? Clearly we cannot account for them or analyse them in terms of further physical parts. The *Theaetetus* is happy just to present the problems. In the *Timaeus*, however, we are given the theory of geometrical atomism, couched in the same letters and syllables terminology. Here though, at the end of analysis by decomposition (into atoms, into atomic planes, into the basic triangles) we have further teleological accounts of the nature of the basic triangles. Clearly this is a contentious area and we will need to look carefully at the nature of geometrical atomism, the receptacle and the sort of explanations envisaged.

Plato also has cosmological reasons for teleology. A central question of cosmology is why the world about us exhibits so much order. The idea that there is one cosmos, and that cosmos organises itself, is in fact remarkably rare in the history of cosmology, and, I argue, is absent from ancient Greek thought altogether. When this idea has been rejected, there tend to be two different types of answer. One can suppose there to be indefinitely many differing worlds/universes, such that the order of our world/universe is explained as an inevitable instance among an indefinite range of possibilities. So the ancient atomists, and so the influential modern view that there may be many universes (the modern term is a 'multiverse'). If one feels that a many worlds theory is ontologically rather profligate, or fails to be a proper explanation (in my own view it fails on both counts; I argue that is Plato's view too), one might argue for a unique cosmos with some extra explanatory principles in play. So Plato, and indeed Aristotle, so various versions of the modern cosmological anthropic principle, and so Christianity. The latter is very important historically. The seventeenth century does not simply revive the views of Leucippus and Democritus, but transforms them within a Christian context. The problems which Plato highlights and solves with teleology are then solved by the existence and actions of a Christian god, who will naturally act for the best. As we shall see, there are certain recurring types of problem in cosmogony and cosmology and certain recurring patterns of strategies for their solution, right down to the current day. Plato, I argue then, investigates a perfectly reasonable option in the face of the perceived difficulties of a materialist account of a unitary cosmos or the multiple worlds account of the atomists. I also contend that Plato, contra the atomists and other physiologoi, argues that not only is there a unique cosmos but also that there are a small and well defined number of types of ultimate particle and well defined, permanent biological species.

If this is so, then we can say some rather more positive things about Plato's teleology. First, in the context of fourth-century Greece, this is a plausible and viable project. It is by no means clear in that context that the atomist/physiologoi approach is correct, (indeed it would appear that there were major flaws in it), or that Plato's approach is incorrect. While this at times has been rather tentatively and grudgingly recognised,[11] one of the aims of this book will be to argue that the ideas of the atomists and physiologoi were rather less plausible and viable than is generally supposed, those of Plato rather more so. Secondly, Plato is addressing serious problems in the philosophy of science, problems which are still under debate today. While modern realists would see Plato as overly teleological, they would also recognise many structural similarities with their own views in the way that he attempts to solve these difficulties.

III. Plato and scientific realism

I argue that the most interesting way of looking at Plato is as someone who is struggling with the foundations of scientific realism. The *Dictionary of the History of Science* defines realism as follows:

> Scientific realism is the thesis that the objects of scientific knowledge exist and act independently of knowledge of them. More generally 'realism' asserts the existence of some disputed kind of object (e.g. universals, material objects, scientific laws, propositions, numbers, probabilities).[12]

Or more bluntly, as Musgrave puts it:

> Realists think that scientific theories are true or false descriptions of reality.[13]

Such realism may be contrasted with nominalism (either specifically denying the existence of universals or more generally denying the existence of disputed objects) and instrumentalism, where theories are instruments for calculation but no question arises as to whether such theories are true or false, as many means of calculation may be devised to produce the same result (see below).[14]

Realist positions require that we make and are able to justify certain assumptions about the nature of the world and the nature of investigation. Whether these assumptions can ultimately be justified is not a question I shall be concerned with here, nor shall I argue that Plato was himself a realist, as such labels are unhelpful when dealing with the ancients. There are significant differences between Plato's views and those of modern realism, and ancient debates around these issues are not set up in quite the same way nor conducted in the same terms. Rather, I shall argue that Plato is concerned with exploring some of the assumptions that are required to get realism off the ground, and is perhaps the first person to do so in a comprehensive manner. The point of this approach is not to assimilate Plato to modern philosophy of science, but rather to narrow what some commentators have argued is a yawning chasm between the two, and to try to understand why Plato approaches certain questions in what might otherwise seem to be an inexplicable manner. If, in a positivist frame of mind, reckoning science to be a matter only of empirically verifiable propositions, we dismiss Plato's discussions of some of the central questions for a realist view of science as irrelevant or metaphysical, we risk missing much of philosophical and historical interest.

If it is the case that Plato is a proto-realist, then it cuts against view that Plato was a (proto-) instrumentalist. Instrumentalism can be defined in the following way:

> Instrumentalists assume that theories have the status of instruments, tools, or calculating devices in relation to observation statements. In this view it is assumed that theories can be used to relate and systematise observation statements and to derive some sets of observation statements (predictions) from other sets (data): but no question of the truth or reference of the theories themselves arises.[15]

Duhem has argued that Plato began a tradition of instrumentalism in astronomy.[16] His view is that:

> The hypotheses of astronomy can be viewed as mathematical fictions which the geometer combines for the purposes of making the celestial motions accessible to his calculations.[17]

Thus when Plato issues the challenge of 'Which hypotheses of regular, circular and ordered motion are able to save the phenomena of the planets?', Duhem asserts that this is an instrumentalist move which begins a tradition whose sole aim is to save the phenomena.[18] I shall argue that this view is quite wrong, at least as far as Plato is concerned.

Another issue on which I believe Plato comes down very firmly on the realist side is that of laws and regularities. One problem here is that Plato, and indeed the Greeks in general are lacking a proper conception of physical law. Again I think we can see Plato struggling with the foundations of a realist account here as he has, at various times, three different accounts of what generates regular behaviour. An added complication here is that certainly as far as the heavens are concerned, Plato's views on whether there is perfectly regular behaviour change between the *Republic* and the *Timaeus*. It is perhaps important to note here that Vlastos believes that:

> Most of us have been brought up to think that the laws of nature are in the last analysis radical contingencies – *de facto* uniformities which we must either exhibit as special cases of still more general *de facto* uniformities or else accept as things for which no further reason can be given.[19]

This is perhaps partly why Vlastos sees Plato's efforts as a 'metaphysical fairy tale'. But modern realists, and in my view Plato, would disagree with

Vlastos on natural law. If we see Plato as someone who is struggling with the foundations of realism though, we can begin to appreciate why Plato requires teleology. An important issue here is why the heavens move in a regular and orderly manner. Plato's answer is to attribute life and intelligence to the heavenly bodies and to the cosmos itself. From the modern point of view, the lines of criticism here appear evident, and Vlastos is critical of Plato's 'psychokinetic' forces. Yet this too bears some deeper analysis. A further effect of the lack of a conception of gravity or force in Greek thought is that they struggle to give a mechanical account of the motions of the heavens. Allied to that is the lack of a conception of physical law. Civil law requires people to use their intelligence to act for the good. So Plato employs the alternative of biological and civil law analogies for both the motions of the heavens and for the cosmos as a whole. Intelligence exhibits regularity, and will choose the best, souls are the source of motion, living creatures have a coherence and organisation, so perhaps the heavens have intelligent souls and the cosmos is a living creature. It is important to note that this is all Plato requires and all he supposes. I shall argue he strips away all superfluous animistic features of these analogies to be left with motion, intelligence and coherence. If so, then these explanations look rather different, and again need to be looked at in the light of Plato struggling with some of the problems of giving a realist account. It is not enough for Plato that the heavens move in a regular manner, or that the cosmos has coherence and organisation, there must be real reasons why these things are so. So perhaps Plato has nomological reasons for his teleology as well as epistemological and cosmological reasons. Recognition of this is important in understanding why he approaches problems in the investigation of nature in a certain manner, and for appreciating the full import of his criticisms of the atomists and the physiologoi. It also allows us to examine what resources were available to Plato in answering these questions, and how that may have helped to shape his replies. A further issue to be discussed here is whether it is evident to the Greeks of the fourth century that one should use only mechanical and not biological or other metaphors to describe the cosmos. I argue that it is not, that this is a piece of hindsight available only after the seventeenth century (if at all), and that biological metaphors may have appeared at least as plausible and viable as mechanical ones.

IV. Observation and experiment

Perhaps the most common complaint against Plato has been either that he denigrated observation or even that he banned it entirely from the sort of enquiries he envisaged. Knorr comments:

> In view of the subordination of the phenomenal and the empirical to the

> abstract and ideal, one would not expect Plato's thought to be the foundation for a robust programme of study.[20]

The evidence for such a view is supposed to be strongest in the *Republic*, and at 530b in particular, where Plato is often interpreted as saying that we should treat astronomy in the same manner as we do geometry and so we should downplay the importance of, or even ignore observations of the visible heavens, in the same way that geometers treat diagrams as imperfect illustrations of the true realities. So Hetherington comments:

> Plato's disdain for observation, his near contempt for observation, is readily apparent in his comments on astronomy.[21]

However, I argue that the allegories of line and cave suggest a much more dynamic approach to investigation, one that begins with the world about us and involves the interrelation of sensibles and intelligibles, the investigation of each being necessary but not sufficient for us to master a given discipline. One important implication of this is that, given a two worlds (TW) view of the *Republic*, there is no demarcation for investigation between the two realms. I take the TW interpretation of Plato to be the idea that four distinctions are co-extensive, and divide what there is into two disjoint sets. On the one hand there is that which is intelligible, stable, knowable and incorporeal, while on the other is that which is sensible, changeable, opinable and physical. The first set of entities, the forms, are said to be, while the second, comprising the natural world, are in a state of becoming.

If there is no demarcation for investigation between the two worlds, then this serves to undermine the notion that Plato's science deals solely with intelligible objects and so eschews observation, and the view that it deals solely with physical entities and so can aim no higher than opinion. I argue that the passage at *Republic* 529d ff. illuminates a distinction between doing and teaching astronomy, and assumes that doing astronomy necessarily involves an empirical investigation, or at least an acquaintance with the data. It would seem from comments in the *Epinomis* and *Timaeus* that Plato was indeed acquainted with Egyptian astronomical data. I also argue that Plato's middle period attitude to the regular behaviour of physical bodies does not entail that he would have been dismissive of detailed observational work.

If these arguments are correct, we need to assess the *Republic*'s views on investigation (as opposed to the education of the guardians) in terms of a necessary initial empirical approach followed by a move to the consideration of the relevant intelligibles as part of the same discipline. This gives it a potentially much more reasonable methodology, observation followed by abstraction and mathematisation, than has been allowed for in some of the literature. We might be critical though of the way in which hypotheses are subjects for discussion, not empirical testing, even if they are initially based on observations. One might well argue though that while Plato does not at any stage ban observation from investigation, his attitude to observation gradually becomes more positive, and that is perhaps shown by his comments on sight and astronomy at *Timaeus* 47a ff. This approach helps to solve an important problem, which Musgrave formulates as follows:

> How are we to reconcile a Plato concerned to save the astronomical appearances with the anti-empiricist Plato of philosophical legend?[22]

I hope to show that Plato's alleged anti-empiricism in the sciences has been exaggerated. There is also the much quoted passage of *Timaeus* 68d where Plato has Timaeus say, in the context of his theory of colours, that it would be useless to make any experimental test of 'these things'. If these things are the way that mixes of pigments make colours, then the position looks odd and perhaps damaging. However, if the reference is to the proportions of the various types of fire particle which constitute a colour, we get a much better sense for the passage and the results are sensible and mundane. This interpretation also accords with Plato's consistent conception of the *technai* in other works where one can learn some basic things by trial and error, and the supposed attack on experiment is unparalleled elsewhere in the *Timaeus* or Plato. The nature of the *technai* will also be an important issue, particularly as Plato draws a distinction between what can be learnt by empirical knack and a form of trial and error, but is still necessary for further investigation, and the proper analysis of the subject matter along with the use of mathematics. It is difficult to give a precise translation of Plato's use of *techne* as the English 'art' or 'skill' do not give a sharp enough contrast between knowing the nature of something and merely having some experience, while 'science' is rather presumptive, not least because the *technai* are broader than the modern connotation of science, though the German 'Wissenschaft' might be closer.

Does Plato believe that worthwhile results will come from the investigation of nature? In the *Timaeus*, the answer to this has to be yes. An analogy is developed between the world soul and our own, where the revolutions of the world soul are stable and clear where ours are chaotic and confused. One key way of stabilising these revolutions and so becoming like god is by doing astronomy, and more generally by gaining control over our perceptions and emotions. What sort of account we can have of the physical world? Even if we take the epistemology of *Timaeus* 27c-29d at face value, it is still possible to make out a good case that the investigation of nature will produce some worthwhile results. As reason has won at least a partial victory over necessity, the world has some order and comprehensibility to it. While we may only have opinion of the physical world, *Timaeus* often claims the account to be the best possible, and of course the physical world itself is the best possible. *Timaeus* 68e ff. also tells us that we must seek the necessary explanations (*aitiai*) for the sake of the divine,[23] for without the former we will not be able to come to know the latter, so the investigation of the physical world has some purpose and worthwhile result. The fact that Plato attempts a thoroughgoing teleological treatment of the physical realm should convince us of his seriousness in this matter. So too the dynamics of ascent and descent of the line allegory would suggest some degree of seriousness and optimism towards the physical world. Later in this book I will argue that the *Timaeus* has rather more to say on the subject of epistemology and that we should at least be suspicious of attributing the views of 27c-29d directly to Plato. If the heavens are stable, and amenable to precise mathematical description, then in astronomy at least some more optimism may be justified. Clearly the status of the likely, mythical story of the *Timaeus* will be an important matter here. It is important to point out though that a good deal of the astronomy and cosmology of the *Timaeus* is repeated in the *Laws* and *Epinomis* without the likely account qualifiers.

V. Geometrical atomism

The view that regular and stable celestial motion is fundamental in the *Timaeus* might be thought to run contrary to some other themes in the *Timaeus*. One might argue that because of the action of the 'wandering cause', and because reason only scores a partial victory over necessity, there could never be completely regular behaviour. I argue though that there is a perfectly viable reading of the relation between reason and necessity which allows such regular behaviour. One might also question how the heavens can be stable given what the *Timaeus* has to say about flux. Here we come to what will be a central issue in the latter part of his book, for I will argue that there is a reading of the *Timaeus* which allows for stability at a fundamental level.

Plato puts forward the theory of geometrical atomism. Clearly these atoms are not stable, as they can be decomposed into two-dimensional faces, and the faces in turn can be decomposed into the two basic sorts of triangle. I argue that there is a perfectly viable reading of the *Timaeus* whereby these basic triangles do not undergo any change other than bonding with each other in to complexes and being dissolved from complexes. If so, then there is a fundamental level of stability, and varying levels of stability of complexes of the basic items, from the most transient of phenomena through to, as I argue Plato makes clear, the eternal stability of the heavens. If that is so, then there are further important issues to investigate. First, if the basic triangles do have this sort of stability, how does this affect certain other passages in the *Timaeus*, most notably those to do with the receptacle? I argue that it is quite possible to get viable readings of such passages which fit with the idea of the stability of the basic triangles. What though of the idea of a radical Heraclitean flux as mentioned by the *Theaetetus* and *Cratylus*? It is by no means clear that is Plato's position in those works, and the geometrical atomism of the *Timaeus* may be an attempt to account for the apparent flux of the phenomenal world (and the apparent stability of the heavens) in terms of some basic and stable items which will allow a coherent theory of the world about us. While the phenomena may be in a state of flux, the imperceptible basic entities of geometrical atomism may not be. As indicated earlier, we may not be able to give an account of the basic entities of geometrical atomism in terms of anything more basic, but we can have a teleological account of them.

I take a fairly radical line on the question of the relation of geometrical atomism to the receptacle. Most literature on the question of the ultimate constituents of the world in the *Timaeus* either treats the receptacle passages (48e-53b) as standing alone, or attempts to interpret geometrical atomism in terms of the results of an analysis of these passages. I believe interesting results may be obtained by a different approach, that of looking first at the possible advantages of geometrical atomism and then rethinking the passages on the receptacle in these terms. I argue that there are paradoxes and other difficulties with the receptacle which can be solved by paying close attention to the way in which geometrical atomism is introduced and subsequently employed. It is notable that Plato expends considerable time and effort in explaining the structure of the world, our bodies and our perceptions in terms of geometrical atomism.

One theme of this book will be Plato's criticism of the atomists and physiologoi either for being indeterminate in their views or for not being able to justify their views properly. The atomists, for instance, are not determinate on the issue of the shapes of fundamental particles and do not justify their claim that a vortex will be formed. Other theories (perhaps Empedocles) postulate basic entities which we can see changing and so are not stable. The task of geometrical atomism in alliance with teleology may be to postulate stability at a level below that of human perception, to be able to justify the nature of the fundamental particles and to be able to account for how the cosmos gained its order without recourse to chance and multiple cosmoi.

This may bear on the seriousness of the *Timaeus*. Plato clearly has many philosophical objections to the atomists and the physiologoi. The *Timaeus* may set out to show that a teleological geometrical atomism is both necessary and adequate for the explanation of physical phenomena. If so, the *Timaeus* would then be an important work where Plato objects to and seeks to replace the views of his predecessors.

There are of course other things to say about the nature of geometrical atomism. Vlastos is quite enthusiastic about the possibilities for quantification and establishing, at least in outline, equations for changes. I would prefer to say that Plato geometrises matter (no small achievement in itself) and makes it amenable to mathematical treatment without ever exploiting that with a quantified physics or chemistry, or indicating that such an exploitation should be made. It is often said that it is the atomism of Leucippus and Democritus that such seventeenth-century thinkers as Descartes, Gassendi and Boyle revive. That stands in need of considerable qualification, however. As noted, there are several issues on which the ancient atomists are either indeterminate or produce no justification for their views. Plato employs teleology to come to a satisfactory resolution. Those who revive ancient atomism in the seventeenth century do not accept indeterminism or a lack of justification. Rather, they invoke (the goodness of) a Christian god to solve these problems. Clearly there will be similarities and differences with Plato's means of dealing with these problems, but in employing such solutions, the seventeenth century distances itself from ancient atomism.

It is also worth discussing how Plato's atomism is rather more like modern atomism than that of Leucippus and Democritus. Plato is strong on the idea that there are a small number of types of ultimate particles, which are mathematically well defined, against the indefinite number of shapes and sizes of the atomists. The discussion of geometrical atomism stresses that matter has deep structure in the sense of ultimate particles forming structures which in turn themselves form structures and so on. Although Plato is not specific about how bonding between particles occurs, he is right that this is not an accidental matter based on mechanical interaction but happens in a specific and well defined manner. This of course is a somewhat Whiggish approach to assessing Plato's geometrical atomism, but given the praise that the ancient atomists have received precisely along those lines it is worth pointing out what we might be able to say about Plato too. If we are going to give the ancient atomists credit for in this manner, then we must also give credit to Plato. Perhaps more importantly, Plato's atomism, like his cosmology, is perfectly viable in the context of fourth-century Greece. It is aimed at solving certain problems with the accounts of the atomists and physiologoi, and is no less plausible as an account of the physical world.

1

Cosmology, Cosmogony and Teleology

The task of this chapter will be to argue that Plato had reasons concerning cosmology and cosmogony for employing teleology. Important here is a question which recurs throughout the history of science, indeed right down to the present day. Can one explain the apparent order of the cosmos solely in terms of matter and motion, or matter, motion and forces? Clearly Plato's answer to this is no, and we shall need to look at precisely why he adopts that position, and whether this is a reasonable position to adopt in the context of fourth-century Greece. As the modern answer to this question is also no (though for different reasons) it is also useful to compare the ancient and modern debates. As we shall see, at least thematically there are sets of recurring answers too, and some interesting parallels between modern and ancient answers to this question. This is important as I want to argue for two main points in Plato's favour here. First, the strategy which Plato uses to approach certain questions in cosmology and cosmogony is still relevant to modern debates. Secondly, not only does Plato have important criticisms of the atomists and other physiologoi, but his strategy is at least as plausible and viable as these alternatives in the context of fourth-century Greece.

I. Plato and modern cosmology

A comparison between a modern view of the evolution of the cosmos and Plato's view of the ordering of primordial chaos is instructive. According to modern cosmology, the universe began with the big bang around 15 billion years ago.[1] After the initial expansion there is a period of rapid inflation, followed by a calmer period of expansion. At this time there are the fundamental particles and there is a great deal of radiation, but none of the elements have formed as yet. This 'chaos' gradually sorts itself as the universe expands and cools. First hydrogen and helium nuclei form, and then matter and radiation decouple, with electrons binding to the nuclei of the light elements to form the first atoms. It is from this period, about 300,000 years after the big bang, that we can find evidence of the cosmic background radiation. While this is in remarkable accord with predictions, there are slight inhomogeneities. These small variations in the intensity of the cosmic background radiation, 'ripples', 'wrinkles in time' or whatever one wishes to call them, are the seeds for the future development of the cosmos. Clusters of matter form from these ripples and eventually create stars and galaxies. Stars burn their fuel of hydrogen and helium by nuclear fusion creating the even-numbered elements up to and including iron, and when these stars go supernova the other odd-numbered elements and the elements heavier than iron are formed. Our solar system then forms out of the detritus of such stars. When sufficiently complex organic chemicals have formed, life gets under way and evolves, which brings us to the present day.

At least at first examination (as we shall see, the situation is somewhat more complex), this account is one that relies solely upon 'chance and necessity', as Plato would put it, and requires no other factor to account for the current order of the cosmos. It is critically dependent upon the attractive force of gravity which leads to an accretion of order around the initial inhomogeneities. Plato's view of the transition from chaos to order is of course very different. In places his view of matter in motion prior to the ordering of the cosmos is rather like our modern conception of the motions of gas molecules if we assume that there are no attractive or repulsive forces between them. The motions of the molecules are random, or as Plato puts it at *Timaeus* 30a:

> For god desired that all things should be good, and nothing paltry (*phlauron*) as far as was possible, he took over all that was visible, which was not at rest but in discordant and disorderly motion (*alla kinoumenon plemmelos kai ataktos*), and led it into order out of disorder, judging the former to be entirely better than the latter.

The key assumption for Plato from a cosmological perspective is that chaos is non-progressive. Left to its own devices chaos will continue to be chaos and order will not grow out of it. If some order should come about by chance, it is likely to dissolve immediately rather than breed further order as it does in the modern view of the evolution of the cosmos. Therefore something else is required to explain how the cosmos acquired that order. I take that to be a perfectly plausible position to adopt in an ancient context. This needs to be distinguished from the modern view, where it is possible for an ordered state to come about with randomly moving particles and no forces, but highly unlikely. Indeed as there are vastly more disordered states than ordered ones, the system will tend towards maximum disorder. Plato's view of chaos would seem to be that it neither progresses to greater order nor regresses to greater disorder.

The *Timaeus* gives us a like-to-like principle, whereby similar sorts of entities, primarily the elements of earth, water, air and fire congregate together. So at *Timaeus* 63c we are told that:

> Standing on the earth, and separating off types of earth, and sometimes earth itself, if we draw them into the unlike air by force and contrary to nature, both cling (*antechomena*) to their kind. The smaller is more readily forced into the dissimilar than the larger.

This is repeated in a more generalised form at 63e:

> The thing to be recognised in all of these cases is that it is the motion of each towards its own kind that makes the moving thing heavy, and the place into which it moves 'down'.[2]

In addition to the like-to-like principle, there are two sources of motion. At *Timaeus* 52d ff., we are told that in the pre-cosmos chaos the receptacle, because it is filled with powers that are not balanced or evenly distributed, is itself shaken by these things and by this motion shakes them in turn. It thus acts like a winnowing basket (*plokanon*),[3] and just as the process of winnowing sorts the corn from the chaff, so the four elements are separated out, the most unlike furthest away and the most like closest together.[4] At *Timaeus* 57c and 88de we are told that this shaking continues in the cosmos. At *Timaeus* 58a ff. we are told of the compressive effect which the rotation of the cosmos has on the elements, making them intermingle and denying the possibility of a void on a large scale.[5] While the like-to-like principle and the sources of motion will be used to account for terrestrial gravitational effects, this is very far from anything like a modern theory of gravitation. There is no universal attraction of matter to all other matter. Earth will congregate with earth, but not the other elements. Nor is there any force which operates at a distance. Similarly at *Timaeus* 80bc electricity and magnetism are explained in a general manner as due to contact action and mutual replacement, and there is an outright denial that any attraction (*holke*, 80c3) is involved. Thus it is wise to refer to a like-to-like principle for Plato rather than to forces.

The important matter here for cosmogony is that a combination of the shaking of the receptacle and the like-to-like principle will not produce a coherent, well ordered cosmos but will simply produce a sorting of the elements. The idea that some physical principle, on its own and undirected could produce the sort of order that the cosmos currently displays is an alien one to Plato. That is quite reasonable given usual analogues for these processes. Winnowing corn does not produce any interesting order, but merely the separation and congregation of the entities involved. There is also an important point to make about the modern view, which I shall expand upon below. This is that gravity is able to create the sort of order we see in the our cosmos only because of an extraordinarily delicate (and extremely fortuitous?) balance between itself and the other fundamental forces, allied to a balance with other fundamental constants. Without this balance, the cosmos would collapse very quickly (gravity too strong), or matter would disperse too rapidly in the expanding universe for anything interesting to form (gravity too weak).[6] Only in what appear to be highly exceptional circumstances do we get order from the initial chaos with forces alone.[7] As we have no reason for why there is this balance, one might argue that this merely shifts matters back one stage. Plato might well ask, as many modern cosmologists have, why does gravity have this apparently fortuitous balance with the rest of nature?[8] Finally, the following passage at *Laws* 889b seems critical to Plato's cosmogony and his criticism of the atomists:

> Let me put it more clearly. Fire, water, earth and air all exist due to nature and chance they say, and none to *techne*, and the bodies which come after these, earth, sun, moon and stars, came into being because of these entirely soulless entities. Each being moved by chance, according to the power each has, they somehow fell together in a fitting and harmonious manner, hot with cold or dry with moist or hard with soft, all of the forced blendings happening by the mixing of opposites according to chance. In this way and by these means the heavens and all that pertains to them have come into being and all of the animals and plants, all of the seasons having been created from these things, not by intelligence, they say, nor by some god nor some *techne*, but as we say, through nature and chance.

A 'fitting and harmonious' ordering will not be generated by like to like principles, as that ordering has a mixture of types, and indeed, as Plato emphasises here, of opposites. The only way that Plato's opponents here can explain how such an ordering came about is by chance, and that with opposites looks extremely unlikely. Here again we see a problem generated by a lack of a conception of gravity. While gravity will bring together things that (for the Greeks) are unlike, principles such as like-to-like will not. Plato's comment here that the elements require some *techne* leads into another important area for Plato's cosmogony.

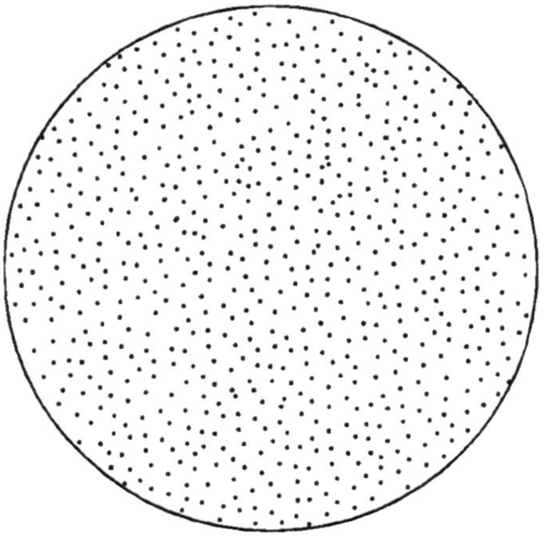

Fig. 1. Random motion With no forces acting between particles in random motion, the modern view is that they will tend towards the most disordered states, Plato's view is that chaos will remain as chaos.

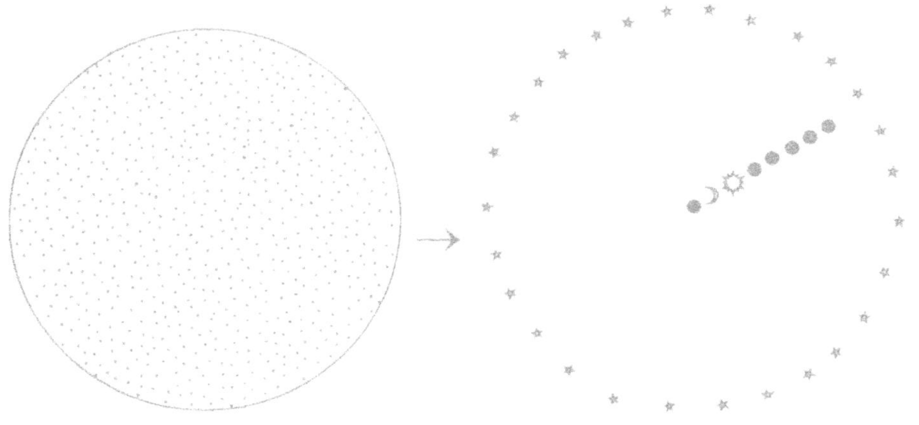

Fig. 2. Random motion to order (Greek cosmos) With no forces acting it is highly unlikely that an ordered state such as the Greek cosmos should come into being, even more unlikely that such a cosmos will stay ordered. One might find it even more unlikely that humans could be produced out of this. According to modern cosmology though, gravity in balance with other forces can produce this order.

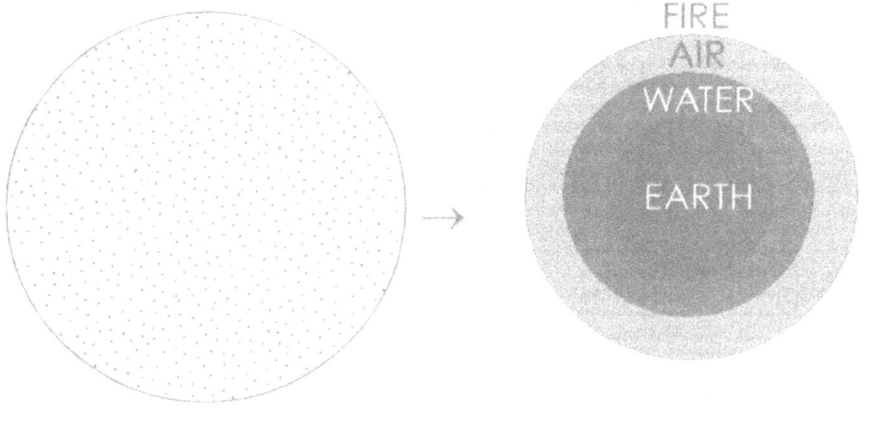

Fig. 3. Random motion to concentric spheres. Like-to-like principles on their own will not produce interesting order, but merely clumps or spheres of like entities. Any mixing can happen only by chance, and the likelihood of that producing the sort of order we see in the cosmos, is, according to Plato, extremely remote to say the least. Plato's position appears a perfectly reasonable one, given the lack of sophistication of Greek conceptions of gravity and force, and the extraordinarily delicate balance of gravity with the rest of nature that is required to produce our cosmos.

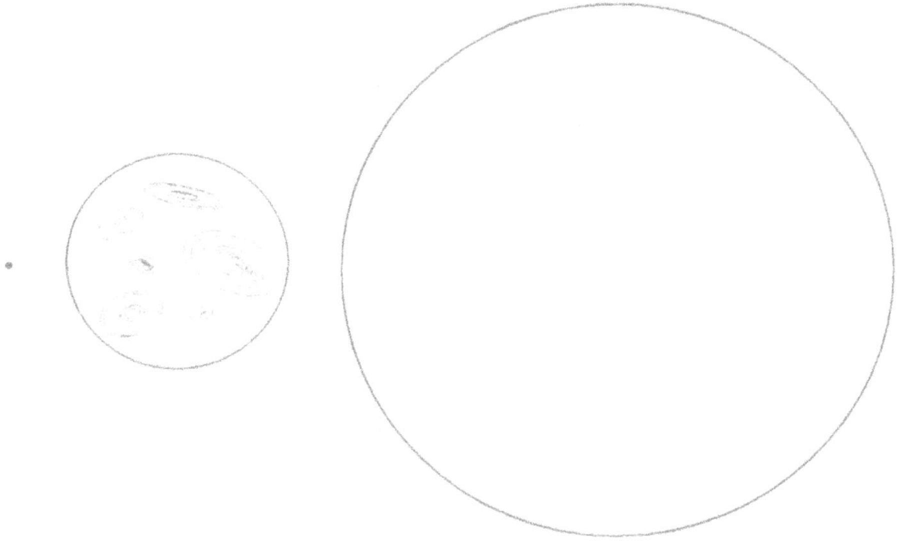

Fig. 4. Modern cosmology The question for modern cosmology, on the other hand, is why, after the big bang, the universe does not recollapse to an uninteresting black hole (left), nor expand so fast that nothing of interest forms as matter is too diffuse (right), but expands at just the right rate to form clusters, galaxies, stars and planets and then life (middle).

II. The chaos of the elements

Plato's vision of chaos is more radical than I have suggested so far. The implications of this are not generally recognised, but are highly important for Plato's cosmology and cosmogony. It is not only the motions and arrangement of whatever Plato thinks exists prior to the ordering of the cosmos which are chaotic, but the elements themselves do not have order. At 53a, Timaeus tells us:

> Before this, all things were lacking in reason and measure (*alogos kai ametros*). When the task of ordering the universe was attended to, fire, water earth and air all possessed traces of their own nature, but were so disposed as would be likely in the absence of god, and such being their nature at that time, god then firstly fashioned them by means of shapes and numbers (*eidesi te kai arithmos*). They were organised by god to be as fair and beautiful as possible, a state which they were not previously in.

We are given a reprise with some further comments on naming slightly later at *Timaeus* 69c:

> As was said at the beginning, all these things were in a state of disorder until God created in them commensurability with respect to themselves and each

> other, so far as it was possible for them to have proportion and commensurability. For at that time nothing partook of these things except by chance, nor was there anything deserving to be named as we now name things, such as fire and water and the others. But all these he first ordered, and then out of them he constructed the universe.

Not only are the elements not properly ordered, but the elemental chaos is non-progressive as well. I shall refer to this sort of order as the intrinsic order of the elements, and the arrangement of the elements with respect to each other as the extrinsic order. With the modern conception of the big bang there is indeed an early phase before the elements in the sense of members of the periodic table are formed, this stage is neither without reason nor without measure. This lack of intrinsic order has some important consequences for the application of a like to like principle to the primordial chaos. Such a principle will not work to produce any sort of ordering in the primordial chaos as we do not have elements. If we interpret 63c-e more strongly as a general like-to-like principle, still this will fail to produce any sort of ordering unless there is some likeness in the initial chaos. Certainly Plato is aware of the idea of a chaos of complete unlikeness. The *Politicus* myth gives us a vision of the deteriorating cosmos, and at 273e a god who eventually intervenes because he is:

> concerned that it should not be storm driven by confusion and broken up into an endless sea of unlikeness.[9]

One might argue though that this somewhat overstates Plato's position on elemental chaos. Just prior to *Timaeus* 53ab, we are told that the receptacle has a winnowing effect, such that the four kinds of element are separated out, the most dissimilar farthest apart, the most similar nearest. So at 53ab, while the elements only have traces of their nature, perhaps they have sufficient traces for the winnowing effect to take place,[10] otherwise the passage at 53a ff. looks strange. However, *Timaeus* 53a is quite specific that it is the four kinds that are sorted:

> In the same way at that time the four kinds were shaken by the recipient, whose motion was like that of an instrument for shaking, and separated the most unlike kinds furthest from each other and forced together the most like kinds, so that the different kinds had different regions, even before the universe was organised out of them.

However, if this winnowing effect does take place, there seems to be no limit on it so eventually we would reach a situation where the four elements are separated out. There is no mention of this winnowing effect or its results at *Timaeus* 30a or 69c, both passages seeming to indicate that there is ordinary chaotic motion. There would appear to be several inconsistencies in Plato's vision of chaos, or perhaps he presents us with differing positions. This leads us to the question of how we should treat the cosmogony of the *Timaeus*, the usual options being a literal or a metaphorical reading.[11] I want to suggest an alternative to this dichotomy, which treats these passages as a framework for thinking about cosmological issues, in particular cosmogony and some epistemological issues related to the order of the cosmos.

The literal interpretation would have it that the *Timaeus* gives us Plato's view of what the primordial chaos was like, and the emphasis then appears to be on cosmogony, and the severity of the problem of producing an ordered cosmos from the initial state. In order to get the current cosmos, some teleological ordering is required. In this one can see the depth of Plato's objections to the atomists in cosmogonical terms. From the initial chaos, there has to be an intrinsic ordering of the elements by chance, which seems radically improbable, added to which there must be an extrinsic ordering by chance, as a like-to-like principle is insufficient, which also seems radically improbable. Taken together, it seems utterly implausible that both the intrinsic and the extrinsic ordering of the cosmos should come about by chance alone.

Plato's actual position on the primordial chaos is difficult to nail down though. That might be due to some indeterminacy or inconsistency in the writing of the *Timaeus*, or may be deliberate. Perhaps he is looking for something more general here, which one might formulate as follows. If the current cosmos has come into being from a less ordered state, one that is to some extent chaotic, then that cannot be explained solely in physical terms, whatever the extent of the disorder. This may be the point of the unquantified comment on 'traces' of order. It covers many positions from no order onwards. If so, Plato has a rather stronger and more flexible position against the atomists. They would otherwise have the option of simply pointing to a disagreement over the nature of the primordial chaos, their view being less radical than Plato's and so perhaps being more likely to result in an ordered cosmos. So perhaps Plato is not stating a specific cosmology, one which he would appear to be confused about, but is rather enunciating a general cosmological principle about the transition from any more chaotic state to the current cosmos. If so, the literal/metaphorical question about his cosmogony may not be appropriate.

III. Chaos and cosmogony

A second approach to the cosmogonical passages, complimentary to the first, is to take them as a counter-factual analysis. An important example of such an approach in Plato is *Laws* 897b. Here, if the cosmos moved in a 'mad and disorderly' manner then we would conclude a bad soul was in charge. It does not though, in fact it moves in an entirely rational manner so we conclude a good soul is in charge. The emphasis in the *Timaeus*, if it too can be interpreted as giving a counter-factual analysis, then appears to be more epistemological. If the cosmos has no order, or only partial order, and no teleology is involved, what would it be like? If the answer to that is nothing like our current cosmos, but a chaos, and a non-progressive chaos at that, then we have the problem of what order we ought to assume our cosmos has, and why. Plato's analysis here is thoroughgoing. The reason why the resulting chaos is so radical is that Plato questions any assumption of order at all. That may well be intended to be a critique of the atomists and the physiologoi, as their conceptions of chaos are by no means as radical as Plato's. Can they justify assumptions of aspects of order within their ideas of chaos?

Their view that the vortex in alliance with a like-to-like principle will produce order seems unjustified, as any mixing of unlikes, as is critical for an ordered cosmos, will come about only by chance, as the *Laws* emphasises, and the like-to-like principle will produce only a clumping of the elements, as the *Timaeus* emphasises.

The atomists may assume there to be all shapes and sizes of atoms, but they also assume that they are intrinsically ordered such that they have determinate shapes and sizes. A radical chaos might lack such an intrinsic ordering. If there were no intrinsic order of the elements at all, then there would be a complete (and non-progressive) chaos, as a like-to-like principle would be unable to function. How does one justify the presumption of intrinsic order at all, and of any specific type of intrinsic order?

Plato's questioning of assumptions of order may go deeper still. The use of the analogy of the winnowing basket would seem to pick up directly on a theme in Democritus. The relevant passage from Sextus Empiricus is as follows:

> There is an ancient opinion which, as I have said, has long been current among the natural scientists to the effect that like recognises like. Democritus is thought to have produced confirmation of this opinion and Plato to have touched on it in his *Timaeus*. Democritus bases his argument on both animate and inanimate things. For animals, he says, congregate with animals of the same kind – doves with doves, cranes with cranes, and so with the other irrational animals. Similarly in the case of inanimate things, as we can see from seeds that are being riddled and from pebbles on the sea-shore. For in the one case the whirling (*ton dinon*) of the sieve (*tou koskinou*) separately arranges lentils with lentils, barley with barley, wheat with wheat; and in the other case, by the motion of the waves, oval pebbles are forced into the same place as oval pebbles, and round pebbles as round as pebbles, as though the similarity in things contains some sort of force for collecting things together. That is Democritus' view.[12]

Plato of course would deny that we can get an ordered cosmos from a like-to-like principle and a vortex. There may be another worry though. At *Timaeus* 52d ff. the motions of the four kinds are chaotic, they transmit this to the receptacle which then shakes the kinds. If we are supposing there to be no order, then there is no reason for the receptacle to move in an orderly manner. This may be an implicit criticism of the atomists' assumption that a vortex would form. Why a vortex as opposed to some other form of motion? The atomists ought not to help themselves to the idea of a vortex without any justification. This is perhaps why Plato chooses the analogy of the winnowing-basket for the receptacle, which is shaken in a fairly random way, as opposed to a sieve which is whirled in a specific manner. This adds another layer of implausibility to the atomists, account. Why should we believe that a vortex, as opposed to anything else, should form in the initial stages? It is interesting that Aristotle too is critical of the atomists on these points. At *Physics* 196a24 ff. he attacks those who make the initial rotation a spontaneous event but do not give this any serious consideration.

There is then an epistemological problem attendant on cosmology. We need to assume that the world is ordered in some fashion so as to be able to explain the current state of the cosmos. What order ought we to suppose though, and why this rather than any other? Plato's answer to this is clear. We should be guided by the principle that the world has been constructed for the best. Why Plato believed that to be the appropriate way to address such problems is a question I shall leave to Chapter 3. In what follows in this chapter I will point out that where Plato is faced with an indefinite field of possibilities, he gives a teleological account of his choice of one, or a small number of those possibilities. That is so for the extrinsic order of the elements (there is one cosmos and that is as well ordered as possible), and for their intrinsic order as well (to account for the multiplicity of phenomena we need to assume two basic entities, the half-equilateral and the half-square triangles of *Timaeus* 53d ff.).

Plato's concerns then are twofold. There is a concern with what order we need to assume to be able to account for the current order of the cosmos. There is also a cosmogonical concern about how the current cosmos could have come about from any previous chaotic state. Physical processes are inadequate to explain this, so there must have been some form of teleological ordering. For Plato, the physical universe is incapable of ordering itself. It is clear then that Plato has cosmogonical reasons for advocating teleology.

IV. Atomists, physiologoi and cosmoi

It is important to recognise that Plato was far from being alone in the view that matter, motion and forces/principles cannot account for the apparent order of this cosmos. Obviously there are other ancient teleologists, but so too the ancient atomists agree on this point, even if they have a different means of resolving the difficulty. Leucippus and Democritus, as is well known, eschew any form of teleology. Fragment 2 tells us:

> Nothing happens randomly, but everything from reason and necessity.

For Leucippus at least the development of the universe begins when many bodies of all shapes move in the void and form a vortex.[13] These bodies then separate like to like, and when the number of bodies prevents equilibrium being maintained in the vortex, the smaller bodies move into the void while the rest stay together, become entangled and create the first spherical structure. All this happens by necessity, as:

> Everything happens according to necessity, for the explanation of the genesis of all things is the vortex, which he calls necessity.[14]

Leucippus and Democritus believe there to be many, in fact an indefinite number of, cosmoi. Certainly that is so synchronically:

> Democritus holds the same view as Leucippus ... There are innumerable cosmoi, which differ in size. In some of these there is no sun or moon, in some they are larger than ours and in some more numerous. The spaces between cosmoi are not equal, in places there are more and in others less, some are growing, some are in their prime, some declining, some are coming to be and others failing. They are destroyed by falling into each other. There are cosmoi bereft of animals and vegetation and all moisture.[15]

This is almost certainly so diachronically as well:

> Leucippus states that the whole is indefinite ... some of it is full and some void ... hence an indefinite number of worlds come from these and are dissolved into these.[16]

If there are an indefinite number of cosmoi, then one would expect some of them to be such as to dissolve back to their original elements, and if the original vortex is indefinite, then more cosmoi will always being made. Whatever principles Leucippus and Democritus use for the workings of necessity to explain the development of the cosmos, it is clear they are not strong enough to generate this cosmos uniquely out of the initial chaos.

Nor does Anaxagoras believe that matter and motion alone, or in alliance with some principles or forces are able to generate our cosmos uniquely out of the primordial chaos. Rather he believes that *nous*, mind or intelligence, plays the critical role in the ordering of the cosmos:

> Nous is the finest and purest of all things, it has all knowledge concerning everything and it has the greatest power. Nous controls all things that have life, both the greater and the smaller. Nous controlled the revolution of the whole, such that it revolved in the beginning. At first it revolved in a small region, but now it revolves in a greater, and will revolve in a greater still. The things which are mixed and separated and divided are all known by mind. Nous ordered (*diekosmese*) them all, this revolution in which the stars and the sun and the moon and the air and the aether which are being separated off.[17]

Plato's objection is well known, that Anaxagoras makes no use of his nous in explaining the nature of the cosmos.[18] The first problem here is that nous does not appear to have any ongoing role in the ordering of the cosmos:

> After mind initiated motion, it began to withdraw from all that was moved, and all that Mind moved was divided. And as this motion and division went on the revolution caused it to be divided much more.[19]

Anaxagoras' position has some interesting similarities with those adopted by some mechanical philosophers in the seventeenth century. There a Christian God created the world in a ready ordered state (in line with the book of Genesis) and then took no further part in the running of the universe. Here nous provides the initial impetus and ordering before withdrawing, and it is interesting to note that like a Christian God, nous knows all and is all powerful.[20] The critical difference here is that in the seventeenth-century conception God also creates a framework of physical law and forces to which matter is subject. Given the initial ordering, these forces and laws are supposed to be sufficient to explain the ongoing order of the world. As in the ancient world, there was a religious debate around this idea. Newton and his supporters argued that removing god from any part in the running of the world would lead to atheism. Leibniz on the other hand argued that to suggest that God was an incompetent craftsman and would produce a mechanism that needed 'winding up and cleaning' was derogatory of the power of God and would lead to atheism.[21]

It is at least a plausible view that without the direction of intelligence and without some framework of laws and forces the cosmos will not be able to keep its initial order, especially in a Greek context. Some have taken the view that even with such a framework the cosmos will not be able to maintain order. Newton, for instance, believed that without the active intervention of God, the stars would move into one giant fireball under mutual gravitational influence,[22] and that the gravitational attraction of the planets for each other would gradually disorder the solar system until it was no longer a system.[23]

Anaxagoras though provides no such framework for the explanation of why the cosmos has continuing order. It is not the ongoing work of nous, yet nor does he give anyone in an ancient context sufficient grounds to believe that matter will be able to sustain or develop the initial ordering. In particular while Anaxagoras allows that nous controls living things, he does not make the cosmos itself a living thing controlled by nous. The *Timaeus* provides a rather different picture, as after the initial ordering the cosmos has a soul, and so do the celestial bodies, which is why they move in an intelligent, ordered and regular manner. That of course is in particular contrast to Anaxagoras' view, of which Plato was well aware, that the sun and the stars were merely hot stones.[24] If nous withdraws its influence, and there is no framework for the maintenance of cosmic order, then either that ongoing order is inexplicable for Anaxagoras, or he needs to rely on the answer of Leucippus and Democritus. If Anaxagoras recognised this difficulty, it may have given him an impetus towards a multiple-worlds view.[25] The evidence for this is cited by Simplicius:

> We must suppose that men and the other animals which have life have been put together just as we have, and that that they have sun and moon and so forth, just as we have … the separation would have happened not only with us, but with others as well.[26]

Simplicius (quite rightly) has his doubts about how far we can push this evidence in the direction of multiple worlds for Anaxagoras. Although not taken up by Simplicius, the point about separation here seems significant, especially if we take it in conjunction with fragment 12 above which might be thought to be suggesting that separation is a continuing process. If so cosmoi might be continually created, and Leucippus and Democritus at least take it as straightforward that many cosmoi separate off from the vortex. One might add that fragment 2 might well suggest that the vortex is indefinite in extent, which could quite possibly support an ongoing process of the separating off of cosmoi:

> Air and aether are being separated off from the great surround, and this surround is indefinite in extent.[27]

There is no suggestion that separating off stops. Against this one must say that most of the more reliable secondary evidence is against multiple cosmoi for Anaxagoras. Simplicius tells us that:

> Those who say that the cosmos is unique, like Anaxagoras and Empedocles, have no need of an infinity of elements, so that Empedocles does better to make them limited.[28]

I shall come back to the question of an infinity of elements for Anaxagoras later when we look at how Plato's geometrical atomism relates to these questions. It is unclear from the remaining fragments of Anaxagoras' writings if he believed nous to be sufficient to generate our cosmos uniquely.

Anaxagoras' nous then cannot explain the ongoing order of the cosmos. There is though a second difficulty, about explaining the initial ordering of the cosmos. Why does nous choose this initial ordering rather than any other state, as surely there were other possible arrangements than the vortex? If there were not, nous would not be required as an arranger, though it might be required as an initiator of motion, though one might again ask, why this motion rather than any other motion? Unless there is some means of differentiating between the possible initial states, then any choice that nous makes seems to be radically underdetermined. If so one might say that the initial ordering is not an act of intelligence at all, if it is not based on any rational criteria. Yet Anaxagoras says nothing about this being the best cosmos. Again, one might say that Anaxagoras is making no use of his nous. The *Timaeus* again produces a very sharp contrast to this. The demiurge produces this cosmos because it is the best cosmos, and so has to be the choice of intelligence.

Anaxagoras then agrees that matter and motion alone are insufficient to account for the initial ordering of the cosmos. Like Plato but unlike Leucippus and Democritus, he considers that some intelligence must be involved in this initial ordering. I will postpone discussion of Empedocles for the moment, as his case is somewhat more complex. While he posits a unique cosmos, despite allowing some role for chance in physical matters, he feels the need to multiply biological entities to give a plausible zoogony where he allows a considerable role for chance.

With the multiple cosmoi of Leucippus and Democritus in mind, we might begin to consider the question of parsimony in ancient cosmological thought. One of the objections levelled against teleological accounts is that they are ontologically profligate. If we can produce an explanation in terms of matter, motion and perhaps forces alone, what need have we of the entities that would be needed to underpin a teleological account? The problem for the physiologoi though is that they cannot produce a plausible account in these terms and need to multiply entities so that there are indefinite cosmoi. As our cosmos is entirely separate from the other cosmoi, and we have no evidence for their existence, they are purely theoretical entities. The alternative, as taken by Plato and indeed Aristotle is to posit one cosmos and some teleological principles. Plato may well feel that the sort of teleology he advocates to be more parsimonious then the physiologoi. One might ask: How much more profligate of entities can you get than postulating an indefinite number of cosmoi? As we shall see, a factor that may tilt this in the teleologist's direction is that this sort question about the order of the world about us and these sorts of answers will recur over the number of species and the number of sizes and shapes of atoms. While the physiologoi again multiply entities, Plato's account is parsimonious over the number of species and the number of sizes and shapes of the ultimate particles while using the same teleological principles.

V. Recurring cosmological problems

Earlier I suggested that the question of whether one can explain the apparent order of the cosmos solely in terms of matter and motion, or matter, motion and forces is a recurring theme in the history of science. Certainly it occurred in the seventeenth century in relation to the mechanical philosophy. The basic principle of the mechanical philosophy was that all physical phenomena could be analysed in terms of the mechanical interactions of matter in motion. Matter was simply extension, and had no active properties whatsoever. Thus matter had no inherent attractive or repulsive forces, and matter interacted by contact action only. Phenomena such as gravity and magnetism were reduced to patterns of matter in motion (vortices of matter give gravity, the motion of screw-shaped particles gives magnetism), but are not inherent properties of matter. So too (for Descartes at least) light and colour could be reduced to pressure waves and particle spin. As the mechanical philosophy was formulated in a Christian context, there was not a problem about the initial organisation of the cosmos. According to the book of Genesis, God created the universe in a ready ordered state. Descartes actually believes that the principles he formulates would allow the generation of the current cosmos from a primordial chaos:

> It may be possible to start from primeval chaos as described by the poets, i.e. total confusion in all parts of the universe and deduce from it, in accordance with the laws of nature, the precise organisation now to be found in things.[29]

However, he backs off from this position because:

> The sun and the earth and moon and stars thus existed in the beginning, and, what is more, the earth contained not just the seeds of plants but the plants themselves; and Adam and Eve were not born as babies but were created as fully grown people. This is the doctrine of the Christian faith, and our natural reason convinces us that it was so.[30]

While Descartes distinguished between *res extensa*, the extended substance, and *res cogitans*, the thinking substance, and proposed a mind/body dualism for humans although not for animals, Hobbes proposed a rather more radical materialism that would reduce mind to matter and motion as well. Against the mechanical philosophy, the Cambridge Platonists (More, Cudworth, Barrow) argued that matter and motion alone could not account for all the phenomena we see around us. Certainly there were religious motivations for this reaction, particularly in respect of Hobbes, but the Cambridge Platonists also focused on what they perceived as weaknesses in the mechanists account. In particular, the mechanical philosophy had problems providing plausible accounts for gravity, magnetism and cohesion. Thus we find More making comments like:

> The phenomena of the world cannot be solved mechanically, but there is the necessity of the assistance of a substance distinct from matter, that is, of a spirit incorporeal.

So too he says that:

> In all nature there can be nothing more certain than or well tested than that the phenomenon of gravity is repugnant to mechanical laws; and further,

> that its explanation cannot be resolved into causes purely mechanical and corporeal; but that it is necessary here to admit certain additional causes which are immaterial and incorporeal.

The seventeenth-century debate was conclusively won by the mechanical philosophy. That has had significant implications for the way in which we explain physical phenomena, and has coloured our thinking about the cosmos right down to the twentieth century. The dominant seventeenth-century metaphors were those of clockwork and mechanism. Descartes, original conception of forces being reduced to contact action was softened by the inclusion of forces acting at a distance (the Newtonian conception of gravity), while retaining the general mechanical conception of the world.[31]

However, these matters do not stand still and our increased knowledge and understanding of the cosmos has resulted in a recurrence of this debate in the twentieth century, in a slightly different form. As outlined above, for modern cosmology the explanation of the development of the cosmos from the primordial chaos of the big bang in terms of matter, energy and known forces alone is not a problem, at least in outline. We have pieced together a series of events within a reasonable timescale which would lead from the big bang to the cosmos as we know it today, and indeed have spectacular confirmation for some parts of this account.

There is a problem though in explaining why the fundamental constants of nature have the values that they do. Examples here are the gravitational constant, the velocity of light, Planck's constant, the electro-weak force, the nuclear strong force, etc. We are aware of no constraints or determinants on these values, and there is no apparent reason for them being at their observed levels. What is more, some argue, it is extremely fortuitous that they are at these levels, as in some cases very small deviations would upset some very fine balances and the universe would no longer be able to create or support life forms like our own. On the grand scale, without this 'fine tuning', the universe might expand too quickly, thus preventing the formation of galaxies, stars and planets, or recollapse long before the life cycle of stars had created the conditions necessary for life. The balance of electro-weak, nuclear strong and gravitational forces is critical not only to the stability of atoms, but also to the conditions required for their generation, the manner of their generation and their relative abundance when generated. It is also critical for the life cycle and nature of stars. Figures for the degree of this fine tuning are typically of the order of one part in 10^{50}. Thus our universe seems very unlikely indeed. So we might well ask, why is the universe like this? Why this universe and not others? Typically, though perhaps not exclusively, there are three types of proposed answers.

First, one can posit some more explanatory principles. God is an obvious one, and one which soon involves some teleology. God selects the values of the fundamental constants in order that life forms of our type can later come about in the universe, and perhaps even in order that the passion of Christ can be enacted. Alternatively, another approach is the cosmological anthropic principle, which has two popular forms:

The Weak Anthropic Principle (WAP)

The observed values of all physical and cosmological quantities are not equally probable but they take on values restricted by the requirement that there exist sites where carbon-based life can evolve and by the requirement that the Universe be old enough for it to already have done so.[32]

The Strong Anthropic Principle (SAP)

The Universe must have those properties which allow life to develop within it at some stage in its history.[33]

The anthropic principles are supposed to lessen our surprise that we live in the sort of universe that we do. Exactly what sort of explanations are generated by the anthropic principles, and what merits those explanations have, are a matter of considerable controversy.[34] Barrow and Tipler are quite explicit about the teleological nature of the Anthropic Principle,[35] and recognise that SAP is more open to teleological interpretation than WAP.[36]

The second type of answer is that one might hold that we do not yet know enough yet about the ultimate laws of physics. When we have the 'theory of everything' (for which the modern acronym is TOE), perhaps from a synthesis of quantum mechanics and general relativity, then it will be clear why the constants are at these levels and indeed have to be at these levels. For some people it is definitive of a TOE that it should explain the values of the fundamental constants.

The third type of answer is to suggest that our universe is only one of many universes. The modern term here is 'multiverse'. If there are an indefinite number of universes, instantiating all the possible levels of all the fundamental constants, and all possible combinations thereof, then it is no surprise that our cosmos, however unlikely it may be, exists. This can work diachronically or synchronically, or both together. When the big bang occurs, it does not send material flying out into pre-existing time and space, but creates time and space as well. The straightforward model of the big bang is that from the initial instant one space-time entity is created which expands to become the cosmos as we know it today. Other models have been postulated, however, where the initial conditions are sufficiently extreme such that the initial space-time not only warps but separates off into many small bubbles, known as a quantum foam or froth.[37] Each of these bubbles then expands in the usual way to form a universe like the one we know. It is possible that the fundamental constants have different values in these different universes.

There are certain models of the cosmos whereby the universe collapses after the big bang and an expansion phase. One possibility with these 'bounce' models is that with each collapse and subsequent big bang the fundamental constants are reset, such that with a greater number of these bounces, the more likely it is that our universe comes into existence. In fact, the latest evidence at the time of writing from the recession velocities of supernovae is that the cosmos is expanding a little quicker than was previously thought, and so standard models do not predict recollapse, though there are of course other models under which this may be still a possibility. The synchronic model suggests that there are an unlimited number of universe in existence now. One general argument in favour of the multiverse is generated from what is known as the 'Copernican principle'. This is an anti-anthropocentric principle which states that:

> We do not occupy a privileged position in the universe.[38]

The Copernican principle has met with a great deal of success. First it was recognised that the earth is not central and stable in the cosmos, but in fact orbits the sun. It was recognised that our sun was not unique, and perhaps not our solar system either, but that are many suns and perhaps many planetary systems. It was recognised that our galaxy was not unique and that ultimately we do not have any privileged position in terms of observing the universe. So perhaps too what we refer to as our universe is not unique either, and there are other, maybe many other 'universes', and we in fact live in a multiverse.[39]

The parallels with ancient debates are evident. Leucippus and Democritus do not feel that they have sufficient principles to generate this cosmos uniquely from the initial chaos, so they postulate an indefinite number of cosmoi, the ancient analogue to the modern multiverse. Plato and indeed Aristotle, on the other hand, take up the alternative of positing a unitary cosmos and some more explanatory principles. It is perhaps indicative of one of the weaknesses of ancient science that no one adopts the second modern position of admitting that there simply is not yet enough knowledge about the fundamental workings of the cosmos. The problem of whether the order of the cosmos can be explained in physical terms alone is then a recurrent one in the history of science. When it is felt that it cannot be, there are also typical recurrent lines of approach, in terms of multiple cosmoi or in further explanatory principles.

VI. Biology and zoogony

If there is a problem for the ancients in this regard for cosmology and cosmogony, then a fortiori there is a similar problem with biology and zoogony, if we hold that life forms have an even greater order and complexity than the physical cosmos. Attempting to explain the body in terms of matter and motion was a job the eighteenth century found impossibly difficult. From the middle part of the seventeenth century onwards, there was a programme known as mechanical biology. This was founded on the mechanical philosophy, which had great success in the physical sciences during the seventeenth century and which people now sought to extend to the study of life forms. The two key principles for the study of life forms were that (1) all physical phenomena were subject to a reductive analysis in terms of matter, motion and the contact action of matter; (2) animals, and the human body up to and including the brain, were entirely physical. The body was no more than matter in motion. Animals, and human bodies too, were no more than machines. Very complex machines, no doubt, but machines all the same, and the principles for analysing mechanisms were to be applied to the body.[40] Some such as Boyle distinguished between watches as the creation of man and animals as the creation of god, while holding that mechanical principles could explain both.

Mechanical biology functioned on two levels. On a macro level, it treated the body as a machine, treating limbs and muscles as levers and ropes, the circulation of the blood as a hydraulic system driven by a pump (the heart), and the stomach as a device for crushing and pulverising food. On a micro level, the ultimate goal was to explain the functioning of the body in terms of the matter and motion of its constituent parts. As matter was conceived to be passive, ultimately all explanation had to be in terms of immediate and proximate causes.

There were, however, certain very important phenomena that mechanical biology struggled to explain in anything like a plausible manner.[41] Some of these were standard ancient objections, some were generated by the use of the newly discovered microscope. The basic problem for mechanical biology was that organisms appeared able to organise themselves beyond anything that mechanical biology could explain in terms of matter, motion and proximate causes. This manifested itself in the question of reproduction, where things as (mechanically) similar as equine and human embryos in things as (mechanically) similar as equine and human wombs would consistently grow into things as different as horses and humans.[42] A further problem lay with the early development of the human embryo, which in its early stages of development looks nothing like a human being at all, and seems to develop due to some internal dynamic rather than being determined by proximate causes. The ability of some animals to regenerate significant body parts (such as some crabs) or reorganise into two after being divided (some worms) or reorganise themselves after being turned inside out (water hydra),[43] also indicated a greater ability to self-organise than the mechanical philosophy could allow. A further blow to the mechanical programme was the discovery of the cell with the new microscopes, as the cell appeared to be a living entity itself, while the mechanical programme required the parts of the body to be mechanical.

So mechanism could not account for the organisational abilities of animals and cells, and their ability to pass on that organisation and organisational ability. Even Aristotelian ideas of form and the actualisation of potential seemed to provide a better conceptual framework for understanding these phenomena than mechanical biology. Thus biology in the eighteenth century adopted vitalist ideas. From a modern perspective the answer to many of these questions is DNA, something that would have delighted the mechanists but was of course far beyond any conception they had at the time. There are then problems, some well known in antiquity which cannot be solved by the resources available to the mechanical philosophy in the eighteenth century. Let us now take this back to the ancient world and Plato's views.

There is very little information on how Leucippus and Democritus thought living beings came about, but we may well presume that this happened by reason and necessity, and the view attributed to Democritus that the first men came from the earth and life-giving moisture would fit well. Whether this process is supposed to occur in all cosmoi that can support life is open to conjecture, but possibly not if this is merely a matter of chance and necessity. There cannot be life in all cosmoi as the physical conditions in some cosmoi (some have no sun, some are devoid of water) will not support life. If there is an infinity of cosmoi though, there must be some which support life and some which have living creatures. Whether other cosmoi would have different forms of life to those on earth, and whether there would be many non-viable life forms before viable ones emerged, as with Empedocles, are also questions we simply have enough information to answer. With Anaxagoras we have a little more information, as we are told that:

> Anaxagoras says that the air contains the seeds of all things and these are brought down together with water and generate plants.[44]

This may well apply to animals too.[45] The difficulty for Anaxagoras' account is that by the time living entities start to form, nous has long since withdrawn its influence. That leaves us with three alternatives. First, nous in some way remains and helps the organisation of life. As nous is active here, as long as it acts with the best in mind then Plato would have no objection. Secondly, despite there being no framework for ongoing order in the cosmos, life just happens, which seems a little hopeful to say the least. Thirdly, Anaxagoras may believe in multiple cosmoi when he could have an account more in line with Leucippus and Democritus.

VII. Empedocles

Empedocles' account of the formation of living things is both fascinating and fantastic. In the first stage of generation:

> And earth chanced in about equal quantity upon these, Hephaestus [i.e. fire], rain and gleaming air [*aither*], anchored in the perfect harbours of Cypris [i.e. Love], either a little more of it or a little less of it among them. From these arose blood and the various form of flesh.[46]

> And kindly earth received in its broad melting-pots two parts of the glitter of Nestis out of eight, and four of Hephaestus; and they became white bones, marvellously joined by the gluing of Harmonia.[47]

One might ask here if the process in the second passage is the working of love but under some other names, or a different process altogether. One might also worry that getting flesh and getting bones is one thing, getting them to combine together into limbs and other body parts is quite another. There are further stages to this process of generating animals:

> Empedocles believed that the first generation of animals and plants were not generated complete in all parts, but consisted of parts not joined together, the second of parts joined together as in a dream, the third of wholes, while the fourth no longer came from homogenous substances like earth and water, but by mingling with each other.[48]

> On the earth there burst forth many faces without necks, arms wandered bare bereft of shoulders, and eyes wandered needing foreheads.[49]

> Many sprang up two faced and two breasted, man faced ox progeny, and conversely ox headed man progeny.[50]

At the fourth stage the creatures are well formed enough to procreate and so there is no need for these processes to continue, though Empedocles does not explain why they cease. Empedocles does not explain all of this in terms of the actions of love and strife. He also brings in an element of chance:

> Wherever everything came together as if it were for the sake of something, there they survived, having been accidentally brought together artfully, but where this was not the case, they died and are dying, just as Empedocles says of his man faced ox progeny.[51]

> These things fell together, encountering each other by chance, and many other things were constantly being produced.[52]

It is important here that many things other than well formed creatures are produced. While Empedocles believes his principles of love and strife to be sufficient to generate this physical cosmos uniquely, and so has no need of multiple cosmoi, it is clear that he does not believe them to be strong enough to generate the current biological species uniquely. Thus we get a multiplication of biological entities (as many of these cannot breed one is wary of calling them species) to explain the formation of the current species. Thus none of the physiologoi give a mechanistic account of the development of the cosmos that does not multiply entities, whether they be cosmoi or species. Leucippus and Democritus, as far as we know, give the same sort of account based wholeheartedly on a multiplicity of cosmoi to explain both the physical and the biological. Anaxagoras' account uses nous, or fails to explain, or multiplies worlds like the atomists. Empedocles' account of the actions of love and strife, however we are to construe those, gives out at a key moment. That is significant, as Empedocles at least seems to recognise a leap in complexity from physical to biological phenomena. It is also important to recognise that chance plays a part in the description of the physical world for Empedocles. Aristotle tells us:

> Empedocles says that air is not always separated off to the highest region,[53] but in a chance manner – he says in his Cosmogony that 'It chanced to run this way at that time, but often it is otherwise' and he tells us that most of the parts of animals came about by chance.[54]

If Plato found these accounts radically implausible, then I think we may well have some sympathy with him. With Aristophanes' speech in the *Symposium*, Plato may indeed be openly laughing at mechanist ideas, in particular those of Empedocles. In the *Timaeus*, he develops an account of the demiurge's construction of humans. First the head is constructed, and then at *Timaeus* 44c ff. we are told:

> In order that it should not roll around on the ground, with its heights and depths of every kind, and be at a loss in scaling these things and climbing out of them, they gave it body as a means of support for ease of travel.

One might also wonder why love produces good order. If love is a sexual metaphor (derived from more ancient mythological accounts?) Plato might well take the view that sexual hedonism will not produce good (cf. the *Philebus*) but needs to be guided by intelligence, which brings us back to the good. Although this is perhaps a more modern concern (cf. my earlier comments on gravity and the cosmos) one might wonder why love has this power rather than being stronger or weaker, when it would not achieve a proper ordering, and what it is balanced so well against strife. As with the value of the gravitational constant and its balance with other forces, this looks fortuitous and unexplained. Why strife begins to gain on love is also unexplained, as Aristotle points out in *Metaphysics* 1000b12 ff. Empedocles also appears to make a significant leap, from the formation of bones and flesh to the formation of well formed biological entities such as arms, legs, necks and eyes. The account of how the biological elements of the body are organised looks significantly underdone.

There is then a further problem for the physiologoi with zoogony and biological phenomena. Here we find that even Empedocles now needs to multiply entities in order to account for biological phenomena.[55] It is important to note that this is a recurrent problem in the history of science, or at least it was until comparatively recently, as now we have a coherent theory of evolution and a sophisticated analysis of the role of DNA. In contrast to the physiologoi, Plato posits a single cosmos and the minimum number of species to account for the phenomena (as indeed does Aristotle). Certainly Plato gives a heavily teleological account of the formation of living beings. We might balance that against the need for multiple cosmoi or for multiple biological entities, for which there is little or no evidence in the ancient world.

VIII. The shapes and sizes of atoms

One can see a similar pattern of argument occurring with the question of the shapes and sizes of the particles in ancient atomic theories, and perhaps with Anaxagoras' theory of matter too. If there is structure, in terms of groups of atoms forming an object, the size and shape of the atoms themselves, or with Plato the ultimate triangular particles, then that structure is something that needs to explained. Why this structure rather than any other structure? One answer from the atomists is that there are an infinite number of shapes and sizes to the atoms, so it is no surprise that there are these particular atoms which have this particular shape and size:

> Leucippus supposed there to be an infinite number of atoms that are always in motion and have an infinite number of shapes on the grounds that nothing is such rather than such (*dia to meden mallon toiouton e toiouton einai*).[56]

If there are no restrictions on the shapes and sizes of atoms, then one might expect there to be some very large atoms, though of course they may only be in other cosmoi:

> One (Epicurus) supposed all atoms to be very small, and on account of this imperceptible, the other, Democritus that some atoms are very large.[57]

If one accepts that 'nothing is such rather than such' then there cannot be an orthodox explanation of why one atom differs from another. That there are an infinite number of worlds is held to explain why there are certain groups of atoms which make up objects. Infinite number of shapes and sizes to atoms explains why there are these shapes and sizes of atoms. Aristotle is quite clear on what he believes are the main differences between Leucippus and Plato at *De Generatione et Corruptione* 325b20 ff. Both hold that the ultimate constituents are indivisible and are figures, but while Leucippus holds that they are solid and infinite in figure, Plato holds that they are planes and are definite in number.[58] For Leucippus coming to be and passing away depends on the contact and motion through the void of these particles, for Plato on contact only.

Whether Anaxagoras posits multiple cosmoi or not, Aristotle certainly considers him to posit an infinity of principles in contrast to the four element theory of Empedocles:

> These two differ, however from each other in that Empedocles imagines a cycle of such changes, Anaxagoras a single series. Anaxagoras again posited an infinity of principles, namely the homeomerous substances and the opposites together, while Empedocles posits only the so-called elements.[59]

Elsewhere, Aristotle confirms this opinion about Anaxagoras, and suggests that for Empedocles the number of elements is six.[60] One must take Aristotle's evidence here cautiously, as he is attempting to fit all of his predecessors into a scheme of those who posited one principle or a limited, or unlimited number of principles. That Anaxagoras is not an atomist also complicates matters, but there is a sense in which he places no limits on the numbers of principles he uses to explain the physical world, though not quite in the sense of Leucippus and Democritus.

Empedocles specifically limits the elements to four earth, water, air and fire. In a later chapter I will argue that Plato objects to these being elements because they are not basic and stable enough, and I will argue that several aspects of Plato's geometrical atomism can fruitfully be seen as an attack on and an attempt to solve some of the perceived problems of Empedocles' account. While Plato and Empedocles would agree on a small number of basic constituents of the world, there is the question of what we should hypothesise about their nature. Why these elements rather than any others? What shape and size of ultimate particles ought we to hypothesise? Plato of course has a very strong answer to this. Rather than subscribe to the *ou mallon* principle of the atomists, which generates an infinity of shapes and sizes of atoms, we have the heavily teleological account both of the two basic types of triangle and of the way that they combine to form complex plane figures and then three-dimensional particles of earth, water, air and fire. Thus we find Timaeus saying at 53d4-54a6 that:

> This we hypothesise (*hupotithemetha*) as the principle (*archen*) of fire and of the other bodies ... but the principles of these which are higher (*anothen*) are known only to God and whoever is friendly to him. It is necessary to give an account of the nature of the four best (*kallista*) bodies, different to each other, with some able to be produced out of the others by dissolution ... We must be eager then to bring together the best (*kallei*) four types of body, and to state

> that we have adequately (*hikanos*) grasped the nature of these bodies. Of the two triangles the isosceles has one nature, the scalene an unlimited number. Of this unlimited number we must select the best (*kalliston*), if we intend to begin in the proper manner. If someone has singled out anything better for the construction of these bodies, his victory will be that of a friend rather than an enemy. We shall pass over the many (*ton pollon*) and postulate the best (*kalliston*) triangles.

There are a considerable number of echoes of the *Phaedo* in this passage. We have a hypothesis, there is talk of something higher (*anothen*), there is hope to produce a sufficient (*hikanos*) account, and of course at all times it is the best that is hypothesised.[61] One can also see why Plato might well feel that the 'nothing is such rather then such' principle of the atomists is utterly inadequate. In Chapter 3 I shall discuss in general how Plato makes choices from an indefinite field of possibilities and how that relates to the way in which he attempts to solve one of the difficulties raised by Meno's paradox.

Plato's approach to the question of the shape and size of the ultimate particles is then consistent with his approach to the questions of the number of cosmoi and the number of species. Plato uses teleological principles to minimise the number of entities here, and eschews the multiplicative approach. Once more we have our parsimony choice, this time between an infinity of shapes and sizes of atoms and some teleological principles. Note that there is no evidence for an abundance of shapes and sizes of atoms.[62]

I would also argue that, apart from being a perfectly plausible line of investigation in an ancient context, Plato's approach to the question of the nature of the ultimate constituents of the physical world has been historically very significant. There are several matters on which some major trends in the history of atomism have gone with Plato rather than Leucippus and Democritus or Epicurus. Perhaps due to the fact that the revival of ancient atomism occurred in the highly Christian context of the seventeenth century, the atomic theory has limited itself to one cosmos. Rather than an infinite number of shapes and sizes of ultimate particles, the tendency has been strongly to hypothesise a relatively small number of types of ultimate particle with specific shape, usually spherical. From the end of the seventeenth century onwards we have pictured atoms, and indeed some sub-atomic particles (certainly electrons, protons and neutrons in some common theories of the atom) as being perfectly spherical. Why should we do so when surely choice of shape here is underdetermined empirically? One influence on this is the supposed existence of a Christian God who would naturally create the ultimate particles of his cosmos in the best manner.

One might argue then that Plato's geometrical atomism has been significantly undervalued in these respects. There is much more to be said in analysing and evaluating geometrical atomism and this is a matter I shall return to in a later chapter. In the meantime, it is important to recognise that Plato's geometrical atomism is not a metaphysical fancy done for amusement. It is an example of a specific type of answer to a recurrent problem in cosmology. As I shall argue in Chapter 3, it is also typical of Plato's strategy for dealing with a problem in the philosophy of science. Where we have an indefinite field of possibilities to choose from, Plato begins with hypothesising the best shapes for the particles. It is also intended to propose answers to problems that Plato sees with the assorted element theories of the physiologoi, and as we shall see later, it is intended to answer certain problems raised in later Plato about the nature of the ultimate physical constituents of the cosmos and the relations between them.

IX. A teleological cosmos vs. multiple entities

The physiologoi then postulate multiplicities, be they cosmoi, shapes and sizes of atoms or species. One might object to this on grounds of parsimony, but I would argue that Plato also objects to this approach on the grounds that the multiplication of entities in this manner is non-explanatory, and that the supposed cosmoi of the physiologoi are not sufficiently ordered. As indicated earlier, Plato is not only worried about the arrangement of the constituents of the cosmos in relation to each other, he is also worried about the order that the constituents have themselves. The shapes and sizes of atoms is unexplained by Leucippus and Democritus who consider them to be infinite in variety and number, on the grounds that 'nothing is such rather than such'.[63] The following passage at *Philebus* 17e5, while it is in the specific context of the analysis of limited and unlimited, might equally well apply to Leucippus and Democritus:

> The indefinite plurality of things and in things makes you in each case indefinite of thought and someone of neither status nor account (*ouk ellogimon oude enarithmon*), since you have never yet examined the number in anything.[64]

There are several word plays in the Greek here which associate allowing indefinite plurality in the world with being indefinite in thought and suggest that if you cannot give an account in either words or numbers then you are a person of no account. Similar sentiments, though perhaps less vigorously expressed, can be found in the *Timaeus* as well. The following passage at *Timaeus* 55c comes directly after the description of how the two basic types of triangles combine to form the complex plane triangles and squares and these then form the three-dimensional elements:

> If in considering all this someone should raise the quite proper (*emmelos*) question of whether the number of cosmoi should be said to be unlimited or limited, he will suppose that the view that they are infinite is that of someone

> who is indefinite (*apeirou*) on a matter on which he should be definite (*empeiron*), though to raise the question of whether in truth it is proper to speak of these as one or five is rather more reasonable. Our view declares it to be one, according to the likely account, but another, looking somehow to other things, might hold a different opinion.

Again we have the play on indefinite in the world and of thought, though here it could be rendered 'inexperienced on a matter in which he should be experienced'. *Emmelos* is also interesting as a contrast to the *plemmelos* (discordant) behaviour prior to the ordering of the cosmos in the critical passage on the ordering of the cosmos at *Timaeus* 30a. The passage at *Timaeus* 55c has long been considered a rather puzzling one.[65] If we take the uses of 'cosmos' here to refer simply to worlds, then it is hard to see why Plato should even begin to consider the possibility of five worlds. However Plato has just had Timaeus describe the organisation of the elements out of their component parts. Each element has *summetria* and *taxis* and so might be considered a cosmos in itself, and Plato is quite happy to use *cosmeo* and its cognates to describe the process of ordering the elements.[66] The passage can then be read in the first instance as an objection to the idea that there are unlimited shapes and sizes of atoms rather than unlimited worlds, something that Plato has had Timaeus object to previously at 31a ff. Whether we speak of one or five cosmoi then depends on whether we are referring to the universe or to the elements from which it is organised. That Plato has had Timaeus describe the construction of five elements prior to this passage as well as after aids this interpretation, as we then have five cosmoi. That then leaves the puzzle of the dodecahedron, which cannot be generated out of the two basic types of triangle, and is not used as an element as with earth, water, air and fire, but 'God used it for the whole writing animal figures on it'. Possibly this is a reference to the zodiac, though it is hard to see how Plato is being consistent here.[67]

For Plato then neither postulating an indefinite number of worlds nor an indefinite number of shapes and sizes of atoms produces an adequate explanation of the nature of the world about us. I have a considerable amount of sympathy with Plato on this matter. To start with, his approach to the ultimate particles is not only plausible in context, but as argued in the previous section, is the direction that science has taken on this issue ever since. On the matter of an indefinite number of cosmoi, it may be the case that just as there are other stars with planetary systems in our galaxy, and as there are other galaxies than our own, so too what we call the universe may be only one of many. One should note though that we believe there to be a finite amount of matter in our universe, and a large but finite number of other stars, planets and galaxies. If there are other universes in this sense, what they are like (and whether we can ever know anything of them) is another matter. It is only a matter of conjecture that they have different values for the fundamental constants to our universe. In the absence of evidence, this approach appears to be enormously profligate of entities. There is also a matter of general aetiological principle here. If I ask why something is this like it is, in that it has some determinate property, does it explain anything to say there are an indefinite number of these with all possible values and this is one of them?[68] I'm not convinced that such an explanation generates any degree of understanding, and nor does such an approach seem heuristically useful. One might also worry about the following difficulty. A multiverse will have a structure to it (universes A, B and C are related to each other so) which could be different. How do we explain that structure? On the principle we employed to explain the order of our universe, are there many multiverses? Or is there some principle which explains the order of the multiverse? If this is regressive, then I do not see that it is explanatory. If it is not, we ought only to posit extra worlds when we have concrete evidence of them, rather than as a means of explanation. One might try to deny that the multiverse has structure or could be treated as a single entity, but that would seem very odd, and I would certainly side with Plato on whether indeterminacy in this sense could ever be explanatory.

There are two important premises in play in some arguments in the *Timaeus*. The first is that what there is constitutes a single, whole entity and not an indefinite plurality. We can take that internally, in that what there is forms a coherent whole, or externally, in that our cosmos is all there is. The second is that there is a unique best. Put another way, there are no ties for first place when we ask what is best, but there is a unique best. As we shall see in Chapter 3, that is important epistemologically, but it is also important cosmologically. There is one, and only one best arrangement for the cosmos, and the demiurge will instantiate that best arrangement as he always acts for the best, as is emphasised at the beginning of Timaeus' main speech at 29d7 ff.

Plato has been accused of a fallacy of division at *Timaeus* 31a1 ff.[69] The passage begins with Timaeus asking whether we ought to speak of one heaven, or many or an indefinite number. The critical point is *Timaeus* 31a3-4, where the answer is:

> One, if it is crafted according to a paradigm (*hena, eiper kata to paradeigma dedemiourgemenos estai*).

Does Plato illegitimately assume that because the paradigm is unique and the craftsman copies it, that the copy will be unique too? We have though a perfectly plausible alternative. The copy is unique because the demiourgos will make the best copy of it and there is one and only one best copy. The uniqueness of the copy is generated by the nature of the demiourgos (acts for the best, no reason to do anything else) and not an attempt to copy the uniqueness of the model.

Does Plato consider the cosmoi of the physiologoi to be properly ordered?[70] If order has come about accidentally, and there is nothing which maintains that order, that may be a very fragile order liable to deteriorate or even collapse entirely at any moment. The next stage of the development of this sort of cosmos may exhibit some disorder. If it does not, then it looks increasingly fortunate we are living in an in ordered cosmos (or one with a great deal of order). Think of this in terms of our randomly moving gas molecules with no attractive or repulsive forces. It is highly unlikely that they would come together as eight clumps of the appropriate size and distribution along with many other clumps arranged in a spherical formation around them (i.e. like earth, sun, moon and planets and fixed stars). It is even more unlikely that they will retain that formation and manage to execute the motions of the heavens. Of course that is so long term for the atomists, as their cosmoi are perishable, but to the extent to which chance is involved it also applies from moment to moment. Leucippus and Democritus do have some mechanism for the accretion and maintenance of order, in their conception of atomic bonding and their principle that like attracts like,[71] but there remains an element of chance for which they need to postulate multiple worlds. One matter that is critical to the account of the *Timaeus* is that it provides not only for the initial ordering of the cosmos by the demiurge, but also for the ongoing order of the cosmos. There is another related issue here, which in modern terms is the difference between regularities and laws. If the order of the cosmos is (to some extent) accidental then (to that extent) there are not laws of nature but only regularities, if we are fortunate enough to live in a cosmos which displays these regularities. Later I will argue that regularities are not enough for Plato, and he requires something stronger. These are not laws in the modern sense, and again teleology is involved, but Plato requires more than the atomists multiple worlds can offer in order to explain the regular and orderly behaviour of the cosmos.

X. Cosmological reasons for teleology

Plato then has cosmological reasons for advocating teleology. There is a recurring question within cosmology, which continues in the present day (though in a slightly different form), as to whether the order we see in the current cosmos can be explained in physical terms alone. When the answer to this is no, there are typically two strategies. Either one can postulate extra explanatory principles for a unique cosmos, or postulate multiple cosmoi. With the Presocratic debate, no one gives a one cosmos mechanical account. Leucippus and Democritus, Anaxagoras and Empedocles all either use teleological principles themselves or need to explain some aspects of the cosmos in terms of multiple cosmoi, multiple shapes and sizes of atoms, or multiple species. No doubt there is a sense in which it is true, as Vlastos puts it, that Plato's thought is:

> The conscious abandonment of the line of thought which had led, in the systems of Leucippus and Democritus, to the first rigorously mechanical conception of the order of nature.[72]

One must take care though to remember the role of chance and multiple worlds in their systems. Plato's general strategy for cosmology can be defended in several ways. It is not at all clear in the ancient world that a one world mechanical account is either viable or plausible, so we cannot criticise Plato for rejecting this. Plato rejects the multiplication of entities (cosmoi, types of particle, species) as ontologically profligate and aetiologically inadequate and has good reasons (cosmological and epistemological) for adopting a teleological approach. Moreover, Plato breaks important new ground in being the first to give a thoroughgoing teleology. To return to my general objection to Vlastos' view, this is no 'metaphysical fairy tale'. Plato's teleology attempts to answer serious cosmological and epistemological problems.

From the modern viewpoint, it has yet to be shown that Plato's general strategy is incorrect. At present there is the problem of explaining the values of the fundamental constants, for which a bare one world account is not enough. Plato's arguments about the explanatory value of the multiple worlds hypothesis still bite, and so postulating extra explanatory principles is still a viable approach. Relative to modern cosmological anthropic principles, Plato's teleology is excessive. While the details of Plato's approach may require historical mitigation (and there is plenty to be had), his general strategy does not.

One might though argue along the following lines. Where science has made its most spectacular theoretical progress, it has done so by rejecting teleology in favour of more or less mechanical processes of evolution. We now have non-teleological accounts of the evolution of the cosmos from the big bang, and of the evolution of life. Thus we should do without our non-progressive teleological prop, and merely consider non-teleological explanations. Again though, there are several lines of defence. Firstly, even if this is so, it is an argument drawn from the experience of two millennia of research subsequent to Plato. Plato as the instigator of this tradition of teleology is in no position to judge its history. Moreover, Plato would have considered himself to have made great progress, and to have laid down a highly progressive programme in astronomy and cosmology. Secondly, proponents of the cosmological anthropic principles would vehemently deny that science has always made progress by jettisoning teleology. They would argue that anthropic principles have made important predictions in modern science, most notably the prediction of (then unknown) resonant energy levels for carbon which are critical for stellar nucleosynthesis. Historically, one might argue that the Platonic programme in astronomy was enormously successful. Apart from the overtly teleological and Neoplatonic approaches of Copernicus, Kepler and the overtly teleological and neoaristotelian Harvey, much of the thinking of the scientific revolution was underpinned by the idea of a benevolent Christian deity creating a good world.[73] A further question we might ask about cosmology is this. Although we can now give a non-teleological account of the evolution of the cosmos, this seems only to push the problem back a level such that we now have the problem of explaining the values of the fundamental constants. Should we succeed in doing this, we might only push the problem back another stage. Ultimately, we may have to leave some problem unexplained or use teleology. Plato then has serious and reasonable cosmological reasons for employing teleology.

2

Astronomy, Observation and Experiment

Plato's comments on astronomy and the education of the guardians at *Republic* 528e ff. have been hotly disputed,[1] and have provoked much criticism from those who have interpreted them as a rejection or denigration of observational astronomy.[2] In this chapter I argue that the key to interpreting these comments lies in the relationship between the conception of enquiry that is implicit in the epistemological allegories, and the programme for the education of the guardians that Plato subsequently proposes. We have, I suggest, been too eager to stress the similarities here, when recognition of the differences may supply us with the tools for a better understanding of *Republic* 528e ff., one that to a large extent disarms the anti-empirical critique. My discussion proceeds in three stages. First, Plato takes great care to place his comments on astronomy in the context of the preceding epistemological allegories. Is there any evidence here, where Plato might be thought to discuss enquiry in general, that he rejected or denigrated observation? I argue that even if Plato advocated a 'two worlds' (TW) ontology, he still envisaged a dynamic process of enquiry, interrelating sensibles and intelligibles, the investigation of each being necessary but not sufficient to achieve the overall aim. Secondly, Plato appears to be deriving how we ought to go about educating the guardians from how we ought to conduct our enquiries. While the two are intimately related there are important differences, recognition of which turns the supposed rejection of observation into an affirmation of the need for an initial empirical approach. I contend that this makes good sense of the internal structure of 528e ff., and of its relations to other parts of the *Republic*. Thirdly, I discuss the *Republic* in relation to the evidence offered by later works on the question of astronomy and observation. I raise the questions of whether Plato's views on the nature of the sensible world necessarily lead to the denigration of careful and prolonged empirical work, and whether his views on the role of observation develop.

A second passage which is though to be indicative of Plato's anti-empirical attitude is at *Timaeus* 68b-d, where he is alleged to say that it is useless to make any practical test of his theory of mixing pigments to obtain various colours, and to try to do so would be to reveal one's ignorance of the differences between man and god. I argue that this passage is actually about the proportions of the various types of fire particle in the visual stream which constitute a colour, and that while the passage is clearly pessimistic about practical tests in this case, there may be some quite sophisticated and reasonable worries about practical tests in play here, and practical tests may well be applicable in many other cases. I also argue that Plato's conception of the *technai* is consistent both with my analysis of *Republic* VII and *Timaeus* 68b-d. For each *techne* there is both the possibility and the necessity of learning something by empirical means before we move on to the higher and more theoretical part of the discipline.

I. Two worlds and investigation

In order to support the view that the comments at *Republic* 528c ff. disparage observational astronomy, one might begin by asking, in which section of the line allegory does Plato place astronomy? As we are told that the 'real objects' of astronomy are 'accessible to reason and thought alone' (529d1-5), and stand in relation to the visible heavens as the objects of geometry do to any diagram which depicts them (529d7-e6), we might well feel that astronomy belongs in the *dianoia* section, L3 (I number the line sections L1-L4 ascending).[3] If so, and we are taking a standard TW view of Plato, then one can see how anti-empirical considerations might easily be generated. I characterise the TW interpretation as the view that for Plato four distinctions are co-extensive, and divide what there is into two disjoint sets. On the one hand there is that which is intelligible, stable, knowable and incorporeal, while on the other is that which is sensible, changeable, opinable and physical. The first set of entities are said to be, while the second are in a state of becoming.[4] If we are dealing solely with intelligible objects in astronomy, perhaps we ought then to 'proceed by means of problems' and 'leave the things in the heavens alone' if we are to pursue the 'real science of astronomy' (530b6-9). The final phrase here is translated in a manner which is sympathetic to this position, and later I will discuss whether this is the best rendering. Here one might also emphasise the abstract definition of astronomy given at *Republic* 528a and 528e, as the study of solids in motion, with its close links to arithmetic, plane geometry and solid geometry in the curriculum for the guardians. Thus one can find Heath commenting that for Plato:

> Science then is only concerned with realities independent of sense-perception; sensation, observation and experiment are entirely excluded from it.[5]

Within the framework of assigning a line section to science in general and astronomy in particular, other views have been canvassed. Lee for example has argued that science belongs in the *pisteis* section, L2, and comments that:

In the *Timaeus*, Plato includes the natural sciences in this subsection, as they can never reach ultimate truth, being concerned with the changing world.[6]

Cornford on the other hand has argued that there is no 'science of Nature' for Plato as there can be no exact truth in our account of the objects of physics, which for Cornford are those of the lower part of the line.[7] Rather than work within this debate, I wish to question whether this general approach is correct. Implicit here is the assumption that science for Plato can be demarcated by its place on the line, and most commonly, that the border of the two realms forms an upper or a lower boundary. My main strategic concern will be to argue that Plato allows and indeed emphasises a continuous progress in enquiry and the cognitive ascent of the philosopher, from sensibles to intelligibles. So too he believes that subsequently the intelligibles may inform us about the sensibles. If so, then there can be no demarcation for investigation between the two worlds, or by position on the line generally. Consider *Symposium* 211b6-d1, where Diotima is instructing Socrates in the ways of love:

This is the correct way of proceeding or being led in matters of love; beginning with particular instances of beauty one must always ascend, as if employing rising stairs, from one beautiful body to two, and from two to all, and from beautiful bodies to beautiful ways of living, and from this to the beauty of learning, and from here to the ultimate higher study, which is concerned with beauty itself and nothing else, so that finally one may know what beauty is.[8]

Let us call this investigation 'erotics'. The key point here is that border of the two realms does not form the demarcation between erotics and non-erotics. The investigation begins with sensibles and ascends to intelligibles as part of the same discipline. This is entirely in accord with the *Republic*. Concerning investigation thought of in terms of the line allegory, at 510b4-9 Socrates says:

In one division the mind is compelled to investigate by using the originals of the visible order in their turn as images, and has to base its search on hypotheses and proceed from them.

The force with which this is expressed is interesting. The mind is compelled, *anankezetai* (510b5) to investigate in this manner, ruling out any other method of accessing the intelligibles. The *Seventh Letter* strongly supports this view (though of course there are doubts about its authenticity). 342b distinguishes name, account, image and knowledge, and states that the first three are the necessary means for attaining the fourth.[9] At 343d we find that we must 'Work between all of these four, passing up and down over each'. Thus we need to grasp the image (i.e. the physical manifestation, cf. 343a3 ff.) in order to access the intelligibles, and as we must pass 'up and down' between all four in investigation, there can be no demarcation at the border of the two worlds. These passages strongly suggest that investigation for Plato is the ascent of the line, including the traversing of the border of the two realms, rather than the exclusive property of any one section. If so, then individual investigations are better conceived of as dynamic vertical ascents rather than static, horizontal positions on the line. It is important not to prejudge the 'horizontal/vertical' debate here when translating the names of the line sections; e.g. by rendering *episteme*, used to refer to the upper parts of the line, as 'science'. Thus science and individual disciplines are demarcated by their subject matter (e.g. astronomy is general enquiry applied to the heavens) rather than by position on the line.

Here one might stress the links between the allegories of line and cave. Our starting point is in the cave and the problem is how we use enquiry to make our ascent. To ask which section of the line a discipline belongs to is to miss this dynamism. Rather we might ask, given the current development of a discipline: What section of the line is it now in, and so how ought we to pursue it in the future? There can be little doubt that the general epistemology and methodology set out in the allegories is intended to be applied to astronomy.[10] There are several other passages which are helpful here. These can be grouped under two headings, those which show that an empirical approach is insufficient, and those which show it to be necessary. It is often possible to pair these within one work, e.g. *Phaedrus* 268a-269a where we find that to possess the skill (*techne*) of medicine one must have more than the necessary 'empirical knack' (*empeiria*), one must in fact understand the nature of what is involved, here the body or the soul.[11] There is then an essential empirical component to investigation for Plato. Obviously there are restrictions on the usefulness of empirical work, for while it may be necessary for the initial stages of the ascent in the line allegory, it is clearly futile for the higher parts (e.g. *Republic* 510b where the investigation disposes of images). Hence we find passages which decry its use for this latter purpose, but not blanket bans. One might also compare *Gorgias* 485c ff. and 500e ff. where identical distinctions are drawn between *empeiria* and *techne* concerning rhetoric and medicine, and one might also look to *Gorgias* 465a and 501a on the requirement that to be in possession of a *techne* one must be able to give a rational account. For necessary/insufficient pairs in later works, one can compare, e.g., *Timaeus* 47a and 91d, *Philebus* 55e and 56a, and *Laws* 719e-720d.

II. The dynamics of investigation

Also important is the subsequent descent of the line. At *Republic* 520c, Socrates speaks as if to his fully educated guardians, telling them that it is their duty to descend into the allegorical cave, when they will:

> see far better than those in that place, and will know each of the images, what it is and what it is of, because you have seen the truth concerning beauty, justice and goodness.

That intelligibles can help to inform us about sensibles shows any supposed demarcation at the border of the two realms to be passable in both directions, 'upwards' from the sensibles and subsequently 'downwards' from the intelligibles. Fine has argued that in the *Republic*, Plato is a coherentist who offers us an interrelation model of knowledge.[12] Instead of justifying our notion of the Form of the Good in terms of anything more fundamental, or in terms of knowledge by acquaintance, we do so in terms of its explanatory power, by showing how well it allows us to explain the nature and interrelationships of Forms and sensibles. That the sensible world is (as far as possible) teleogically structured, and can be explained in teleological terms is a point that Plato is more insistent on in the *Timaeus*, but is still an assumption that underlies the discussion of the *Republic*.[13] If so, as we descend the line in showing the explanatory power of the Form of the Good, we must draw sensibles into our explanatory structure. So if we think of enquiry in terms of explanation, again we pass the border of the two realms in both directions.[14]

The passage at *Republic* 520c also suggests a second strategy by which we might undermine the supposed demarcation, which is to argue that Plato is not in fact committed to TW. In relation to the later dialogues, this may well be a highly profitable line of thought. Whether it can be successful for the *Republic* though is open to doubt. That we can opine intelligibles does not breach TW strictures, if we take it that it is possible to have knowledge subsequent to opinion of intelligibles, but that this is not the case for sensibles. Conversely, we are able to perceive or think about sensibles, but cannot perceive intelligibles. At *Republic* 506c Socrates has belief but not knowledge of the form of the good, while at 523a ff. we reason about our contradictory perceptions of our fingers. Thus we ought to read 'knowable' as 'opinable and knowable' and 'opinable' as 'only opinable'. So too with 'perceptible' and 'intelligible'. This is not to suggest that we can reason about sensibles in the strongest sense, as their deficiencies prevent us having knowledge about them, but surely we can think about them, or bring them before the mind by some route other than perception, perhaps by memory or imagination.[15] This weaker construal of TW meets all the 'unattractive consequences' of Fine's stronger construal, according to which there can be no belief concerning Forms.[16] These consequences lead Fine to reject TW for the *Republic*. Where cognitive states are correlated with objects (e.g. 511de), on my view these are maximal states to which we may progress from lower ones.[17] Thus we can come to know intelligibles in the manner of the Meno by attaching an explanatory account, though the *Republic* has a more sophisticated view of this process. Thus the *Republic* does not 'radically reject' the Meno's account of knowledge.[18] Rather than reading *Republic* 520c as allowing unrestricted knowledge of sensibles, it may tell us no more than that we can know what a particular sensible is (this is an image of Beauty) and what it is of (we can know Beauty itself).[19] Knowledge of sensibles themselves might still be elusive. The analogues for the relation of Forms and particulars that Plato employs in the *Republic* suggest that sensibles are imperfect relative to Forms, and that these deficiencies affect statements about them.[20] In astronomy we are told that it would be foolish to seek truth in the visible stars, as they are physical and so undergo deviations with the result that there is no constant relation between their periods.[21] If there is no exact and consistent truth, then for Plato there can be no knowledge. What might be possible however is knowledge of the ranking of sensibles in order of their approximation to the relevant known ideal, once we know what that ideal is. This would be important in politics, because the guardians with their knowledge of abstract justice could then correctly rank actions as more or less just, and derive their right to rule from this ability.[22] It is also possible that knowledge of the Forms would allow us improved opinions of sensibles and a better understanding of their functioning, without allowing us knowledge of them.

III. Vlastos on investigation

Vlastos has argued that for Plato, while there are necessary empirical preliminaries that must be undertaken, a discipline itself is an a priori matter with intelligible objects.[23] He cites *Phaedrus* 268e1-6, where Socrates (hypothetically) says to a man who believes he has knowledge of harmony because he can strike the highest and lowest notes:

> My good sir, anyone intending to become skilled in harmony must know these things, but nothing prevents someone at your stage from being without any knowledge of harmony. You know that which of necessity must be studied before harmony, but not harmony itself.

If we were to accept this argument, we would need to distinguish the discipline name, e.g. 'astronomy', referring only to the intelligible part of the investigation, from the whole investigation, including the necessary empirical phase. The nature of the investigation would be unaffected, as merely naming its intelligible part does not establish a demarcation. One would then have to discuss which of these we ought to refer to as Plato's astronomy. What is clear, however, is that enquiry considered as a whole does indeed have an essential empirical component.

Should we accept Vlastos' view though? First, in general Plato refers to the whole investigation by the discipline name. We have seen that at *Symposium* 211b erotics covers a broad investigation, and passages at *Republic* 526de and 530d which we shall examine in the next section will help to confirm this. Secondly, I have argued that an important factor in investigation for Plato is the interrelation of sensibles and intelligibles, involving the subsequent descent of the line, giving us some information about the contents of the cave and drawing the phenomena into our explanatory system. Vlastos' model of enquiry for Plato though is very much a unidirectional one, completely discarding the empirical preliminaries once we begin the 'real' discipline. Apart from the question of whether this is a good description of enquiry as envisaged by the epistemological allegories, one might also wonder about its relation to Vlastos' view that the distinction between forms and particulars is an attempt to formulate a distinction between necessary and contingent or a priori and a posteriori.[24] If, as Jordan has argued, we can undermine this view of the forms, then we can undermine any related conception of enquiry too.[25]

A different objection to my position could be based on the association of the methodologies of astronomy and geometry. One might propose the following dilemma. If geometry is to be pursued solely through intelligibles, then so is astronomy, and if astronomy is at least partly empirical then so is geometry. There can be little doubt that there is a close association between astronomy and geometry for Plato.[26] However, we might ask the following question: Ought we to assimilate astronomy directly to geometry, or ought we to consider how both relate to a general theory of investigation? If the latter, and the general theory recognises variables which affect individual disciplines, then there may be important differences between how we should presently pursue astronomy and geometry.[27]

The problem with the dilemma as posed is that it fails to recognise the dynamic nature of investigation for Plato. While geometry is at the moment pursued through intelligibles, the line allegory makes it clear at several points that we can only access this level in by ascending from the sensible via the contemplation of diagrams,[28] and that we must eventually transcend the current level.[29] How we currently ought to pursue a discipline for Plato is a question that is dependent on how far that discipline has developed. Although not directly expressed by Plato, we might allow him the following consideration. Only a few of the most elementary perceptions are required to make us enquire into the nature of number,[30] while geometry requires more, and more specific perceptions for us to investigate the relevant intelligibles, and so on in a continuum via astronomy and harmony.[31] So while astronomy and geometry share the same basic methodology, it would be wrong to equate the two directly as astronomy will require a much broader empirical base. As astronomy may be at a different stage of its development, it may require a different prescription for its future pursuit, such that we cannot without further consideration impose recommendations for geometry onto astronomy. Even given that Plato does not directly assimilate astronomy to geometry, we might be critical here feeling that there is a fundamental division between disciplines that require empirical confirmation and those that do not. Plato though approaches the matter from a different perspective. He seems to have in mind a general account of the genesis and development of disciplines, and the evolution of geometry from land measuring and other 'practical geometry' to abstraction to diagrams and then to intelligibles is something well theorised by the line allegory.[32] This may well be related to the way in which he treats human intellectual development in the cave allegory.

IV. Doing astronomy and teaching astronomy

Let us now consider *Republic* 528e-530d. Plato's critics have made much of this passage, while others have held that it is difficult to produce a coherent interpretation of it as a whole.[33] The context here is crucial, and I must disagree with Crombie who suggests that:

> If there is anything about scientific method in the *Republic*, it will be in the curriculum for the further education of his rulers which Plato lays down in book 7.[34]

The epistemological allegories have already given us a general methodology, which as we have seen, certainly applies to astronomy. At *Republic* 521c there is a marked change of direction. Socrates switches to discussing how the guardians of the ideal state might be produced, comments that what is at issue is the conversion of the mind 'from a kind of twilight to the true day' (*Republic* 521c5), and then says:

> It is necessary therefore to ascertain what sort of studies have the power to effect this.

The question now is how we might employ a discipline for specific educational ends rather than how we might pursue an investigation.[35] This means that we must be very cautious in drawing any conclusions concerning scientific method here. As a shorthand I shall refer to 'teaching astronomy' and 'doing astronomy' respectively. At *Republic* 526d8-e7 Socrates says that:

> It is necessary to consider whether the greater and more advanced part of the subject, if taken further, has the effect of making it easier to grasp the form of the good ... It is useful if it compels the contemplation of being, rather than becoming.

The method of education is then to encourage the guardians to consider the intelligibles. Note that only part of the whole subject has this effect.[36] Socrates also employs a distinction between astronomy as it is used now in philosophical education, and how it ought to be so used. At *Republic*

529bc he criticises the current use of astronomy, and at 529c4-6 Glaucon asks:

> But how is it necessary to learn (*manthanein*) astronomy, as opposed to the manner in which they learn (*manthanousin*) it now, if it is to be useful for the studies which we recommend?

It is important here to translate *manthanein* as 'to learn' rather than as 'to study', which would introduce an ambiguity which is not in the Greek.[37] Study is ambiguous between learn and research (i.e. teaching and doing astronomy) in a way in which *manthanein* is not.[38] It is this sentence, in conjunction with 526de, which sets the terms for the discussion which is to follow. The aim is to determine which part of astronomy deals with the intelligible and so ought to be used in philosophical education, in distinction to the current curriculum. How we ought to teach astronomy is going to be derived from how we ought to do astronomy, and contrasted with how it is taught now. This accords well with the broader relations of the two sections we have examined. How to educate the guardians is being derived from the earlier epistemological discussion.

First, Socrates considers the status of the celestial bodies, which though they may be the best and most precise of visible things, are inferior to the true entities, the orbits which the real speed in true number and pattern move relative to each other, which are accessible to reason and thought but not to eyesight.[39] The stars are likened to a diagram, as well drawn as might be, but still having inherent shortcomings, which we ought to use as in geometry. We do not examine a diagram hoping to discover any precise truths, but we consider the intelligibles which the diagrams express.[40] In line with the programme of 526de and 529c, Socrates separates the sensible and intelligible parts of astronomy. Note that there is no ban on observation here, where investigation is being discussed we are to use the visible heavens as a model, as in the line allegory. Having set up the appropriate distinctions, at *Republic* 530b6-c1 he then draws the concomitant conclusion:

> It is by means of problems, then, that we shall proceed with astronomy as we do geometry, and we shall leave the things in the heavens alone, if we propose by really taking part in astronomy to make useful instead of useless the understanding that is by nature in the soul.

The key question here is whether 'proceed with astronomy' refers to teaching or doing astronomy, or both. Consider the structure, wording and placement of this sentence. It is the culmination of this section, and as such should be completing the programme of 526de and 529c. If it refers only to teaching astronomy, then it gives an appropriate answer to the questions raised. If we are to use astronomy properly for education we must employ the intelligible part of the investigation, and not divert our students with observations, as philosophical educators do now.[41] The 'if' clause is specifically and exclusively educational, and the conditional structure allows us to draw a contrast with how astronomy is taught now and how it ought to be done.[42] Crucially, the attempt to derive how to teach astronomy from how to do astronomy in terms of the part/whole relationship implied at 526e makes sense on this reading. For the structure of the passage to work, doing astronomy must involve something more than teaching astronomy. It may also be significant that Plato uses *ean*, meaning 'to leave alone, to leave be or to heed not', rather than the stronger phrase *chairein ean*, meaning 'to renounce' as he does in other contexts. *Ean* fits the contrasts perfectly. We leave observation on one side for education, but not for other purposes, where one might expect *chairein ean* if Plato were renouncing observation entirely.[43]

The crucial point made by this carefully constructed and placed sentence is that astronomy as it is taught now and as it ought to be done do involve examining the physical heavens – this sentence implicitly affirms rather than explicitly denies this – in distinction to how we ought to use astronomy in philosophical education, when we should leave the things in the heavens alone, and consider only the intellectual problems.

If what I have argued earlier is correct, then this makes good sense in terms of the methodology suggested by the epistemological allegories just ten Stephanus pages earlier, and Plato has taken great pains to place his comments in the context of that methodology. It is thus possible to appreciate the distinction between doing and teaching that Plato is showing us in this passage. If we ignore this, then much of the structure of this passage is otiose. An additional advantage of this reading is that it does not require us to attribute any gross exaggeration or overstatement of position to Plato. As Heath has suggested:

> It may be that, when Plato is banning sense-perception from the science of astronomy in this uncompromising manner, he is consciously exaggerating.[44]

There is no ban and there is no exaggeration, but there is a comparison between teaching and doing astronomy. We ought also to consider the section on harmony at *Republic* 530d6-8 which immediately follows this passage, particularly as Socrates says:

> As our eyes are framed for astronomy, so are our ears for harmony, and these two are in some way kindred disciplines.

This sentence is either heavily ironic, or affirms that sense perception has a role in both of these disciplines, if our eyes and ears are framed for them.[45] Presumably much of what is said about the methodology for harmony applies to astronomy and vice versa. The key to this section is again the question that is being asked. At *Republic* 530e5-7 Socrates says that we must maintain the principles that have already been laid down, and:

> We must prevent those that we shall nurture from ever attempting to learn (*manthanein*) in an imperfect manner, one that does not always bring them to where it is necessary for them to go, just as we were saying concerning astronomy.

The same comments about the translation of *manthanein* apply here as at *Republic* 529c. The primary question then is not how we should do harmony, but what part of harmony is suitable for use in the education of the guardians, although there is naturally some consideration of the former in order to be able to derive the latter, as there was correspondingly in the astronomy section. The comments on harmony are then amenable to the same sort of doing versus teaching analysis as those on astronomy.

V. 'Real' astronomy

A further matter of translation concerns the phrase *ontos astronomias metalambanontes* at *Republic* 520b8. This is sometimes translated with the adverb *ontos* taken with *astronomias*, such that we get 'partaking in real astronomy'.[46] That is clearly favourable to positions which see Plato banning observational astronomy here, and to Vlastos' position whereby 'real astronomy' deals with the problems set up by the empirical preliminaries.[47] The more natural reading though is to take adverb *ontos* with the verb *metalambanontes*, giving 'really partaking in astronomy', allowing some difference between teaching and doing astronomy. This is a particular difficulty for Mourelatos, who bases his discussion around the question 'What exactly is the "real astronomy" discussed in 529c-530c?', taking 530b6-c1 as a pronouncement on how 'real astronomy' ought to be done.[48] He divides the interpreters into two groups, the 'a priori' and the 'save the phenomena' views.[49] The first group sees Plato as pursuing 'pure, ideal, or strictly a priori astronomy',[50] dealing solely with intelligible objects. We have already seen the theoretical difficulties with this position, and the problems with generating it from the epistemological allegories. Here we see that to generate it from *Republic* 530b6-c1 we need to force the translation and ignore the doing/teaching contrast that would seem to be the key to this passage.

Mourelatos' second group sees *Republic* 530b6-c1 as an exhortation to study the real and not the apparent motion of the heavenly bodies, the reported challenge to 'save the phenomena' by regular circular motion perhaps being the problem that we are meant to ascend to. Mourelatos cites Vlastos as the most able supporter of this position.[51] According to Vlastos, the meaning of the key phrase 'we shall leave alone the things in the heavens' (*ta d'en toi ouranoi easomen*) is that (i) we should dismiss the things in the heavens as objects of knowledge, and (ii) we ought to use them to set up the problems that 'real astronomy' then tackles, solutions here amounting to knowledge.[52] Vlastos recognises that 528e-530b offers no direct support for the second part of his assertion,[53] and so he seeks general epistemological support elsewhere. This though is problematic, for as we have seen enquiry for Plato seems to require a closer interrelation between empirical and intellectual work than Vlastos' account allows.

Vlastos' difficulties here, I suggest, stem from trying to read how to do astronomy directly from a passage which deals with how to teach astronomy. There then appears to be a gap between Plato's general conception of enquiry and how we should do astronomy. Vlastos' work can be seen as an attempt to bridge this gap from both ends, by downplaying the interrelation of empirical and intellectual factors in investigation (hence 'empirical preliminaries' and 'real astronomy') and seeking a greater role for empirical factors than might be initially suggested by *Republic* 530b6-c1. However, if we reckon there to be a fairly strong interrelation between observation and the subsequent processes in enquiry, recognise the do/teach contrast and do not reckon 530b6-c1 to be a statement about 'real astronomy', then there are no gaps to bridge. Doing astronomy as opposed to teaching astronomy is entirely consonant with Plato's general conception of enquiry.

If this is so, it will also defuse some of Vlastos' criticism of Plato's programme.[54] Vlastos argues that Plato envisages a sharp division of labour between those 'high grade drudges' who observe the heavens,[55] and the philosophers who think about the problems suggested by the results, and that this will be detrimental to astronomy. This conflates doing and teaching astronomy once more. The guardians, as part of their philosophical education will indeed avoid observation, and merely use the observations of others to set up problems for them to consider. Here the division of labour may be even sharper than Vlastos suggests, but for the purposes of investigating the heavens this division may not exist at all. Dicks' explanation of Plato's comments at *Republic* 530b6-c1 is unconvincing.[56] He remarks that:

> There had been accumulating during the fifth and fourth centuries BC a mass of crude, but none the less practically useful observational material ... what was now needed was for astronomers to sit down and do some hard thinking about astronomical theory ... that is precisely what Plato realised, and this is why he urged astronomers to concentrate on the mathematical side of their subject ... to the extent even of calling a temporary halt to the mere accumulation of more observations.[57]

However, there is no mention by Plato or any of his ancient commentators of any (temporary or otherwise) halt to observation. What is more, the difficulties that Plato raises concerning observation are philosophical worries. The question is not of how much observation we should do before considering more theoretical matters, but whether observation will help at all with the education of the guardians. Again, difficulties are created by failing to recognise the possible differences between doing and teaching astronomy. At *Timaeus* 39c5-d1, we are told that:

> Men have not pondered the periods of the other stars ... thus men do not name them nor investigate the mathematical relations between them.

As Plato gives considerable significance to the periods of the planets and their interrelations in the *Timaeus*, he can hardly have been calling for a temporary halt to observation. Bulmer-Thomas has argued that what we are meant to leave alone at *Republic* 530bc are *ta d'en toi ouranoi*, the 'things in the heavens' and that these are synonymous with *ta en ouranoi poikilmata* at 529c7.[58] According to Bulmer-Thomas, the *poikilmata* are patterns rather than individual bodies, and should be interpreted as constellations. Thus Plato is not rejecting observational astronomy, but only the fanciful grouping of stars into constellations which Bulmer-Thomas argues was happening at the time.[59] However, Plato's philosophical worries would again appear to cut deeper than this. First, the epistemological problem of the compresence of opposites, which not only underlies the important discussion concerning the 'sight-lovers' of *Republic* V and the epistemological allegories, but is also used to set up the discussion of education in *Republic* VII, affects all physical entities.[60] From this perspective, it is hard to see what advantage there might be in studying individual stars as opposed to constellations. Secondly, there is the problem that at *Republic* 530b2 where the heavenly bodies are said to 'deviate' such that there is no constant relation between their periods. The periods mentioned are day, night, month and year which are defined by the relation of single heavenly bodies (earth, sun and moon) to each other,[61] so again it is difficult to see how studying individual bodies rather than constellations is going to help here. Once we recognise the differences between doing and teaching astronomy, the problem becomes more acute, for it is not at all clear how studying individual bodies as opposed to constellations, fanciful or otherwise, is going to help with the education of the guardians. One might then pose the following dilemma. If we do not recognise any difference between doing and teaching astronomy, then, contrary to the philosophical thrust of the passage, the guardians must observe the heavens. If we do recognise some differences, and *Republic* 530b6-c1 contrasts doing and teaching astronomy, then as doing astronomy does not involve leaving alone *ta d'en toi ouranoi*, if the *poikilmata* are fanciful constellations then these are what we must investigate. Plato may well have objected to the fanciful grouping of stars into constellations, but that cannot be the whole story here.

VI. The *Republic* and other works

So far I have concentrated largely on the *Republic*. One reason for this is that if Plato did denigrate observational astronomy, then the purported evidence is strongest here, and it is worth investigating whether this charge can be repudiated on the basis of the *Republic* alone. There are two further reasons for this approach. First, in later chapters I will argue that there is a significant change in the nature of celestial motion between the *Republic* and the *Timaeus*, the heavenly bodies no longer being subject to the 'deviations' of *Republic* 530b2, but following predictable courses with invariable relations between their periods. This is a complex matter with many implications for other positions that Plato might be thought to adopt in the *Timaeus*. If there is regular celestial motion in the *Timaeus* and other later works, as I shall argue, then this may significantly affect the role that observation plays in astronomy. Secondly, there is clearly a serious re-examination of epistemology and the nature of knowledge in the *Theaetetus* and other later works, which may revise some of the theories of the *Republic*. Any evidence from the later works would need to be subjected to detailed scrutiny before it could be adduced in a discussion of the *Republic*'s astronomy.

Bearing these reservations in mind, the later works certainly do not recognise any denigration of observational astronomy. The clearest passage is *Timaeus* 47a ff.:

> Sight, in my opinion, is the cause of the greatest benefit to us as not a word of our current account of the universe could ever have been given if neither the stars, the sun, nor the heavens had been seen. But now the sight of day and night, of months and the cycle of the years has procured for us number and the concept of time, and has also led us to seek the nature of the universe. From these we have derived philosophy, and no greater good has or will come as a gift from the Gods to mortal men ... God devised and gave us sight in order that we might observe the rational revolutions of the heavens and use them against the revolutions of thought that are in us.

This would seem to make it entirely clear that observation is an essential part of the investigation of the heavens, and as the passage goes on to discuss the merits of hearing, is well worth comparing against what we have been told about sight and hearing in *Republic* VII. I would agree with Vlastos that in the *Timaeus*:

> Plato's language here is saturated with the terms of observational astronomy.[62]

This is something we can examine in more detail in chapter five. If Plato did issue a challenge to 'save the phenomena', that too would seem to sit strangely with an anti-empirical reading of *Republic* VII. While Plato was a theoretician rather than a practising scientist, another question here is his attitude to observational studies carried out by others. Let us first consider the case of Aristotle. Aristotle was essentially a theoretical astronomer who took over a tradition from Eudoxus and Callippus.[63] In doing so he would have been absorbing the large body of data that provided the basis for these theories, and then improving the relation between theory and data. Indeed, on more than one occasion Aristotle refers to results obtained by Egyptian and Babylonian astronomers.[64] However, Aristotle does not appear to have made detailed observations himself. If we did not know that he had taken over a large body of data, we might reckon the empirical support he provides for his theories to be sketchy and arbitrary.

We might take roughly the same line with Plato, as he too is a theoretician rather than a gatherer of data. As he is perhaps the progenitor of the tradition the Aristotle later takes on board, we cannot deploy quite the same argument in his case. We need some more direct evidence that he took over a body of data, and this I suggest is provided by the *Epinomis* and *Timaeus*. At *Epinomis* 986e6-987a6 we are told:

> The first to observe these bodies was a non-Greek. Tradition nurtured those who first took these things to mind, due to the excellence of the summer season which both Egypt and Syria adequately possess, and revealed to their sight, so we say, they always beheld the stars together ... These observations have since disseminated everywhere and have been shown to be true by the test of time.

If we are to believe the *Epinomis*, it would seem that Plato was well aware of detailed Egyptian astronomical observations, and was fulsome in his praise of their accuracy.[65] It is worth noting that in the *Timaeus*, while Plato does not refer directly to astronomical records, he praises the Egyptians for having the most ancient and comprehensive records,[66] and at *Laws* 967b ff. he is full of praise for those who have accurately studied the heavens. At *Timaeus* 39c ff. Plato is critical of those who have not studied all of the planetary orbits. Plato is also aware of Hesiod (*Epinomis* 990a) as one who is not a true astronomer, being someone who has only 'observed settings and risings' rather than studied orbits. As the *Timaeus*, *Laws*, and *Epinomis* are all later works than the *Republic*, they may represent a change of mind on Plato's part. However, according to my arguments, they need not; the *Republic* does not condemn or denigrate observational astronomy, but has a firm view of its position as the necessary starting point for investigation, and of the limitations of a purely empirical approach.

One might argue that while the *Republic* has a role for observation, because its physical entities do not behave in a perfectly regular manner, it discourages careful and prolonged empirical investigation. As Shorey comments, however:

> That the stars in their movements do not perfectly express the exactness of mathematical conceptions is no more than modern astronomers say.[67]

It is important to recognise that there are two explanations as to why this should be so. According to the *Republic*,[68] physical bodies are not capable of regular behaviour and so are likely to 'deviate' from any ideal projection. A modern explanation though might state that while bodies do perfectly obey laws, when we deal with a phenomenon like planetary motion there are problems of complexity and isolation.[69] Once we postulate e.g. an inverse square gravitational law, with a force propagating to infinity, then we need to include every massive particle in the universe in our calculations if we want complete precision. At some point we will have to accept 'anomalies' relative to our model of planetary motion as due to extraneous forces on a system which we cannot isolate. So both moderns and middle period Plato agree that there are anomalies. Yet we do not infer from this that modern science disparages careful observation. One might reply that the moderns have had great success in closing the gap between anomaly and theory, and that there is a difference, perhaps of several orders of magnitude, between what the moderns and the ancients would be prepared to accept as an anomaly. That is undoubtedly true, but the ancients had success here as well (especially in astronomy), and we have little knowledge of what they would have regarded as a serious anomaly. One might also argue that modern chaos theories and recognition of the butterfly effect actually paint a somewhat bleaker picture of what investigation might achieve than middle period Plato or the ancients in general might have accepted, but again we do not take these ideas as a discouragement to detailed observational work. We need not agree with Farrington then who argues that Plato's view that the heavenly periods bear no exact relation to each other was thoroughly reactionary, as it was likely to discourage precise observation and theorising.[70]

After the comments on the regularity of the heavens at *Republic* 530b3-4 concerning the heavenly bodies we are told that 'It would be absurd to seek in every manner to grasp the truth of these things'. Now, if this is merely a point about the philosophical status of the heavens, we might well agree. However, some, including Vlastos have construed it as a methodological point, disparaging prolonged and detailed observational work. We may use the heavens to set up the problems that 'real astronomy' subsequently investigates, but perhaps this involves no more than a cursory examination.[71] Given that the doing/teaching distinction and the dynamic interrelation model of Platonic enquiry will undermine any theoretical support for this position, is there any warrant for this interpretation? This seems doubtful. The point of this passage, as we saw in section IV, is to determine the philosophical status of the objects of astronomy in order to determine what part of astronomy the guardians should study, rather than to make methodological prescriptions for doing astronomy. As for the evidence elsewhere in Plato, passages in the *Timaeus* and *Epinomis* that we have seen earlier would seem to indicate that Plato had no objection to prolonged and detailed observational work, though as a theoretician this is not something he would carry out himself, nor is it something that he would recommend for the philosophical education of the guardians.

This is not a crucial matter though, for as I commented earlier, by the time he composed the *Timaeus* Plato had changed his views on the regularity of the heavens. Thus the question of whether Plato would have ignored perturbations in the orbits of planets as real anomalies, and so would never have discovered Neptune and Pluto become redundant.[72] A common view of Plato's middle period attitude to the imperfection of the sensible world does not necessarily result in the disparagement of detailed observational work. One might argue another line though, that if Plato thought that astronomy had now risen to line section L3, he may have thought the empirical work to be complete, and so did not encourage further observation. While it may well be the case that Plato believed astronomy to have progressed as far as L3, *Timaeus* 39c ff. quoted above indicates that Plato thought the empirical investigation of the heavens to be far from complete, a point which tells against Vlastos' sharp division of 'real astronomy' and its preliminaries as well.

VII. Celestial motion in the *Republic*

As I shall argue that there is a significant change in Plato's attitude concerning the regularity of celestial motion from the *Republic* to the *Timaeus*, let us examine exactly what the *Republic* has to say on this matter. The key passage in the *Republic* is at 530a ff., where Socrates asks:

> Won't the true astronomer be similarly persuaded when he gazes upon the movements of the stars? He will hold that the architect of the heavens and what is in them has constructed these things as beautifully as is possible for such works. But concerning the proportions (*summetria*) of night to day, of these to month, of month to year, and of the other stars to these and each other, wouldn't he consider it absurd to expect these things always to behave in the same manner and never to deviate (*paralattein*)[73] in any way, as they are physical and visible, and absurd too to seek by every means to grasp the truth about them?[74]

The position of the *Republic* seems quite clear on whether there is a constant relationship of the periods of the stars to each other, and hence whether there is regular celestial motion and whether the visible heavens are amenable to precise mathematical description.[75] There are important criticisms to be made of this sort of position, though whether they apply to the *Timaeus* as well is another matter. As Sorabji notes:

> The *Republic* passage is particularly interesting, because Plato suggests that, if the heavenly motions do not conform with mathematics, this shows that there is something wrong with the bodies, not the mathematics. This is contrary to the spirit of those who believe that nature obeys exceptionless laws, and who would seek ever to improve their mathematical formulae, to approach closer to the laws.[76]

As to why Plato believes the heavens behave in this manner, I have two suggestions, one metaphysical, one cosmological. At *Republic* 529cd Plato has Socrates say:

> The things that are decorations in the heavens, since they have decorated the visible, are regarded as the best and most exact things of their type, but they fall far short of the true entities, the orbits which real speed and real slowness, in true number and all true figures, move in relation to each other, and carry the things in them; these are accessible to reason and thought, but not eyesight.

One might propose two quite different analogues for how either mathematics, or the 'real speed and real slowness', relate to the motions of the visible heavens. First, there is the relation between an abstract geometrical entity and any diagram of it that we might draw. Secondly, there is the relation between a number, and any group of things that has that number. The key difference between these two is that while a geometrical diagram cannot perfectly instantiate an abstract geometrical entity (I cannot draw a perfectly straight line, nor can I draw something that has length but not breadth), a group has its number precisely. Possibly it is the first relation that is in play here and is causing the problems. This interpretation has the advantage it can be applied quite generally, which seems to be the intention of the epistemological allegories, and pays due attention to the fact that Plato pushes the diagram analogy fairly hard in the line allegory. If we look to the line allegory, it may well be that this analogue is used quite generally for the relation of the intelligible to the sensible. *Republic* 510d ff. is important here as this analogue is used as the model for the relation between sections L2 and L3 of the line. One might argue though that this looks rather like the old (mis)reading of the sticks and stones passage of *Phaedo* 74a ff. On this view, Plato does have a concern about diagrams, though not because diagrams are inherently inaccurate, but because, in any diagram, while A may be equal to B it is not equal to C, and while A may be double C it is not double B. Diagrams are inherently particular, and suffer from compresence. Thus we find Socrates saying at *Republic* 529de:

> One must use these decorations of the heavens for the sake of that study, just as if one had chanced upon some thoroughly well drawn and perfected diagram by Daedalus or some other skilled craftsman or artist. Anyone acquainted with geometry, seeing such things, would consider them beauti-

> fully crafted, but would think it laughable to examine them seriously, with the hope of grasping in them the truth concerning equals or doubles or any other ratio.

So this passage does not require anything imperfect about the physical world, one just has to recognise that particulars are always subject to qualification in a way in which equal and double themselves are not. The problem with this interpretation though is that Plato then goes on to tell us that the periods of the planets are subject to deviation, as they are physical and visible, and it would seem difficult to explain why Plato should do that if the compresence of opposites/qualification of particulars is the only difficulty in play here. Nor will it be able to deal with a passage often cited in relation to the *Republic* in this context, *Politicus* 273b ff. where it is the bodily element of the cosmos which is responsible for its gradual degeneration.

An alternative diagnosis of the problems of *Republic* 528 ff. might come from the myth of Er. Here there is a pivot, the spindle of necessity (616c3) about which the cosmos rotates. While the fixed stars may be an integral part of the spindle, the hemispheres of the sun, moon and planets which sit in the bowl of the great spindle cannot be as they are said to rotate slowly in the opposite direction (617a6-8). The spindle anyway is said to 'turn upon the knees of necessity' (617b4), so each moving part of this model has a pivot. The critical passage for our current purpose is at 617c5-d1:

> Clotho with her right hand grasped and turned the outer revolution of the spindle, leaving intervals of time between, while Atropos with her left hand did similarly with the inner. Lachesis grasped each in turn with each hand.

It would appear that some active intervention is required to keep the cosmos in motion.[77] Here Clotho and Atropos take hold of the rims of the spindle and of the hemispheres and turn them intermittently. A spindle seems a particularly apposite model here, as this is just the sort of thing that gradually slows down and has to be kept moving when in use. So without the actions of the daughters of necessity the cosmos will gradually grind to a halt. However, if their activity is intermittent, then as *Republic* 530ab has it, the relations of the periods of the various heavenly bodies will not be constant.[78]

With this view, one might argue that there are two difficulties that run through *Republic* 529c-530d, which are the compresence of opposites (so *Republic* 529de) and a cosmological problem epitomised in mythical form with the myth of Er concerning astronomy only (so *Republic* 530b). The problem here is that the epistemological allegories seem quite general, and one must worry that we are told that the periods of the heavens deviate because the heavens are physical and visible, not because of something specific to the heavens. So too astronomy is being treated as a general case for the purposes of the epistemological allegories.

A third possibility is that Plato has not yet fully sorted out his model of participation and his worries over the nature of the world, and so runs the compresence of opposites problem in parallel with an imperfection of the physical world view in the *Republic*. One might argue that he does something similar in the *Phaedo*, if we compare the sticks and stones passage with the 'fish's eye' view which is put forward at *Phaedo* 109b ff., which compares the relation of the submarine to the terrestrial and the terrestrial to whatever Plato envisages here as beyond the terrestrial. A complicating factor here is that in some works (*Phaedo*, *Republic*, *Politicus*) Plato may countenance a degenerating cosmos as I shall discuss in greater detail later.

Whatever the diagnosis of why there is not regular celestial motion in the *Republic* I shall argue that it is clear that the *Timaeus*, and indeed the *Philebus*, *Laws* and *Epinomis* adopt a substantially different position. Perhaps Plato moves to the second analogue for the relation of mathematics to these motions, one that would certainly better underpin such a cosmology. Admittedly direct evidence of such a change is sparse, but the contrast that the *Philebus* at least is interested in is between the arithmetic of the philosophers and that of the masses, the former working with pure numbers, the latter 'with two armies and two bulls and two extremely small or extremely large things'.[79] The *Philebus* also suggests that this analogue is the appropriate one for many other fields of study.[80] If this second analogue is in place for the relation of mathematics to the motions of the heavens in the *Timaeus*, then that would explain the move to regular motion, and if it is in place quite generally, it would explain the *Timaeus*' constant insistence that events occur *kat'arithmon*. Certainly it is the case that there are cosmological changes from the myth of Er in place in the *Timaeus* such that the cosmos no longer requires a pivot to turn on, any external bonds to hold it together or any external interference to keep it in motion.

VIII. *Timaeus* 68b-d

There is a passage in the *Timaeus* which some commentators have taken as showing that Plato believed the idea of experiment to be problematic. *Timaeus* 68d tells us that:

> If some investigator were to make a practical test of these, he would show his ignorance of the difference between human nature and god's, in that god has sufficient knowledge and power to blend the one into many and dissolve the one back into many, but there is not now, nor will there ever be a man capable of these things.

It is important here to take due care in translating *basanon* at 68d3. Although experiment is a tempting translation, it is critical to recognise that *basanon* has broader resonances and of course ought not to be assimilated to any modern notion of experiment. LSJ give 'a test to try whether something was genuine or not ... enquiry, esp. by torture'. On the interpretation of this passage, Sedley comments:

> Plato knows well enough that his own formulae for colour mixture would not be confirmed by experimental test. But that, we are told, is because the very idea of experimental science is flawed. To try to reproduce natural processes in the laboratory is to play god, in ignorance of the fact that in this respect we can never duplicate god's work.[81]

Sedley takes 68b-d to be about the mixing of pigments, but there is an alternative view held by Cornford, and by Lloyd who comments that:

> The blending in question appears to be not a matter of mixing pigments, but one of combining fundamental atomic particles.[82]

Whether 68d is about mixing colours or pigments is important for assessing Plato's views on learning by making practical tests. Let us put this passage in context first. I take it that this section on sight and colours runs from 67c to 68d, and is the fourth part of the piece running from 65c ff., which wants to analyse taste, smell, hearing and sight in terms of geometrical atomism. At 67c ff. we are told that there are several types of fire particle, small, medium and large, which contract, leave unaffected or dilate the visual stream. It is the mixture of these which create the sensation of 'colour' (with due caveats about Greek conceptions of colour/hue/intensity, etc.).[83] So we might expect the main conclusions here to be about the way that geometrical atomism relates to perception.

At *Timaeus* 68b5-9 we are told that it would be foolish to say what proportions they are mixed in, as there is 'no necessary or likely account' (*mete tina anangke mete ton eikota logon*). Does this refer to mixing pigments (the immediate context) or mixing fire particles (the more general context)? It would be extraordinary if we could not have some sort of account of how many drops of black and white make a shade of grey. What proportion of large to small fire particles are involved though might well appear to an ancient to be something that is beyond us, and is likely to stay beyond us. Fire particles are far below our perceptual thresholds. So there will be problems for humans both in gaining information about them, in terms of identifying or quantifying types, and manipulating them. Those might well appear, quite reasonably, to be intractable problems to an ancient. These are not problems which god has, however. When he orders the cosmos, earth, water, air and fire only have traces of their proper nature, as is emphasised just after our section at *Timaeus* 69bc. Clearly he is capable of operating at that level in a way that humans are not.

There is also a question about the status of Timaeus' account here. At 68d2, the results of mixing pigments are those which will 'save the likely tale' (*diasozoi ton eikota muthon*). However, if 68b5-9 is about pigment mixing, there is no necessary or probable reason, and if 68d2-7 is about pigment mixing too, any practical test will be useless. Better then that we take 68b5-9 to be about mixing fire particles as then *Timaeus* can have his likely account of pigment mixing, while being able to say nothing about the mixing of fire particles. If 68d2-7 is about fire particles too, then he can have learned the probable account through experience. One might try to avoid this by arguing that while 68b5-9 is about proportions of pigments, the rest of this section is about simple qualitative results of mixing pigments. That though would leave us in the distinctly odd position of being able to have some account of what colour occurs when we mix pigments, but being unable to have any account of the relative quantities which would give different shades, hues or intensities. If 68b5-9 is about mixing fire particles, and one translates accordingly, this alters the balance of the passage such that it is less odd that the final section refers to mixing fire atoms rather than colours.

How does Plato treat similar activities to pigment mixing in other contexts? There is a group of practical activities in which Plato would say ability was acquired by *empeiria kai tribe*, experience or empirical knack, and grinding away. Typical examples here are cooking,[84] learning a musical instrument,[85] base rhetoric and crude medicine.[86] Plato is of course highly critical of the limitations of learning in such a manner. As we have seen, such empirical preliminaries may be essential for learning or investigating a subject thoroughly, but they are very far from being the whole story. Unless we analyse the nature of what we are dealing with, or introduce some mathematics we will never be able to progress to *techne*. Nowhere though does Plato suggest that we cannot learn something, however basic and limited, by *empeiria kai tribe*. I take *tribe* in particular to be quite pejorative, with overtones of 'grinding out', and learning by trial and error, but one does not show one's ignorance of the difference between man and god by attempting these activities. So it would be odd if the results of mixing pigments, which Plato might be expected to say can be learned by a painter by *empeiria kai tribe*, were singled out for extremely strong criticism at *Timaeus* 68b-d.

What is it that god has sufficient knowledge and power to do (*hikanos epistamenos hama kai dunatos*, 68d6) that we do not? Surely not mixing pigments, as it is evident we at least have the power to do that (many into one at least),[87] and Timaeus has just been talking about the results (and it is those which on the pigment mixing interpretation are problematic, not the ability to do mixes). We have seen that here is a considerable difference between what god and humans might do with fire particles. Another possibility here is suggested by *Sophist* 265c-266d. *Sophist* 265c gives us the standard idea that we and other living things, along with earth, water, air and fire are all products of divine workmanship. 266bc though tells us that dream images, and those which occur in the day, and are said to be natural, such as shadows and the interaction of the two light streams, are in fact divine works too. So try to investigate how fire particles mix to create images might well be to show ignorance of the difference between man and god, but then this is a special case restricted to the way that fire particles mix. There is no indication of this view in the *Timaeus* though, either in the discussion of mirrors or of what god does.

Does Plato anticipate failure for his account of colour mixes if tested? Although some of the terms used appear odd to us, do we know enough about Greek colour terms to be sure of this? According to Aristotle, bright (*lampron*) seen through a dark medium or reflected in a dark surface is seen as red,[88] and he has some very odd things to say about the colours of the rainbow.[89] If Plato does anticipate failure, why give this account at all? All he needs to say for the structure of this section is that variations in colour are due to variations in fire particles in the visual stream.

On an initial reading, the references of *Timaeus* 68b5-9 and *touton* at 68d2 would appear to be to mixing pigments. There are several difficulties with this view though, which also gives Plato a pretty poor position, and one that is not only contrary to his consistent conception of *techne* in many other works would allow us to learn about pigment mixing by *empeiria kai tribe*, but also is unparalleled in the *Timaeus* and elsewhere. Here we see the importance of the previous analysis of *Republic* VII for interpreting Plato. One might try to cite passages there as support for a radical anti-experiment reading of *Timaeus* 68b-d. So I would rather take the (on the face of it) less likely (but perfectly possible) references of 68b5-9 and 68d2-7 to be to fire particles. Doubtless Plato could have been clearer as to exactly what he meant at *Timaeus* 68d, as Lloyd comments,[90] but I don't see that we need to conclude he was hostile in general to experimental science from this passage.

IX. Plato and experiment

If *Timaeus* 68b-d is about the mixing of fire particles, and we cannot by practical test find out the proportions of those fire particles which make up a colour, to what extent are the worries here generalisable to other situations where we might hope to make some progress by the use of practical tests? One might worry that this will affect all cases which involve either blending the many into one or dissolving the one into many, as only god is sufficiently wise and powerful to do this. However, it is evident that we can mix or dissolve some ones and manies, both physically (mixing pigments) and philosophically.[91] So *Timaeus* 68d must only be about ones and manies that only god can blend or dissolve, and not all ones and manies. I have two suggestions on this matter.

If we take our cue from *Sophist* 266a ff., then what god can do but we cannot is produce living creatures (animals and ourselves), organise the elements (earth, water, air, fire) and produce images, either those that occur in dreams, or those which occur by day and are said to be natural (*autophue*).[92] If so, then the extent of *Timaeus* 68b-d is severely limited. There is something special about images which means we cannot find out about them by practical test. Perhaps too there is something special about the production of living creatures that we could not find out about by practical test either. However, that would not affect the study of the constitution and function of animals and ourselves, and indeed *Timaeus* will have a great deal to say about both of these later on. If this is the problem, it is important to note that it would be still quite possible to learn (in a limited manner) many things by *empeiria kai tribe* as Plato's account of the *technai* has it, something I take to be important for both the suggestions I put forward in this section. There is a down side to this view which is that it would have god intervening in the world to produce images virtually constantly, which would run contrary to the general tenor of the *Timaeus*. It also seems slightly odd to put images on a par with the production of living creatures. As noted in the previous section, the *Sophist* talks specifically of dreams and mirror images in relation to divine works, and while the *Timaeus* has a discussion of both dreams and mirror images at 45d ff.,[93] and talks of how god designed the eye and eyelids, etc., it makes no mention of dreams or images or mirror images as being due to divine work.

My second suggestion on what god can do but we cannot is that god does have the knowledge and the ability to manipulate the various individual fire atoms but we do not. As god has the ability to do the intrinsic organisation of atoms, he clearly can manipulate the atoms themselves, while it is clear that (in the context of fourth-century Greece) we cannot. This immediately looks to be a more widespread problem, though one must note that there is no difficulty here if we have the knowledge and ability to manipulate, so practical tests on larger scale agglomerations of atoms (perhaps at the perceptible level) would be perfectly possible, as would learning by *empeiria kai tribe*. While the *Timaeus* is pessimistic about finding out the proportions of types of fire atom that produce colours, it is happy to give an account which is in accord with empirical considerations of other phenomena at the atomic, and even the sub-atomic level. With the discussion of digestion at *Timaeus* 80c ff., for instance, we are given an account of the incoming atoms of food being cut up and reconstituted in order to replenish the body, and we can even have an account of the way in which the triangles go to make up the elements. Why can we have this account? It is important to note that in relation to the fire particles, there is no necessary or probable reason (*Timaeus* 68b7) for the proportions of the sizes constituting a colour. This is not something which is said of the other geometrical atomism analyses of perception which precede that of sight, nor is it said anywhere else about geometrical atomism analyses of the body or anything else. So if fire particles are too small for us to have the knowledge and power to manipulate properly, and there are no plausible hypotheses to be had, then perhaps there is indeed no point in practical tests in this particular case. In other cases though there may well be 'necessary or probable' hypotheses which we can test the consequences of against experience or against practical test, even where we lack the knowledge and ability to manipulate the relevant entities.

This may indicate that Plato is worried about a more sophisticated difficulty with practical tests. In modern parlance this is known as the underdetermination of theory by data. In the next chapter I shall look at the relation between Plato's views and the modern underdetermination problem in detail, in particular the way that Plato's worries are generated by Meno's (the learner's) paradox. This is a problem which worries him in many places in the *Timaeus*, and is arguably one of the driving forces behind the highly teleological account. The aspect of underdetermination that may be in play here is that no amount of data will determine which theory we ought to adopt, but extra (non-empirical) criteria are required. As we shall see, Plato does have a means of resolving underdetermination and the non-empirical criteria he adopts, involving ideas of the good, have interesting relations to modern realist resolutions. The problem with fire particles though appears to be that in this case these criteria are inadequate as there are no necessary or probable reasons to prefer one hypothesis over any other, and as we might devise many hypotheses to cover the results of any practical test, we cannot learn of the proportions of sizes of fire particles constituting a colour by means of a practical test. This gives us a general problem with experiment which Plato believes to be soluble, and a highly specific one to do with fire particles which is not, and the possibility of learning by *empeiria kai tribe*.

Clearly *Timaeus* 68b-d contains some pessimism about what practical test might be able to achieve. How general that may be is another matter. *Sophist* 266a ff. would suggest the problem is highly specific, but it is not clear that Plato has these considerations in mind in the *Timaeus*. If the problem is that we do not have the knowledge or the power to manipulate microscopic entities (and never will have), then although this problem is not general (we can do practical tests where we can manipulate), then one must feel that Plato was unduly pessimistic. Criticism on this matter though ought to be tempered by the context of fourth-century Greek science, where without any technology of optical aids or delicate and intricate experiment, or progress in these areas, access to the world of the atom may well have seemed impossible. If underdetermination is the worry, then there is a particular and unresolvable problem with fire particles and a general but resolvable problem with practical tests. Even if colours were the issue at *Timaeus* 68b-d, these considerations if duly reformulated would also limit the extent to which there is a doubt about practical tests in general.

None of this is to deny that there are weaknesses in Platonic conception of science in relation to experiment, just as there are with virtually all Presocratic conceptions of science.[94] All are too optimistic about what theorising can achieve and all fail to constrain or test their theories properly by experiment, even if they have commonplace observations to support their theories. What I do wish to deny here is that we can single out Plato as an exception in having a highly anti-empirical attitude. There is no ban on or denigration of observation in *Republic* VII, nor is there a general ban on experiment at *Timaeus* 68b-d. This is important, for as we shall see, these passages are much quoted in diverse circumstance as being indicative of Plato's anti-empirical attitude. Nor can we cannot support an anti-empirical reading of either by reference to the other, or anti-empirical readings of other passages by reference to these critical passages. Plato has a coherent conception of *technai* which is consistent with both of these positions. There is both the possibility and the necessity of learning something by empirical means before we move on to the higher and more theoretical part of the discipline. Thus Plato does not, as Farrington puts it, 'substitute mathematics for physics',[95] nor does he exclude observation and experiment from science.[96] The anti-empirical charges against Plato's conception of science, as far as they single him out as an exception against the background of the Presocratics, are ill founded.

3

Meno's Paradox and Underdetermination

The main aim of this chapter is to argue that Plato had epistemological reasons for employing teleology. There are considerable similarities between some of the problems posed by Meno's paradox and a problem in modern philosophy of science known as underdetermination. While Meno's paradox is broader and more general, it can be seen to have the following problem is common with underdetermination. Which hypotheses, from an indefinite field of hypotheses, should we adopt for further consideration, and which one should we finally accept? The modern problem gives us the difficulty of which criteria we ought to employ when experiment underdetermines theory. Which solution ought to be adopted, and indeed whether there is a solution here are hotly debated issues.[1] There are those who hold that while scientists do employ non-empirical criteria in theory selection, the choice of these criteria cannot be justified in any absolute manner. It is a purely pragmatic, instrumental or conventional choice, to provide simpler theories and easier calculations. These theories may be more convenient for humans to work with, but there is no question of whether they are true or false. Some hold that theory choice is determined by sociological rather than intellectual factors and so argue that we should pursue a sociology of science, or more radically a sociology of knowledge. Such positions are in general associated with some form of instrumentalism, anti-realism or relativism.[2] Those on the other hand who believe that some form of non-empirical criteria for theory choice can be justified tend towards some form of realism. Plato believes that he has an answer to Meno's paradox. This is important, as it affects the way in which Plato thinks about certain scientific problems, especially in cosmology. In the *Timaeus* Plato several times faces the problem of choosing from an indefinite field of possibilities (which cosmos/ what is the nature of order/ the ultimate particles/ celestial motion) and in each case proposes an answer which is shaped by his solution to Meno's paradox.

I. The modern underdetermination problem

The modern problem of underdetermination in the philosophy of science, in its simplest form, can be stated quite baldly. However much empirical data one has, that on its own is insufficient to determine which theory, law or explanation to adopt. Thus this problem is often referred to as the underdetermination of theory by data.[3] The modern form is perhaps best illustrated by the graph drawing problem.[4] Here we have a set of data from an experiment, and we would like to determine the mathematical relationship between x and y.[5] We can draw a straight line through the data, but the difficulty is that there are many relationships which we can postulate which will all account for the data. It should be evident that not only can one fit many curves to the data, but that one can generate indefinitely many such curves. The equations for those curves will be more complex than for a straight line, but with no further criteria than simply accounting for the data, they cannot be ruled out. While one can draw a straight line through the data, there are indefinitely many other possible ways of accounting for the data (see Figs 5 and 6). How can we justify a preference

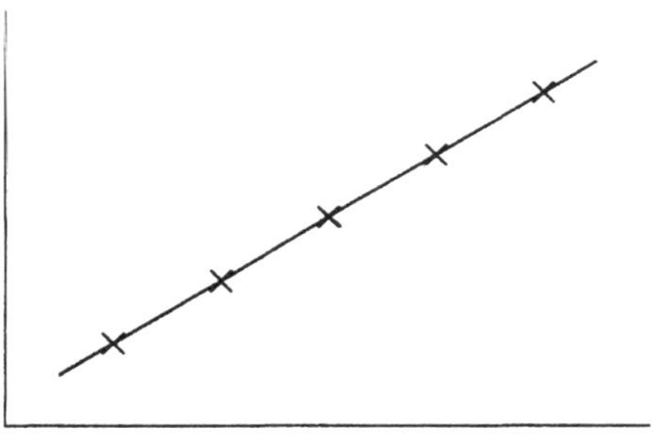

Fig. 5. Data plot.

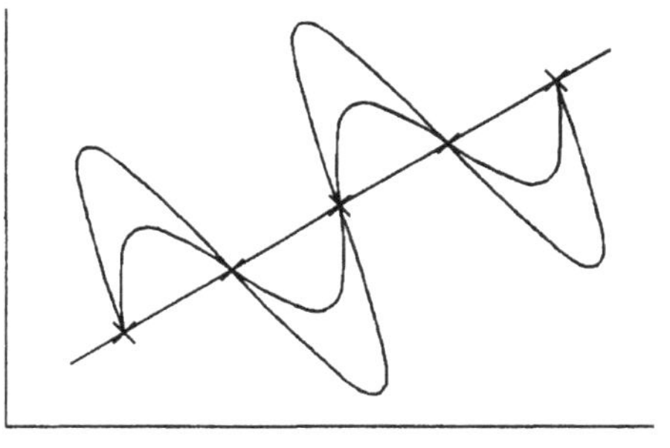

Fig. 6. Data plot with curves.

for any one way? This is a critical problem, for we might hope to generate laws of nature in this manner. Figures 5 and 6 could represent F=ma or V=IR or similar relationships thought to be fundamental to science. More data will not be any help here, for while it might rule out some sets of curves, there will still be indefinitely many left. There is a deliberate ambiguity in these diagrams, with lines and curves meeting at each datum. One could account for the data not only by straight line or regular curve, but by any combination of these, i.e. (between successive points) straight line, then small amplitude curve, then large amplitude curve. With no criteria other than accounting for the data, nothing says we have to use continuous functions (Fig. 7). Simply by varying the amplitude of any sine-type curve which passes through the points (and with sufficient ad hoc manipulation, which again we have no grounds to rule out, there

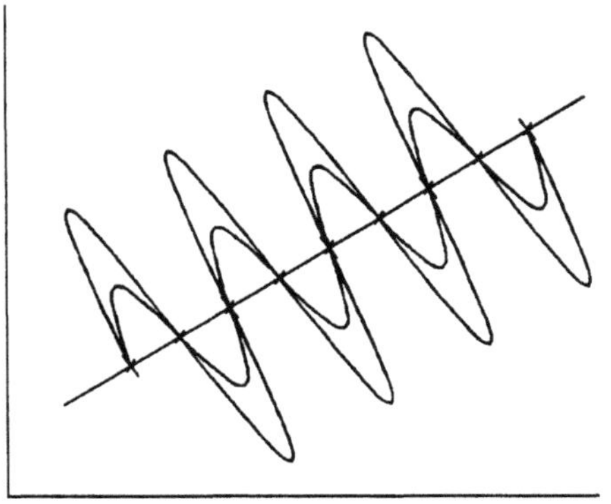

Fig. 7. More data, still underdetermination. More data does not help with underdetermination, as there will still be indefinitely many ways of accounting for the data, even using continuous functions.

always is such a curve) one can generate an indefinite number of functions relating x to y. In order to determine which is the correct relationship then, we need some more criteria. However, if science is conceived as being a purely empirical exercise dependent only upon the data, then there are no further criteria.

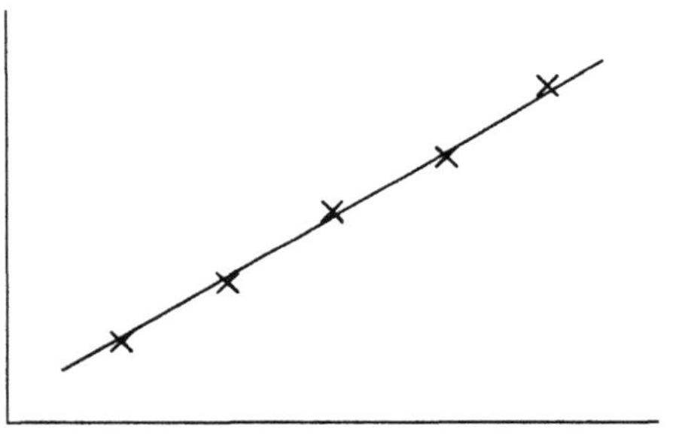

Fig. 8 Data plot, not saving the data. Anyone producing experimental results like these would be very happy to draw a straight line here, even though the 'best' straight line does not go through any of the data points.

Practically, scientists are not often bothered by this philosophical difficulty. In fact, they are quite happy to do something worse (Fig. 8). Given this rather fine set of data, no practising scientist would have a problem with drawing a straight line through the data.[6] However, the straight line does not go through any of the points and so does not 'save the phenomena' at all. The use of error bars, while they may soften some of the practical difficulties, will not solve the philosophical problems. Error bars indicate the supposed amount of experimental error in a measurement and give a spread of values around the measurement instead of a point (Figs 9 and 10).[7] Even if error bars soften the problem about drawing straight lines which do not go through all the points, they produce a further underdetermination problem allowing many straight lines to be drawn through the error bars.

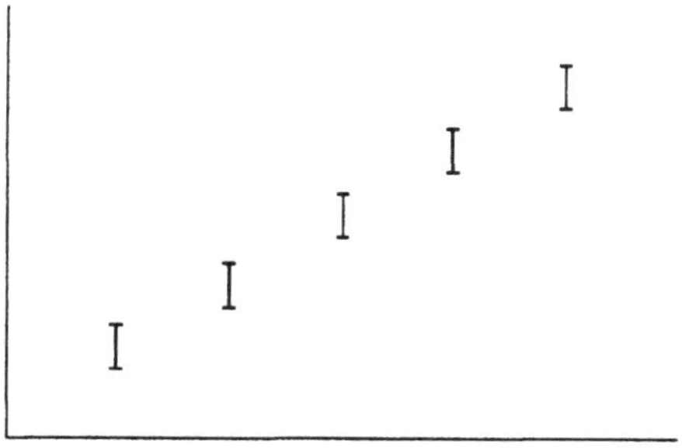

Fig. 9. Error bars plot.

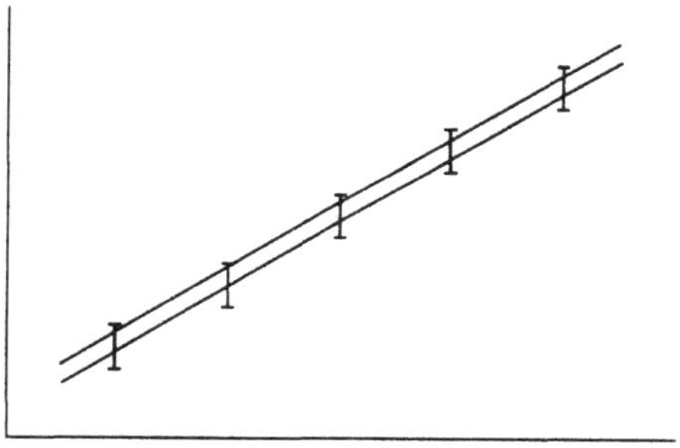

Fig. 10. Underdetermination with error bars.

One still has the problem, however, that one can draw indefinitely many curves through the data. While in this situation one can draw a straight line (in fact indefinitely many straight lines!) through the error bars, one can easily conceive of a situation where that would not be possible.[8] Another theoretical, if not practical difficulty is that of the 'rogue' result. Anyone who has ever tried to get an experiment to work will recognise data like that shown in Fig. 11.[9] Such results are always likely to occur, especially where small external factors (e.g. dust) have a critical effect. Given underdetermination though, what grounds do we have to rule out 'rogue' results, as is commonly done? It is easy to recognise with the rogue result that something has gone wrong with the experiment. It is less easy to formulate and justify criteria for excluding such results.[10] Again, if science is entirely data driven then we have no criteria for excluding such results.

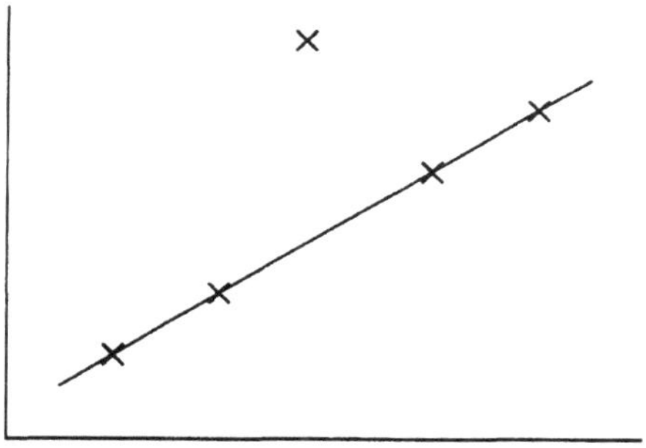

Fig. 11. Poor data plot.

So far we have looked at simple mathematical relationships between two variables, but it should be evident that the basic problem here will apply to more complex theories, and indeed purely verbal explanations as well. For any theory or explanation, by making sufficient ad hoc modifications, or even by simply adding material, one can generate an indefinite number of theories or explanations which will cover the facts.[11]

How troubling is this form of underdetermination? This depends on what resources one believes are available. If for instance one takes the position (at least within science) that our only proper guides in theory choice are experimental data and logic, then theories will remain underdetermined. In practice, there are of course further criteria which scientists use in theory selection. Here one might cite simplicity, unity, elegance, beauty, and relation to other theories, particularly at a higher level.[12] Unfortunately this is not a straightforward matter as it is not easy to capture exactly the sort of simplicity or unity that ought to be employed, nor do straightforward accounts of simplicity or unity give a good description of scientific progress. The nature and status of these criteria is then a matter of considerable debate.

II. The Duhem-Quine thesis

There is another form of underdetermination, known as the Duhem-Quine thesis. It is sometimes held that while Hume has shown that trying to prove a theory with positive instances is futile,[13] the Duhem-Quine thesis shows that trying to disprove a theory with negative instances is similarly doomed. The central tenet of the Duhem thesis is:

> An experiment in physics can never condemn an isolated hypothesis but only a whole theoretical group.[14]

If the group is falsified, we do not know which theory in the group to reject and which to retain, and so theory choice will again be underdetermined. Quine too is a holist in this sense:

> The failure of a prediction falsifies only a block of theory as a whole, a conjunction of many statements. The failure shows that one or more of those statements is false, but it does not show which.[15]

As I argue that Plato believes he has a solution to the problem of positive instances, I shall not be greatly concerned with negative instances in what follows. The problem of negative instances is specifically a twentieth-century one, largely in response to Popper's falsificationism, attempting to show that no theory can be conclusively falsified. However, Gillies has argued that what is often taken as the Duhem-Quine thesis are actually two separate theses, and Quine's views are distinctly more radical.[16] Duhem is concerned only with physics, is concerned with groups of theories, and offers an account of scientific good sense.[17] Quine is concerned with the whole of science, believes the 'unit of empirical significance to be the whole of science'[18] (and not individual theories or Duhem's groups) and offers no systematic account of theory choice in the face of underdetermination. In support of underdetermination, Quine notoriously held that:

> Any statement can be held true come what may, if we make drastic enough adjustments elsewhere in the system.[19]

Of course anyone can hold any statement, but as Laudan has argued, the key matter is whether it is rational to do so.[20] In support of his view Quine suggests that we can arbitrarily redefine terms in the theory,[21] abandon the laws of logic or plead hallucination.[22] Plato has ample resources to deal with the question of the rationality of any of these procedures, especially in a dialectical context. He also has important resources to deal with the resulting position. There are certain parallels between Quine's position and that of Protagoras. Elements of the strategy Plato employs against Protagoras in the *Theaetetus*, most notably the generalised application of theses, including reflexive application, and the consideration of how the resulting positions can deal with the assertion of the negation of the main thesis or themselves can also be deployed against Quine.

At first reading, one might get the impression that Quine is referring only to scientific statements when he holds that any statement can be held true. In fact though there is no restriction on the application of Quine's underdetermination principles and so they can be applied quite generally to what others might distinguish as philosophical and scientific problems. As Quine rejects the analytic/synthetic distinction,[23] gives us the image of 'the totality of our so-called knowledge or beliefs ... is a man made fabric',[24] and indeed explicitly states that 'ontological questions are on a par with questions of natural science',[25] that would have to be so. Indeed, in places Quine is quite explicit about this.[26] So any statement, not just scientific ones, will be underdetermined. This has some unfortunate consequences.

Quine may make comments about simplicity ('total science ... has as its objective the simplicity of laws),[27] conservatism (as to why we 'adjust one strand of the fabric of science' rather than another),[28] and pragmatism (theory choice where rational is pragmatic),[29] which would appear to give him a reasonable position. He appears to have some criteria for choosing between systems despite empirical underdetermination. But these criteria too are underdetermined, if the underdetermination principles apply generally, and so there is no choice between criteria and so no choice between rival systems. Secondly, the underdetermination principles themselves (failure of negation to determine theory, redefining terms, altering logic, hallucinating etc.) are underdetermined (by each other or reflexively). Finally, if x = 'any statement can be held true come what may, if we make drastic enough adjustments elsewhere in the system', then the negation of x can be held to be true, resulting in paradox.[30] The resulting position is indefensible (there are no longer any criteria to distinguish it from any other position). In such a situation, as Plato is fond of pointing out, there can be no experts and no learning, and enquiry is hamstrung.[31] As Plato is well aware from Meno's paradox, and later discussions as well,[32] one has to be very careful not to undermine the possibility of enquiry and learning altogether.[33] Whether Quine can mount a defence against this sort of attack is another matter. I merely indicate Plato's strategy against such positions, to show that it is still viable in modern debates.

This approach is also important because of other modern trends, notably within the sociology of science and the sociology of knowledge. There are those who would argue, either on grounds of outright relativism or on grounds of underdetermination,[34] that as scientists employ non-logical, non-empirical criteria in theory choice, then theory choice should be explained by 'social rather than logical factors'.[35] Thus there ought to be a 'sociologising of epistemology'. Plato would reject this relativism (what is the point of discussion if all theories are equally true?), and so too underdetermination. Epistemology and methodology remain as philosophical rather than sociological matters for Plato. Indeed the reflexive argument pioneered by Plato in the *Theaetetus* remains one of the main weapons in the battle against factual relativism.

III. Meno's paradox

If one wishes to be a realist about science, then one must be able to solve the problems created by underdetermination. Effectively this means being able to justify the use of some non-empirical criteria. Plato, I contend, recognises a similar, although more generally expressed problem, which he initially presents as Meno's paradox. This paradox is set out as follows, at *Meno* 80d:

> *Meno*: How will you search for this thing, Socrates, not knowing at all what sort of thing it is? For what sort of thing that you do not know will be proposed in your search? Or even supposing that you should meet with this thing, how will you know that it is this thing which you do not know?
>
> *Socrates*: I understand the sort of point you make, Meno. But don't you see what a contentious argument you introduce here, that it is not possible for a man to search for either that which he knows or that which he does not know? It is not possible for him to search for what he knows, as he knows it, and there is no need to search for such a thing. Neither can he search for what he does not know, as he does not know what sort of thing he seeks.

In context, the matter under debate is virtue but Socrates' reply to Meno is quite general, and can be taken as a problem for all kinds of enquiry. Meno gives us two problems. How does one begin an enquiry (what do we propose in our search?), and how does one close an enquiry (how do we recognise the answer when we meet it?). One of the problems in each case is that of theory choice. There are a range of theories one might adopt, either as candidate theories at the outset of an enquiry or as final theories at the close of an enquiry. What criteria do we have for theory choice? If we accept Socrates' distinction between knowing (fully) and not knowing (anything), then either we have no need for enquiry or we have (as yet) no criteria for distinguishing between the merits of theories, and so theory choice is radically underdetermined. There are of course differences here with the modern underdetermination problem. Meno's paradox is more general, dealing with all types of enquiry, is not specifically set up as a problem about experimental evidence, and there is no mention of choice from an indefinitely large number of theories. Meno's paradox can also be seen as being regressive. If we do not have the criteria for judging theories, then we must search for them. But how do we begin and end that search? It does though give us the same problem about how to finish an enquiry – what criteria do we use to select our final theory? So too Meno's paradox sets a problem for starting an enquiry, in terms of how we select candidate hypotheses for testing, whether the testing be philosophical or experimental. This is a significant but perhaps underrated problem for modern philosophy of science, particularly where theoretical assumptions need to made in order to facilitate experimental design.

One might argue that modern science is spared some of the more radical consequences of Meno's paradox, as we now know which sort of theory or explanation we are looking for, even if there is an underdetermination problem with the final form. While that may be so for the majority of science, an exception to this may be cosmogony, and in particular the big bang. A good deal of our mathematical physics becomes inapplicable, as the zeros associated with the initial singularity generate infinities,[36] and as neither time nor space are considered to exist prior to the big bang, causal explanations seem inappropriate as well. If there was literally nothing prior to the big bang, it is hard to see what determined it (especially if the laws of physics did not exist). Thus we do not know what sort of explanation to propose or to accept for the big bang. The typical reaction – 'Is that *really* an *explanation*?' – to some modern cosmogony would seem to be a healthy one, and an indication that Meno's paradox still has some bite beyond the more standard problem of underdetermination.

So how does Plato answer Meno's paradox for the investigation of the natural world? In the *Timaeus*, we have seen choices from indefinite fields. So we have the choice of two sorts of fundamental particle, one cosmos, and the slightly less evident choice about the origin of humans. The procedure here is clear enough. We must, in the first instance, hypothesise what we believe to be best, as is made clear at *Timaeus* 54ab in relation to the fundamental particles. At this point the *Timaeus* is quite tentative, and says if anyone can hypothesise better particles, he will be treated as a friend rather than an enemy. The next task is to see how good a theory these particles can generate, whether this theory coheres with others and whether we can have plausible explanations of the phenomena. This I take to be one of the important points of giving an extended discussion of what we can do with the theory of geometrical atomism in the latter part of the

Timaeus. The nature of the good that Plato invokes will be an important concern here, but first I want to look at how Plato's application of the good develops. If we go to the *Meno*, Plato's first answer to Meno's paradox is the theory of recollection.[37] For the purposes of this book, I take this to be a relatively short-lived and uninteresting answer. If we go to the *Republic*, we find a fully fledged epistemology with hypotheses leading to the cognition of the form of the good. The *Republic* perhaps reflects one of the key differences between Plato and Quine. Quine believes that:

> We have no reason to suppose that man's surface irritations even unto eternity admit of any systematisation that is scientifically better or simpler than all possible others. It seems likelier, if only on account of symmetries or dualities, that countless alternative theories would be tied for first place.[38]

Plato supposes just the opposite, that there is one and only one system which is best. Whether he is justified in doing so is another matter, though it is worth noting that at least within modern science this is a matter of contention, as many would hold that there is a unique unified ultimate physical theory, a theory of everything (TOE) which physics is heading towards, and indeed may be relatively close to achieving.[39] Meno's paradox bites here too, as there is considerable debate about what the general form of such a theory would be. Instead of the *Republic* though, I propose to begin with the *Phaedo*. The *Phaedo* has some interesting points to make balance of explanation and ontology, and highlights differences with the physiologoi.

IV. Socrates' autobiography

At *Phaedo* 96a ff., Plato has Socrates recount his experiences of enquiring into why things should come to be and pass away. Socrates first tells of his youthful infatuation and subsequent disenchantment with *peri phuseos historian*, the 'enquiry concerning nature'.[40] He then hears of the works of Anaxagoras, and hoping to find teleological explanations, rushes to read his books, but is disappointed. Unable to find or be taught the sort of explanations he requires, Socrates sets out on his *deuteros plous*, a second voyage or perhaps second best way. The *protos plous*, the 'first voyage' has been identified with the views of the *physiologoi* which interested Socrates in his youth. This interpretation runs into a major difficulty though, as the *deuteros plous* involves both the method of hypothesis and the theory of forms. It would seem unlikely that Plato would have Socrates describe these as a second best to the physiologoi. There are two standard strategies for dissolving this tension. The first is to suggest that the *deuteros plous* description is ironic and so is not to be taken seriously, the second to postulate that the *protos plous* is some unattainable ideal, relative to which the *deuteros plous* is a second best.[41] What I want to suggest instead is that Socrates' autobiography is primarily a discussion of ontology and explanation.[42] The question I believe is being posed is: What is the minimum number of entities we must postulate to have a theory that will do all the explaining we require of it? The central issue is then a balance between ontological parsimony and explanatory richness. The *deuteros plous* may then be a second attempt at this balance, explanatorily richer, but requiring a greater ontological commitment.

One way of proceeding with this programme is to begin with a minimum number of entities, to examine how they are explanatorily deficient, and then to build a richer ontology specifically designed to remedy these inadequacies. This is indeed what we find in Socrates' autobiography. The puzzles which Socrates cannot solve at *Phaedo* 96d ff. may seem a rather odd collection. Undoubtably, part of the function of these puzzles is to help us formulate what is an adequate explanation. Some criteria of adequacy here are that the possession of a property ought not to have two opposite explanations, as when both division and addition are cited as reasons for something being two, nor should a single explanation explain the possession of opposite properties, as when 'a head' is cited as the reason for one person being small and another being large, nor should the explanation cite the opposite of the property. Something cannot be large by virtue of something small.[43] Why, though, does Socrates mention some clearly inadequate explanations in each case? If we are restricted to a parsimonious ontology these may be the only sorts of explanation available. It is also interesting to note the nature of the account of mind that is on offer here. At *Phaedo* 96b, we find that among the questions that Socrates asked himself in his youth are:

> Is it blood with which we think, or air, or fire, or is it none of these? Is it the brain (*ho engkephalos*) which grants the sensations of hearing, sight and smell, are memory and opinion produced from these, and is it from memory and opinion acquiring stability that knowledge is produced?

Note the use of *ho engkephalos*, brain, rather than Plato's usual *he psuche*, mind/soul here, and the external and causal explanation of the acquisition of knowledge. Knowledge here is a state that occurs due to the causal effect of perceptions eventually creating stable opinions in the brain, rather than being due to any activity of the mind. The discussion then takes a detour via Anaxagoras. Socrates is excited by the prospect of teleological explanations on hearing that Anaxagoras believes that intelligence orders everything.[44] He is disappointed though, and at *Phaedo* 98b says:

> My wonderful hopes came to ruin, since proceeding with my reading I perceived a man neither making use of (his) intelligence, nor assigning to it any reason for the order of things, but giving the reason as air, aether, water and many other absurd things.[45]

I take it that Plato has two objections here. First, Anaxagoras gives explanations couched in the same inadequate, entirely physical terms as those of the physiologoi. Secondly, he postulates an entity which does no explaining. Anaxagoras is reputed to have remarked that though all men have intellect not all of them use it.[46] This might be seen as a proto-Ockhamite objection, that Anaxagoras is multiplying entities beyond necessity. This has an important function as a contrast to the lessons of the physiologoi's ontology. The reader has now been armed with the criteria with which to judge any future proposals. Are they too parsimonious and explanatorily deficient, or are they too profligate, with entities that are inefficient or even non-explanatory? Closing the first part of his autobiography at *Phaedo* 99c, and having bemoaned the fact that no one has given him proper teleological explanations, Socrates says:

> They do not truly believe that it is the good (*agathos*) and proper which binds and holds everything together. I would most gladly become anyone's student concerning such a reason and how it prevails; but since I was deprived of this, neither able to find it myself nor to learn it from any other, would you like, Cebes, for me to demonstrate how I worked out and created for myself a second voyage in search of explanation?

What one would expect if Plato is following the programme I have suggested is that some more entities will now be postulated, designed to remedy the deficiencies of the views already examined, and this that we find right at the beginning of the *deuteros plous*. At *Phaedo* 100b Socrates says:

> I hypothesise there to be something beautiful itself by itself and similarly a good (*agathos*) and a large and all the others.

That these new entities are designed to help solve some puzzles is uncontroversial. At 100c ff., Socrates goes through the difficulties introduced at 96d ff. and shows how we can know have some adequate explanations which cite forms as explanatory entities.

Now consider the relation of the *deuteros plous* to the proposed *protos plous*. The *deuteros plous* is more profligate of entities than the *protos plous*, but is also more explanatorily sufficient. There is a sense in which it is a second best relative to the *protos plous*, in requiring greater ontological commitment, but also a sense, in terms of explanatory adequacy, in which it is better. The *deuteros plous* is also secondary in terms of exposition in terms of the programme that I suggest Plato is engaged in here, and secondary in terms of the chronology of the metaphor of Socrates' autobiography, without necessarily being a second best.

We can also consider the *deuteros plous* as an epistemological metaphor. One meaning of the phrase is to take to the oars when there is no wind to move the ship.[47] Thus when the easy methods which we hope will bring us to our destination without doing any work ourselves fail, we must make headway by our own active efforts, 'row our own boat' if you like. One sense might be that when we cannot be taught, we must actively seek knowledge ourselves. This has deep resonances throughout Plato's work, from the relation of the Socratic elenchus to the experts, to the *Apology*, to Meno's paradox, through to the *Theaetetus* and the criticism of Protagoras. Like using the sails, being taught is easier, but should this fail to provide understanding, then we must take to the oars and play an active part ourselves. Note the relative ontological commitments here. Being taught requires only a passive account of the brain and a causal account of the acquisition of knowledge, like that of the physiologoi at *Phaedo* 96b, while the *deuteros plous* requires a more active account of mind.

One can also consider this metaphor in the following manner. If sense perception is inadequate for what we require, then we must have recourse to reason, again requiring a more active account of the mind. Doubtless the *deuteros plous* description is ironic, but Platonic irony is not a binary, on/off phenomenon and the dichotomy of the serious Socrates giving us Platonic doctrine against the ironic Socrates who is not to be believed is too simple.[48] Rather, Plato's irony can focus our attentions on ambiguities, forcing us to do some reasoning for ourselves in order to arrive at a consistent account. The *deuteros plous* is presented as simultaneously a best and a second best, and the irony focused on this description forces us to sort out in exactly which separate respects the *deuteros plous* is best and second best. We are led to consider the ontological cost of explanatory adequacy.

V. Teleology and the *Phaedo*

Does the *deuteros plous* give us any teleology? At *Phaedo* 97d Socrates makes it clear in discussing Anaxagoras that he is interested in teleology:

> Considering these things, I was delighted to think I had found in Anaxagoras a teacher of the reason for things in accordance with my own way of thinking. I thought that firstly he would tell me, whether the earth is flat or round, and having told me this, he would explain the reason and the necessity, stating the nature of the better and why it is better for the earth to be as it is.

Earlier, I suggested that Plato hypothesises the entity required for such explanations, the *agathos*, as the *deuteros plous* gets under way. One might argue that if Plato makes no use of this *agathos*, then he is prey to the same objection that he deploys against Anaxagoras, that of postulating an entity which does no explanatory work. However, it may be that just as the puzzles of 96b ff. were given solutions in terms of the newly hypothesised entities, so the puzzle of 97d is resolved at *Phaedo* 108e ff. Socrates says that:

> I am now persuaded that, first, if the earth is round and in the centre of the heavens, then it requires neither air to prevent it falling nor any other necessitation of this sort, but the uniformity of the heaven itself in every way and the equipoise of the earth itself is sufficient to restrain it. For something which is equipoised and is placed in the middle of something homogeneous cannot yield to being moved aside in any way, but will remain steadfast.

That Socrates discusses the shape, position and stability of the earth, the three things mentioned for explanation at 97d ff. can hardly be a coincidence, especially as Socrates also comments that his theory has no need of air to prevent the earth from falling, perhaps a reference back to *Phaedo* 99b where he attacks the vortex and 'kneading-trough' theories:

> One man makes the earth stay in position by means of the heavens, employing a vortex, and another has it flat like the lid of a kneading trough,[49] resting on the air for support. But as for the power by which things are now situated in the best way the could be, they neither search for it nor attribute any special force to it. They believe they will find some Atlas stronger, more immortal and better able to hold everything together, and they do not truly believe that it is the good and proper which binds and holds.

Specific targets here may be Empedocles and Anaxagoras, though the criticism is quite general and Plato may have others in mind too.[50] Concerning the *Phaedo* 108e ff., Sedley has argued that Plato gives a teleological explanation:

> The earth is stable because its stability makes it an integral part of the best possible overall arrangement of the cosmos.[51]

According to Sedley, the teleological reading he derives here is not evident, but Plato has left enough clues for working out the true explanation of the earth's stability. Perhaps the main objection to the idea that the *Phaedo* countenances teleological explanation is that he has Socrates say that he was 'deprived' (*esterethen*) of it, 'neither able to find it myself nor to learn it from any other'.[52] I would agree with Wiggins against Vlastos on the translation of *Phaedo* 99c.[53] The aorist *esterethen* 'I was deprived' of teleology does not prejudice any future venture. The reading Vlastos argues for, that Socrates is for all time deprived of teleology requires a perfect tense, i.e. *esteremai*. So while the *protos plous* was devoid of teleology, that is not necessarily so for the *deuteros plous*, especially if the *deuteros plous* is being more generous in terms of ontology so allowing fuller explanation. What is different about the *deuteros plous*? It appears to be more active. Socrates has failed to find out about teleology by being someone's pupil, now he will try to find out for himself. But does he not try to find out for himself in the *protos plous*? In a sense yes, but it is important to recognise how he tries to do this. When Socrates refers back to his youthful experiences at *Phaedo* 99e he says:

> I was afraid that I might totally blind my soul (*psuchen*), looking at things with my eyes and endeavouring to grasp them with each of my senses.

He then goes on to say that in the new enquiry he 'took refuge in theories', which marks the beginning of the method of hypothesis. So instead of a merely empirical enquiry or being taught (both of which would require only a passive account of the brain, note that *psuche*, soul, returns in place of *engkephalos*, brain, here) we are to have a theoretical enquiry where Socrates will have to do some active theorising, one in which he certainly theorises the existence of the *agathos*, and may in the myth give an indication of the application of the *agathos*. It would hardly be any great surprise then if we need to do some of this work ourselves, to take a more independent and active approach to the text in order to uncover some teleology, or as Sedley puts it:

> [Plato] expects his reader to work to uncover the deeper-seated teleology of the myth.[54]

Such readings are of course impossible to prove, and as Fine has pointed out, it may be the case that Plato is merely offering us an account he sees as preferable to those of the physiologoi, without yet being teleological.[55] Whatever the position of the *Phaedo* here, clearly there are significant developments by the *Republic* and further developments in the *Timaeus*.

VI. Healthy hypotheses

Whether or not the *Phaedo* gives us any teleology, it does have some important points to make about method. If it is the case that Plato advocates a coherence theory of truth in the *deuteros plous*, then this may shed some light on one of the perennial difficulties associated with the method of hypothesis, the meaning of *sumphonein* at *Phaedo* 100a5 ff.[56] Socrates says:

> Therefore this is the way I have begun, hypothesising in each case whichever account (*logos*) I consider to be the healthiest (*erromenestaton*), and whatever seems to me to agree (*sumphonein*) with this, I posit as being true, and whatever does not, as not true, whether concerning the reason for or any other entity.

The problem here is that the logical relation which *sumphonein* denotes will work well for one arm of this procedure, but not the other, whether we take this relation to be entailment or consistency. With entailment, all that is entailed by a true hypothesis is true, but many things not so entailed are also true. Further, as Gentzler points out, the set of propositions not entailed by our hypothesis will contain pairs of negations, and it cannot be reasonable to posit both members of these pairs as false.[57] With consistency, all that is inconsistent with a true hypothesis is false, but many things that are consistent with it are also false. It may also be the case that while B and C are consistent with our hypothesis A, they are not consistent with each other. Socrates also says that he can demonstrate the immortality of the soul using this method. Consistency though will merely leave us with a set of assertions, not a proof.[58] Such considerations have led some to conclude that we cannot read *sumphonein* univocally through Socrates' account.[59] Robinson comments:

> Both our interpretations of the metaphor of accord and discord have run into grave paradox ... There is no third interpretation. We have to choose between consistency and deducibility as the meaning of 'accord'. The better is consistency, as the paradox to which it leads is much less than to which deducibility leads.[60]

However, if Plato has in mind a coherence theory of truth then the problem dissolves. *Sumphonein* simply means 'coheres', and it is definitionally true in a coherence truth system that whatever coheres is true and whatever does not is not true. Where Plato wants to express logical consequence he usually employs *sumbainein*.[61] Gentzler gives the following definition of coherence:

> A proposition P coheres with Q if and only if P is consistent with Q and stands in a suitable inductive or deductive relation to Q.[62]

The broader the conception of that which has to do the cohering, perhaps the easier it is to see how such coherence could work. While it is possible to argue that propositions may cohere, it is perhaps easier to see how accounts might generate coherence. The question then is how we should translate *logos* at 100a4. The narrow translation, given by Robinson, is 'proposition', where I have favoured the broader 'account'. Some points in favour of the 'account' view. If one thinks in terms of the *Meno*, then knowing something means being able to *logon didonai*, generally rendered 'to give an account', which suggests more than a proposition. Similarly, at 101d6 in the *Phaedo*, when one has to give an account (*logon didonai*) of a hypothesis, one would expect more than a single further proposition. It may well be that hypotheses are complex too, involving more than one proposition. If so, then *Phaedo* 101d, where we are meant to examine the consequences of an hypothesis, whether they agree with one another or not makes more sense. While one proposition cannot have contradictory consequences, complex hypotheses can.[63]

Gentzler raises the worry that there will be an infinite set of propositions that are consistent with P but which do not stand in a suitable inductive or deductive relation to P, and so must (unreasonably) be reckoned false,[64] and that:

> Perhaps Plato did not assume a clearly defined decision procedure for restricting them to a manageable number; rather he took it for granted that his audience would have an intuitive, if rough idea of the difference between the logical consequences of our hypotheses that were worth considering and those that were not.[65]

While this is a different problem from formulating hypotheses in the first instance, again one can see how reference to the good might be important for Plato here. The clearest passage on the later decision making process comes at *Epinomis* 991d ff. We are told that there is an important matter, which if conceived correctly is a great boon to men, and if not we can only call upon the gods. To those who learn in the proper manner, every diagram, every system of number and harmony, and the ordered motions of the stars will reveal themselves to be part of a unitary and coherent (*homologian*) whole, as there is a single bond between these things, and we must keep this in mind while investigating. The conclusion is that:

> If one goes about this in some other way, it is necessary to call upon luck, as we say.

One might compare this with other passages, most notably *Gorgias* 503d ff. where those who are properly skilled work with the best in mind, and do not do things at random or produce random effects. Those who do not aim at the best, and are not in possession of the proper skills can only learn by *empeiria kai tribe*, an empirical knack and grinding away, that is effectively by luck and trial and error.[66] We might also look to the *Republic* where knowledge is structured around the good (or perhaps the structure of knowledge/the world is the good), and it is notable that we have the first appearance of the demiurge ordering the heavens in best way possible way at *Republic* 530a. It is also important that the demiurge, the skilled workman god, creates *taxis* and *cosmos* in the universe just as skilled craftsmen (e.g. doctors) create the same attributes in their subject matter. This of course is the key link for Plato. If the world has been structured for the best, then we can have an appropriate epistemology which uses reference to the good in order to overcome such problems as underdetermination.[67] One interesting aspect of this is the contrast of chance and the good in both cosmology and epistemology. Without the good any structure in the universe is a matter of (highly improbable) chance, and if we do not look to the good then learning too will be a matter of chance. If one believes that learning is not a matter of chance, and that there is such a thing as expertise, then we might turn this the other way round. If we want an epistemology which will be able to give us realist answers, then we must suppose the cosmos to have good order. Thus Plato has epistemological reasons, rooted in Meno's paradox/underdetermination for teleology. One might also trace this back to the original Greek idea of the cosmos as an ordered whole that is understandable by human beings. For Plato, the cosmos is an ordered whole because it is teleologically ordered, the accounts of the atomists and physiologoi being inadequate. The cosmos also needs to be teleologically ordered so that human beings can come to understand it. As we saw in Chapter 1, Plato takes the order/chaos analysis to a fundamental level, of the intrinsic as well as the extrinsic order of the constituents of the cosmos. Plato does not betray the original idea of the cosmos, but he does see some of the difficulties associated with it and attempt some solutions. One might see here considerable affinities with modern cosmological anthropic principles.

How much of this does the *Phaedo* have? Here an important question is how to translate and interpret *erromenestaton* at 100a4. Robinson here translates 'strongest', which working within a correspondence theory of truth is fair enough. However, *erromenestaton* also has connotations of 'in good health, vigorous, stout', and I have translated 'healthiest'. Within a coherence theory of truth based on relationship to the *agathos*, this has important connotations, of selecting the 'healthiest' account most likely to cohere with 'for the best' explanations. Perhaps a good example of this is *Phaedo* 108e and the question of the earth's shape and position. Plato does not have Socrates propose any random shapes or positions here, but ones which are likely to produce coherence, 'healthy' hypotheses framed with 'for the best' explanation in mind. If Sedley is correct about teleology in the *Phaedo*, then we see the beginnings of a matching of a teleological structure for the world with an orientation of epistemology towards explanations citing the good. One critical difference between the *Phaedo* and the *Republic* of course is that in the *Phaedo* the *agathos* is just one form among many, and not as in *Republic* the key form, though one must note that at 99b ff. Socrates is critical of the physiologoi who do not attribute any 'special force' to the good, nor do they recognise that it is the good which binds and holds everything together.

VII. Teleology and epistemology

Plato then has epistemological reasons for employing teleology. In order to solve Meno's paradox, and in particular the difficulty of which hypotheses from an indefinite field to adopt for further consideration (how to start an enquiry) and which one to accept ultimately (how to end an enquiry), he gives us an account where 'healthy' hypotheses are selected for their likely coherence and relation to the good. Vlastos is critical of Plato's use of teleology in the following manner:

> Suppose we could have asked the physiologoi – any of them from Anaximander down – the following question: 'In your inquiries into nature, when you have to decide whether or not something is thus and so, would you think it right to settle the issue by arguing, "It would be better, more beautiful, if things were thus and so; ergo, they are thus and so?" ' There can be little doubt that we would have got an emphatic 'No' from each.[68]

This is not quite Plato's position on healthy hypotheses, but the criticism could be easily reformulated to take account of that. If I may suggest Plato's reply, it might be 'Fine. But now tell me how you are going to resolve such matters. How are you going to justify a choice from an indefinite number of possibilities? If you don't have an answer, your position is indeterminate.'[69] This will remain a standard implicit criticism of the atomists and the physiologoi on several issues for Plato. Just how do they make a choice about conceptions of order/chaos, the nature of matter, the behaviour of matter, etc., and can this be justified? To return to some issues from Chapter 1. How can the supposition of a vortex, rather than any other form of motion, be supported? On what grounds does Anaxagoras' nous choose the ordering of the cosmos? There we saw that Plato has teleological answers to epistemological problems in cosmology and cosmology. One point of this chapter has been to explore the grounds that Plato has for his method in approaching certain problems in cosmology and cosmogony.

How does this account relate to modern realist positions in the philosophy of science? Here we need to look more to the *Republic* than the *Phaedo*. One modern view is that the ultimate goal of science is the final theory or theory of everything (TOE). This would sit atop a pyramidal structure of scientific knowledge, with other theories of physics, then chemistry and biology supporting it and being in turn supported by it as part of the structure. Some would argue that we are beginning to have quite a good idea of what such a theory would look like.[70] Certainly it will involve a unification of quantum mechanics and general relativity, and it is hoped that such a theory would predict the observed values of the fundamental physical constants. The ramifications of such a theory might well be mathematically highly complex, but it is thought that the foundations will be conceptually simple and elegant. One might well argue that this is the case for general relativity and modern theories in particle physics. As Einstein famously put it, 'God is subtle, but not malicious.'[71] For this sort of position, there are several criteria for theory selection in order to be able to meet underdetermination. One looks for conceptual simplicity, coherence with other high level theories, and the ability to unify existing theories at a higher level. Proponents of this position would argue that it gives a good description of scientific progress (the standard problem being how science can 'progress' if one incorrect theory follows another) as more general and unified theories supersede theories that are less so approaching the TOE.

We might well compare aspects of the *Republic* with this conception of science. The *Republic* too has a pyramidal structure, with theories interdependent on one another. The *Republic* is of course more general, dealing with all of knowledge rather than just science, but there is a similar emphasis on unification and coherence with higher level theories. Relative to the modern realist position I have sketched, Plato's view, as presented in the *Republic* appears overly reliant on the good and ethical and aesthetical judgements. In later works though the conception of the good appears to be somewhat more technical, involving unity, mathematics and structure. In this context Aristoxenus' comments on Plato's lecture *On the Good* may be of some considerable interest:

> Each came expecting to learn something about the things which are generally considered good for men, such as wealth, good health, physical strength, and altogether a kind of wonderful happiness. But when the mathematical demonstrations came, including numbers, geometrical figures and astronomy, and finally the statement, good is one it all seemed to them, I imagine, utterly unexpected and strange.[72]

Sayre has argued that the comments attributed to Plato in his *Lecture on the Good* can be understood with reference to the views he puts forward in the *Philebus*.[73] It is evident early on in the discussion of the good and the good life that neither can be equated with any one thing exclusively, such as intelligence or pleasure, but must consist of a mixture. As the discussion draws to a close and the good is brought into view, and at *Philebus* 65a Socrates comments:

> If it is not within the power of one idea to hunt down the good, let us grasp it with three, beauty, proportion and truth, and let us say that all these as a unity among the elements of the mixture are most correctly held responsible, and it is on account of the goodness of these that the mixture itself has become good.

At 66a ff. Socrates and Protarchus proceed to rank intelligence and pleasure against the three of this unity. First come 'measure, the mean and what is appropriate' (66a7), second are 'proportion, beauty, completeness, sufficiency' (66b1), third come 'wisdom, intelligence and understanding' (66b6), fourth are 'knowledge, skill and right opinion' (66b8) and finally come the most pure forms of pleasure. Now we know from earlier in the *Philebus* that measure is achieved by the imposition of limit, which is equivalent to participating in unity. So if measure is the primary constituent of the good, then as measure is participation in unity, so in one sense unity might be equated with the good. Gaiser summarises the main features of Plato's principles as preserved by the doxographers as follows:

> The goodness (*arete*) of a thing is shown by its permanence, beauty and form. These qualities depend on order (*taxis, kosmos*); that is, on a well proportioned arrangement of parts within the whole. The basis of order is then unity, and thence unity or oneness is the cause of all good, or the good in itself.[74]

This conception of the good, found in the *Philebus* and the *Lecture on the Good* differs from the position of the middle period, and presents a more technical and mathematical view of the good. When we then talk about the good structure and arrangement of the universe, we are more likely to refer to mathematical notions such as proportion and due measure rather than ethical or aesthetical criteria. The latter may still be relevant though (aesthetical considerations may be important for modern scientists too; there is certainly a disposition to believe that the most beautiful, elegant theory ought to be correct, including theories about how the cosmos is structured). One might here consider that when *Timaeus* describes the construction of the world soul (see *Timaeus* 35e ff.), he does so primarily in mathematical terms of proportion and due measure with other consideration subservient. One would like to be considerably more specific about the increased role for mathematics in investigation and explanation, but unfortunately Plato is very vague here, just giving us theoretical outlines of a greater mathematical orientation. How much of a greater orientation is another question. The curriculum of the guardians is structured to lead to cognition of the good, and contains mathematics, astronomy and harmony,[75] and indeed what is good for the state is that which binds it into one (*Republic* 462b).[76] Perhaps then there is only a change of emphasis, but that in itself may be quite significant.

If this general view of Plato's later conception of the good is correct, then it considerably narrows the gap between Plato's teleology and a realist conception of modern science, which would want to cite unity, simplicity, elegance etc. in order to be able to answer underdetermination. It does not close it, of course, and one would not expect that. It does show though that Plato's teleology is no mere 'metaphysical fairy tale'. Vlastos is also concerned that:

> The insistence on circle and sphere being the most beautiful and most perfect shapes gave undue importance to aesthetics.[77]

This criticism though seems misplaced, if the basis of Plato's theory of the good lies deeper in ideas of unity, comprehensibility and order.

VIII. Duhem and instrumentalism

Duhem has argued that Plato began an instrumentalist tradition of 'saving the phenomena' in astronomy, one which was highly influential in antiquity.[78] Duhem's views have been highly influential themselves, though Lloyd and Musgrave have now seriously undermined many of his claims about later astronomers.[79] Here I shall consider Duhem's views on Plato in detail. Duhem draws on Simplicius' commentary on Aristotle's *De Caelo*, saying that:

> We find the Platonic tradition formulated in the following terms: Plato lays down the principle that the heavenly bodies' motion is circular, uniform, and constantly regular. Thereupon, he sets the mathematicians the following problem: What circular motions, uniform and perfectly regular, are to be

> admitted as hypotheses so that it might be possible to save the appearances presented by the planets? ... The object of astronomy is here defined with utmost clarity: astronomy is the science that so combines circular and uniform motions as to yield a resultant motion like that of the stars. When the geometric constructions have assigned each planet a path which conforms to its visible path, astronomy has attained its goal, because its hypotheses have then saved the appearances.[80]

Saving the appearances is then the end point for astronomy done in this tradition. Duhem contrasts this instrumentalist tradition with another and sums up the differences between the two approaches:[81]

> The hypotheses of astronomy can be viewed as mathematical fictions which the geometer combines for the sole purpose of making the celestial motions accessible to his calculations; or they can be viewed as a description of concrete bodies and of movements that are actually realised. In the first case, only one condition is imposed on hypotheses, namely, that they save the appearances: in the second, the intellectual freedom of the astronomer turns out to be much more limited, for if he is an advocate of a philosophy which claims to know something about the celestial essence, he will have to reconcile his hypotheses with the teachings of that philosophy.[82]

Duhem places Plato firmly in the first of these categories, and considers the position of the physicist to be like the prisoners in the cave allegory:

> A slave to positive method, the physicist is like the prisoner in the cave: the knowledge at his disposal allows him to see nothing except a series of shadows in profile on the wall facing him; but he surmises that this theory of silhouettes whose outlines are shadowy is only the image of a series of solid figures, and he asserts the existence of these invisible figures beyond the wall he cannot scale.[83]

For Duhem, however, such an assertion is invalid. It exceeds the legitimate methods of the physicist, who must remain a prisoner in the cave.[84] Plato does make some comments on the prisoners which would suggest the best they could do would be something like instrumentalism, in particular *Republic* 516cd, where there is competition among the prisoners to predict which shadows will appear next. But for Plato we are not stuck in the cave, and we can, and indeed ought to escape.[85] As commented earlier, if the guardians did not derive some benefit from escaping from the cave, such that they can do better than the permanent inhabitants on their return, they would have no intellectual mandate to govern. As *Republic* 520c says that they will 'see far better than those in that place, and will know each of the images, what it is and what it is of' when they descend, it seems that they will be far better off then the prisoners, to an extent incompatible with instrumentalism. Duhem may be an instrumentalist on the basis of this allegory, but Plato most certainly is not. To make this more specific to astronomy, there is no suggestion in the discussion of astronomy as part of the education of the guardians that seeking the 'true realities' is not part of the task of the true astronomer.[86]

The dynamism of the enquiry in the *Republic* is important as it relates to Duhem's treatment of *hupothesis* and its cognates in the passage from Simplicius is also highly suspect. The Platonic challenge asks for hypotheses, but Duhem's use of 'hypotheses' in the subsequent discussion, and in other works makes it clear that he does not recognise any significant distinction between Plato's use of *hupothesis*, and his own use of hypotheses, and so he can make a case that Plato was an instrumentalist, at least as far as astronomy is concerned.[87] However, Plato's use of *hupothesis*, both in the line allegory and in the *Phaedo* is surely significantly different from Duhem's.[88] For Plato, hypotheses are something we seek to go beyond as part of a dynamic enquiry, something we perhaps even return to destroy as hypotheses. They are certainly not the end point of an enquiry as they are for Duhem, but somewhere near the beginning. So too standard Greek usage, while not as loaded as Plato's use, would also have hypotheses are something assumed for further discussion. So when we come to the Platonic challenge, we have no reason to suppose that we should rest satisfied with our hypotheses of regular, circular and ordered motion, and feel that we ought to go no further. One might add that in the *Phaedo* the aim is to reach *ti hiknanon*, something adequate, and in the *Republic* it is to eventually to come to the *arche anupothetos*, the unhypothesised principle.[89]

According to Duhem, the only constraint on hypotheses should be that they should save the phenomena. The intellectual freedom of the astronomer should not be restricted by having to make his hypotheses conform to some conception of the 'celestial essence'. Is this the case with Plato? Again, surely not. In the passage from Simplicius Plato demands that hypotheses should posit regular, circular, and ordered motion. In view of what Plato says at length in the *Timaeus*, *Laws* and *Epinomis*, it should come as no great surprise that Plato limits hypotheses, and limits them in this manner. These are some quite considerable constraints on the sort of hypotheses Plato will consider. Let us be clear what those are to start with. Duhem translates the passage from Simplicius as:

> Quels sont les mouvements circulaires et parfaitment réguliers qu'il convient de prendre pour hypothèses, afin que l'on puisse sauver les apparances présentées par les astres errants?[90]

That I suggest significantly underplays the constraints. While Simplicius *in De Caelo* 488.23 has Plato asking for motions which are *homalon kai tetagmenon*, 492.3 has him asking for motions which are *homalon kai enkuklion kai tetagmenon*. Presumably *tetagmenon* asks for something more than *homalon*, rather than merely reiterates it, and the idea of *taxis* is an important one for Plato, especially in the *Timaeus* and *Laws*. Here

Laws 898ab is important. First it speaks of regular and uniform motion, and then of motion which is according to *hena logon kai taxin mian ampho*, one reason and order. *Laws* 898b contrasts motion which is not regular or uniform and also motion which is *med' en kosmoi med en taxei mede en tini logoi*, 'not arranged or ordered or according to some reason'. On the evidence of the *Laws* then, it would seem then that Plato asks for regular, circular and ordered motion. At 991e, the *Epinomis* is even more emphatic on this matter. Every aspect of the analysis of the heavens (diagrams, numbers, harmonies) must be brought into one system.

The critical question here is, Are the restrictions which Plato places on hypotheses merely means to make calculations of the motions of the heavens simpler? The answer to this has to be no. Why do the heavenly bodies move in this manner for Plato? Because in addition to their bodies which are 'mainly fire',[91] they have souls and are intelligent, and so choose the appropriate best available motion.[92] As *Laws* 893b ff. makes abundantly clear, high on the list is regular and circular motion. Plato restricts hypotheses to combinations of this sort of motion because of his view on the nature of the heavens. So too Plato wants ordered motions, which I take that to mean that we must consider individual motions in relation to the ordered whole of the heavens. As I will argue in greater detail in a Chapter 5, the model of the heavens which Plato gives us in the *Timaeus* is too simple to account for the known phenomena, because of the insistence on regular, circular and ordered motion. In the following passage Duhem is highly critical of Aristotle, but everything he says here is equally applicable to the conditions for celestial motion which Plato sets:

> There were so many restrictive conditions that he imposed upon the hypotheses of the astronomers, and he would not have hesitated to reject a combination of motions that presumed to dispense with any of them.[93]

It is also important that Plato's idea of *techne* goes beyond instrumentalist limitations. *Gorgias* 500e tells us that the *techne* of medicine investigates the nature of those it treats and the reason for what it does, and *Phaedrus* 270b tells is that it is necessary to analyse the nature of the body (for medicine) and the soul (for rhetoric), if you intend attain the *techne* and not merely have practice and experience (*me tribe monon kai empeiria*).

That the *Timaeus* presents a myth and a likely story is a problem for the interpretation of that work. Quite how we ought treat such an account I shall return to later. Even if we take the likely account at face value though, that will not help the instrumentalist view, on which all accounts which save the phenomena have equal merit, and we adopt the simplest. The likely account of the *Timaeus* is the most likely account, more likely than any other. It is also important to recognise that much of what the *Timaeus* has to say on astronomy and cosmology is repeated elsewhere, most notably in the *Laws*, and indeed the *Epinomis*, without the likely account qualifiers.

The most fundamental matter here, from which much of the above flows, is quite simple. This is that Plato believes he has a solution to the problem of underdetermination for the formation and selection of hypotheses, and is not in the least concerned about underdetermination in relation to negative instances. Nor do I see that Plato would be worried by other anti-realist arguments. Van Fraassen's argument that the merits of explanations are context relative,[94] such that explanatory merit cannot decide between theories. It is open to Plato to argue that relation to the (form of the) good and the structure generated around it as in the *Republic* is the critical context. The historical argument that so many past theories have been superseded that we should not trust current ones fails to trouble many modern realists and was unavailable anyway in the context of fourth-century Greek science.[95] Arthur Fine has argued that any procedure for choosing between hypotheses cannot itself be justified by that procedure, on pain of vicious circularity, and must be justified by something more stringent.[96] Plato has at least two options here. The first would be to argue that his decision-making procedure is derived from something more fundamental, possibly considerations about how to conduct any investigation at all, and here Plato would be less worried about justifying scientific procedure by extra-scientific means than Fine might be.[97] Secondly, given some of the epistemological considerations examined earlier, Plato might well argue that any circularity here is virtuous rather than vicious when we consider the broader picture.[98]

Thus it is hard to see that Plato has any motivation towards instrumentalism. I see no reason to suppose that Plato was an instrumentalist, either in general or in astronomy, or that he began a tradition of instrumentalism in astronomy.

IX. Saving the phenomena

Some doubts have been raised about the nature and authenticity of the challenge that Simplicius ascribes to Plato. Let us look at the full passage. Simplicius tells us that:

> First of the Greeks [to hypothesise regular, ordered, circular motion] was Eudoxus of Knidos, (1) as Eudemus states in the second book of his History of astronomy, and so too Sosigenes having taken this from Eudemus, (2) and as Sosigenes says, Plato had posed the following problem for those engaged in such studies: 'Which hypotheses of regular and ordered motion are able to save the phenomena of the planets?'[99]

Some commentators worry that apparently we have the testimony of Eudemus and Sosigenes for (1), we only have the testimony of Sosigenes for (2).[100] It seems extremely doubtful whether we can set great store by

Simplicius' 'characteristic pedantry' here.[101] At 488.23 the motion asked for is *homalon kai tetagmenon*, but when he gives a recapitulation of the passage above at 492.31 ff., at 493.3 the motion asked for is *homalon kai engkuklion kai tetagmenon*.[102] As Vlastos comments, it may simply be the case that Sosigenes failed to acknowledge Eudemus, or that for stylistic reasons (not wishing to repeat himself) Simplicius failed to acknowledge such an acknowledgement.[103] Even if (2) was not in Eudemus, Simplicius repeats (2) at 492.31 ff. without mentioning either Eudemus or Sosigenes, and raises no doubts about attributing the challenge given in (2) to Plato. Whether Simplicius had access to the works of both Eudemus and Sosigenes is a question here, though almost certainly it is undecidable on the current evidence. That he quotes from Eudemus' *History of Geometry* (at length)[104] and his *Physics* (very often)[105] does not show he had access to the History of Astronomy, and that he quotes extensively from the first two but very little from the third may be good evidence that he did not. His praise of Eudemus' style may also be second hand.[106] Nor do I see that any great mileage is to be had from worrying how Eudoxus was truly the first with these hypotheses if Plato had set such a problem. That Plato set a problem and Eudoxus was the first to attempt a solution is a perfectly sensible reading here.[107] There is some evidence that Plato was not the first to hypothesise regular circular motion, as there is a passage from Geminus which tells us that:

> For the Pythagoreans, who were the first to apply themselves to investigations of this kind, assumed the movements of the Sun, the Moon and the five planets to be circular and uniform. They would not admit, with reference to things divine and eternal, any disorder such as would make them move at one time more swiftly, at one time more slowly, and at another time stand still ... with the imperishable stars it is not possible to adduce any cause of swiftness or slowness.[108]

If we accept this and Simplicius, I would agree with Van der Waerden that the easiest resolution here is to suppose that Eudoxus was first to address the problem as stated by Plato.[109] Depending on what sort of relationship the Pythagoreans envisaged between this model and the actual heavens, Plato may have been the first to believe that the phenomena may be saved in this manner.[110]

This is not to say that there are not legitimate worries about whether Plato actually issued the challenge reported in Simplicius. As Zhmud has argued, there was a tendency to portray Plato as the architect of science which can be traced back to the early Academy.[111] It is odd that if the challenge was in Eudemus that we do not hear about from other sources, and the challenge is not mentioned by either Theon or Geminus. It is possible then that we have only have Sosigenes' testimony, and that he may be embellishing other similar accounts. As Zhmud has argued, there are similarities between Philodemus' History of the Academy and the passage in Simplicius:[112]

> At this time the mathematical sciences (*ta mathemata*) were also greatly advanced, with Plato being the architect of this development; he set problems for the mathematicians, who in turn eagerly studied them.

Whether that shows a great deal is another matter. There is nothing in the challenge which could not have been said by Plato, as problems and hypotheses can be found in the *Republic*, regular, circular and ordered motion in the *Timaeus* and *Laws*. Sosigenes could quite possibly have lifted all the required language from Plato himself, independent of any other tradition. It is unlikely that we can come to a definitive answer on whether Plato actually issued the challenge or not. Unless we can unearth some new authoritative text there will always be some doubts and even if Zhmud is correct, that does not preclude the possibility that Plato issued the challenge and that Sosigenes' report is in fact true.

Some more tractable questions are whether Plato could have issued such a challenge, and whether, given the tenor of his later view on astronomy and cosmology, he was likely to issue such a challenge. So too we can consider the relation of Plato and Eudoxus independently of the challenge. There are some standard objections to Plato issuing the challenge and subsequently influencing Eudoxus. One we have already dealt with, that on the evidence of *Republic* VII Plato rejected or denigrated empirical astronomy and so could not have influenced someone such as Eudoxus. Neugebauer for instance comments that Plato could not have positively influenced Eudoxus:

> His advice to the astronomers to replace observations by speculation would have destroyed one of the most important contributions of the Greeks to the exact sciences.[113]

Two other objections need detailed rebuttal and will be dealt with later in this book. If Plato did not himself advocate regular, circular and ordered motion how could he issue the challenge and so influence Eudoxus? It is clear that in the *Republic* he did not advocate such celestial motion. We can look here either to 530b, where we are told that we cannot expect the periods of the planets not to deviate (*paralattein*) in any way, or to the myth of Er and the intermittent actions of the daughters of necessity. Arguably, that is the case in the *Phaedo* and the *Politicus* as well. However, I shall argue at length in later chapters that this is not the case in the *Timaeus*, and that the *Philebus*, *Laws* and *Epinomis* agree with the *Timaeus*. The later Plato may well have issued this challenge and in doing so influenced Eudoxus.[114] I have argued that Plato as not an instrumentalist, and in the next few chapters we will see how he struggles with the foundations of scientific realism. That might be seen as a barrier to any serious influence on Eudoxus, if Eudoxus was, as some have asserted, an instrumentalist, interested only in a geometric construction of the planets, paths and disinterested in cosmology. Again though, I shall argue that attributing such a position to Eudoxus is quite unwarranted on the evidence we have, which, if anything, points in the other direction. One cannot then argue against the authenticity of Simplicius' report on the grounds that it would have been impossible or highly unlikely for Plato to have issued such a challenge or for Eudoxus to have accepted it. Nor is there anything in Simplicius' report itself that might lead us to have serious doubts over it.

4

Celestial Motion in the *Timaeus*

The main aim of this chapter is to argue that in the *Timaeus* the motion of the heavenly bodies is regular, stable and amenable to precise mathematical prediction, contrary to Plato's view as expressed in the *Republic*. This is important both as a matter of Platonic scholarship, where it has been controversial, and as an episode in the history of science.[1] Plato's belief in the regular, circular and ordered motion of the heavenly bodies arguably sets the stage for astronomy for the next two millennia, so the development of his views is of some interest. It is also significant that we can find this conception of celestial motion in the *Laws* and *Epinomis*, and arguably the *Philebus* as well. If so, then later Plato is perhaps immune to some of the criticisms that might be aimed at the *Republic*. One of the key implications here is that there is a considerable strengthening of the connections between events if Plato gives up the idea of the inherent irregular behaviour of physical entities. The criticism that Plato cannot have exceptionless laws because of this inherent irregular behaviour is then inapplicable for the later works. I also propose an interpretation of the relation between reason and necessity which accords with regular celestial motion and the new ideas about the behaviour of bodies. The change to regular and stable celestial motion perhaps also gives us a new criterion for the relative dating of Plato's works. I shall argue that the last expression of the *Republic* view comes in the *Politicus*, and that those works which share the *Timaeus*' view, the *Philebus*, *Laws* and *Epinomis* might all be dated later (with due caveats on dating). I also argue that it is possible to discern a change in Plato's attitude to the inevitability of political, social and moral degeneration, and that this may be linked to the cosmological changes. I argue that my group of later dialogues disown the inevitable degeneration of the earlier works.

I. Disorderly and degenerating cosmologies

We have seen that according to the *Republic*, the periods of the heavenly bodies deviate and do not conform to regular periods. The *Phaedo* provides a framework which would support similar conclusions. There the myth draws a parallel, which we might term the 'fish's eye' view of the world.[2] Just as a fish might look up and see the sun and stars through the water, and believe the sea to be the true heaven, only to be disabused if it were to put its head above water, so we, living in the hollows of the earth, live in a murkier world than the true heaven. Exactly what lesson we are to draw about the nature of the heavens and of astronomy is unclear, though it may well be that we see an analogue of the blurred and unstable image of the heavens as a fish would see of the world if it were to look up through turbulent waters.[3] If so, then the visible heavens might well defy precise mathematical prediction and the periods of the planets would deviate.

The myth of the *Politicus* is also interesting in this respect. We are presented with a cosmos which, on the face of it, is subject to two cycles. First, there is the reign of Cronus where God guides the rotation of the heavens. When, at the appointed time, he relinquishes his control, the motion of the heavens reverses and the reign of Zeus ensues. It is made clear at *Politicus* 272b that we are currently living under the reign of Zeus:

> You have heard of the life under Cronus, Socrates, but the tale of life under Zeus, that of the current state, you are alive to experience yourself.[4]

The key passage is *Politicus* 273b ff., where the Stranger describes the fate of the world in the reign of Zeus:

> In the beginning it remembered the teaching of its creator and father most accurately, though this eventually dulled. The reason for this was the physical element in its constitution, which had been in it from the earliest stage, and partook of great disorder before the universe came to be in its present ordered condition ... When the world nurtures within itself living things under the guidance of the helmsman, it produces little evil and much good. However, when it becomes separated from him, it fares best during the time immediately after the release, but as time proceeds and it grows forgetful, the old condition of disorder gains sway more and more, and towards the conclusion of time little good and much of its opposite flourishes, and there is danger of the destruction of the world and those in it.

That this change affects the heavenly bodies is attested at *Politicus* 269a. We might also infer this from the Plato's general conviction that the stars have physical bodies and the myth's statement that all physical bodies degenerate. If the cosmos is continually degenerating towards chaos, and as *Politicus* 273b makes clear, this will be a radical and fundamental chaos (unless god intervenes, the cosmos will be 'storm driven by confusion and broken up into an endless sea of unlikeness', *Politicus* 273de), then clearly the periods of the planets will be subject to deviation and the visible heavens will again defy precise mathematical prediction.

One might also argue that in the *Phaedo*, *Republic* and *Politicus* Plato is committed to a degenerating cosmology. The clearest statement of this occurs in the myth of the *Politicus* at 273b ff., but so too the passage at *Republic* 529d ff. can be read as saying that not only do the planets not follow mathematically precise orbits, but that their orbits are also continually decaying. I have translated *paralattein* at 530b2 as 'to deviate' but it is worth noting its resonances as given by LSJ of 'to change or alter a little, esp. for the worse ... to pass aside, turn from the path, to deviate, vary'. Interestingly the key word *paralattein* does not occur in *Timaeus*' account of the heavens, nor in other works (*Philebus*, *Laws*, *Epinomis*) which arguably agree with the *Timaeus*.[5] One might see such a degeneration as an expression of the general principle enunciated at *Republic* 546a that 'All things that come to be must also decay', where it is used to underpin a description of inevitable social and political degeneration. It may not be an accident that Plato mentions Hesiod's theory of metals at *Republic* 546e ff. when discussing his own theory of decay.[6] One can find similar language of corruption and decay in the myth of the *Phaedo*, where at 110a Socrates tells us:

> This earth and the stones and the whole region have been corrupted and eaten away (*diephtharmena ...kai katabebromena*), as things in the sea are by salt water; nor does anything worth mentioning grow in the sea, and practically nothing is perfect, but there are eroded rocks and sand and unimaginable mud and mire, and things are in no way worthy to be compared with the beauties in our world.

One might argue here that as Plato uses the perfect tense one is only justified in assuming that decay has occurred rather than is continuously occurring, though one might counter by saying that is all he requires for his epistemological purposes and the comparison with sea water might suggest a continuous process.

II. The myth of the *Politicus*

How common is the idea of a degenerating cosmos in the ancient world? With literature one could certainly look to Homer and Hesiod. With Presocratic philosophers Empedocles is explicit that we are in a cycle of increasing strife and decreasing love, and certainly at least some and possibly all the cosmoi of the atomists were subject to destruction. With later philosophy, the Stoics held a theory of gradual warming and loss of equilibrium moving towards the cosmic conflagration, and the theme of decay is more evident in Epicurus and Lucretius. Lucretius tells of a farmer who praises the achievements of his forefathers, but:

> He does not realise that everything is gradually decaying and on course for shipwreck, worn out by the long years of old age.[7]

The idea of a degenerating cosmos, or of a degenerating cycle with a cyclical conception of time was then reasonably common in the ancient world. We cannot dismiss the idea that Plato at some stage held a degenerating cosmology on the grounds that such a view would have been unduly pessimistic in the context of other ancient cosmologies, nor if I am correct in my suggestions about the *Phaedo* and the *Republic*, is the *Politicus* alone in Plato's works putting forward a degenerating cosmology.

I have assumed a standard two-phase cycle for the *Politicus* myth, where Brisson and Rowe have argued for three phases.[8] I do not believe that such a move is justified on broader philosophical considerations. Contra Rowe, I don't see that the *Politicus* cosmology is hopelessly pessimistic, compared with other works by Plato, other philosophers or Greek literature. Does the cosmos behave like a rational creature? In many ways, yes it does. It struggles hard to be as much like god as possible, but is eventually overcome by old age and deteriorates towards death. Rational creatures are mortal, even if gods are not, and it is interesting to note that the cosmos of the *Timaeus* is a god as well as a rational creature and does not degenerate. The *Politicus* specifically denies that the cosmos is a god. This consideration is also important in relation to one of the main motivations considering the details of the myth behind the Brisson/Rowe view. At 272e ff., god relinquishes control of the cosmos, and so too the demi-gods let go their appointed spheres of influence. At this point there is a shudder, and great destruction of living things. There is then a period when the cosmos sets itself in order and masters itself and the things in it. It is only somewhat later that the deterioration sets in. A problem with a straightforward reading of the myth, of god's order followed by degeneration, is the chaos left after god leaves and the regeneration before the degeneration. Degeneration ought to follow order, and not chaos, and the regeneration is unexplained. However, if the cosmos is a rational but mortal creature left to its own devices, this is indeed what we ought to expect and there is no need to postulate extra phases of the cycle. While the cosmos is young, it will be capable of such ordering. This is not to say that the mortal cosmos can order itself from any sort of chaos, and certainly not from the 'endless sea of unlikeness' that god has to rescue it from the possibility of, nor even from the state the cosmos is in when god performed this rescue act. The chaos after god relinquishes control is far milder than either of these. There is a shudder, and the death of many living things, but the cosmos is still holding together fairly well. Like any living creature, it cannot order itself from outright chaos, but can to some extent perform repairs on itself and heal its own wounds. Also like any living thing though, it loses this ability as it grows older.

So the second phase begins in some sort of chaos which the young cosmos can put into order. As such ordering is clearly for the best, there will be no problem in having an account of it. As the cosmos gets older though, it begins to lose its faculties, and so is no longer able to sustain the order which it has created, and the cosmos begins to degenerate. As the cosmos begins to age there is no need for a reversal of the cosmos, which might well be thought to bring great destruction with it, as when god relinquished control. It is interesting to compare the *krates* language of

Politicus 273b1 with the *Timaeus*. In the *Timaeus*, if we master our emotions we can become just, if we master our perceptions we can become knowledgeable and indeed like god (as far as possible!). If we look to the account of ageing, in youth our triangles can cut up the incoming ones, in old age the reverse until we eventually die. In the second phase then the cosmos does not actively oppose the rule of god, but imitates it as best it can. In the early stages it does that well, but being mortal, the cosmos cannot do this well for ever. If we can keep the interpretation of the myth simple while coherent, we ought to do so given the simplicity criteria that are mentioned at the beginning of the myth.[9]

I agree with McCabe that it is important that our account of the current phase is such that there can be a debate about the merits of life in each cycle, and that it is plausible that life in the current cycle is better. So too our account of the current cycle must generate some grounds for reconsidering the nature of the statesman.[10] On the accounts of Brisson and Rowe, life in the first cycle would seem preferable, and there seem to be no grounds for the statesman not being a shepherd, as in the first cycle. The theme of independence (*autokrates*) is an important link between the myth and the main body of the work. The cosmos, like the statesman does best when independent and capable of ordering the disparate parts of the cosmos/state into a unified whole. The statesman, however, like the cosmos, is mortal and so cannot do this indefinitely. Here I would agree with Lane that the *Politicus*,[11] in distinction to the *Republic*, lays emphasis on the ability of the statesman to act at the *kairos*, the right moment. If that ability is lost, as both cosmos and any particular statesman will indeed lose it, then we see the unfortunate results. The two-phase cycle then seems to make best sense of the *Politicus* myth, as long as we recognise the cosmos is a mortal rational animal which can regenerate itself and be independent in its youth but degenerates in its old age.

III. Celestial motion in the *Timaeus*

One interpretation of the *Timaeus* has it in agreement with the *Republic* on the question of regular celestial motion. Burkert comments that in the *Republic*, Plato:

> was still of the opinion that real exactitude was impossible in the physical world, so that the true astronomer should not depend on sense perception but busy himself with purely ideal magnitudes and movements. In the *Timaeus*, too, Plato speaks without hesitation of the 'wandering' of the planets.[12]

Certainly there are things to be said in favour of this sort of position. The planets are often referred to by the use of *planes*, 'a wanderer', and its cognates.[13] We might wish to equate this usage of *planes* with that at

Timaeus 48a9, where we are told of *tes planomenos eidos aitias*, the 'wandering cause' which prevents the complete expression of reason in the universe. So while reason might wish the heavenly bodies to move in an orderly manner, the dictates of the compromise with necessity to which all physical entities are subject ensures that they move irregularly. Thus *planes* attributes at least some disorderly motion to the heavens. However, there are considerable difficulties with this view, and I shall argue that celestial motion in the *Timaeus* is entirely regular, and that the *planes* description is used in two separate ways. In the context of astronomy it merely distinguishes the fixed stars from those bodies which move in complex orbits and so 'wander' across the sky, while in other contexts it carries the attribution of disorderliness. First consider the use of *planes* in the context of astronomy. At 34a *Timaeus* tells us that the demiurge gave to the universe:

> a motion proper to its body, that of the seven motions which is best suited to reason and intelligence. Therefore he made it move in a circle, revolving of itself uniformly and in the same place, and he took from it all trace of the other six motions and kept it free from their wanderings (*kai aplanes apeirgaseto ekeinon*).[14]

There is then at least one motion which is *aplanes*. Further, at 40b *Timaeus* distinguishes between 'the stars which do not wander' (*hos aplane ton astron*) (40b4), and the planets 'which turn and as such wander' (*ta de trepomena kai planen toiauten ischonta*) (40b6). This is rather awkward for the position I have sketched, according to which there should be no *aplanes* celestial motion. One might try to argue that the crucial difference between the fixed stars and the planets is that the former are living creatures and the latter are not, but *Timaeus* 38e makes it clear that the planets are alive too. The fixed stars of course have physical bodies, 'created for the most part from fire' (*Timaeus* 40a3). What though of the bodies which do have *planes* motion? Concerning the orbits of the planets, at 39c *Timaeus* tells us that:

> Men do not name them nor investigate the mathematical relations (*summetrountai skopountes arithmois*) between them, so that they do not know that the wanderings of these stars constitute time (*chronon onta tas touton planas*), as they possess immense complexity and form astonishing patterns.

Now, if the wanderings of the planets constitute time, and these wanderings are irregular, then time will be irregular. Thus *planes* cannot attribute irregularity of motion when applied to the heavenly bodies. More likely, it merely distinguishes the mobile hands of the celestial clock (the planets) from the static numerals (the fixed stars). It is illuminating to compare this passage with *Republic* 530a. Both talk of how the motions of the heavenly bodies relate to time, but here the emphasis on the deviations of physical bodies and the consequent irregularity of celestial motion has been lost. Instead, it seems that most men, including perhaps the younger Plato, have been led astray by the immense complexity of the heavens and have not properly investigated the mathematical relationships between the stars. Note for instance the difference between the *summetria*, 'proportions' at *Republic* 530a8 which deviate, and the precision of the *summetrountai arithmois* 'commensurates in number' that we are to investigate at *Timaeus* 39c8. At 529cd the *Republic* emphasises that it is the true realities, those accessible to reason and thought which have 'true number', while the visible heavens are like a diagram. In the *Timaeus*, the emphasis is on the fact that we can apply mathematics precisely to the visible heavens.

If Plato did not at any stage countenance an irregular flow of time, then his conception of what constitutes time may also have changed. In order to judge that celestial time is irregular we need some independent access to absolute time. How this is done is left unspecified in the *Republic*, but it is clear that absolute time there cannot be constituted out of the motions of the planets. It may well be that Plato had not considered these difficulties at the time of writing the *Republic* and had no clear ideas concerning the nature of absolute time or how we might gain access to it. One might compare the clarity of the *Timaeus* on these points, as at 47a we are told that the sight of day and night etc. has procured for us number and the concept of time.[15]

One might try to grasp the nettle here though, and argue that perhaps for Plato, with his view of the imperfection of the physical world, time does not flow in a perfectly regular manner. There is no explicit statement on this matter in the main corpus of his works, though we are told at *Epinomis* 991c that:

> We must grasp the exactness of time, how precisely it marks all the events throughout the heavens.

There are however some points which weigh heavily against the irregularity of time in the *Timaeus*. At 37d Timaeus tells us that the demiurge:

> Intended to create a movable image of eternity, and at the same time as ordering the heavens, he made from the eternity that resides in unity an eternal image moving according to number (*kat' arithmon iousin*), that which we have called time.

So at *Timaeus* 38c we find that:

> From this reasoning and thought by God concerning the generation of time, in order that time should come to be, the sun and the moon and the five other stars, which have the name 'wanderers' (*planeta*), came into being to deter-

> mine and guard the numbers of time (*eis diorismon kai phulaken arithmon chronou gegone*).

Now, although time may be less perfect than that which it is an image of, this does not entail that its flow is irregular. The contrast here between stability in unity and motion according to number is sufficient. All that is required is that time flows while eternity stands still, and the phrase 'moving according to number' would suggest that time flows in an orderly manner. This is reiterated at *Timaeus* 38a9 where we are told that time 'circles around according to number'. It is also worth noting that Aristotle, while he is critical of Plato's views on time, does not comment on this matter, which one might expect if Plato had advocated non-uniform time.[16] A further consideration is the stability of the solar system and the nature of the great year. At 39d *Timaeus* tells us:

> Nevertheless, it is perfectly possible to grasp the amount of time elapsed at the completion of the perfect year, when the relative speeds of all the eight circuits have brought them to a head simultaneously, as measured by the motion of circle of the same and similar. In this manner and for these reasons all of those stars which have turnings on their passage through the heavens were created, in order that this world might be as like as possible to the perfect and intelligent living being in respect of it imitating its eternal nature.

If there is a specific amount of time between the grand conjunctions, then celestial motion must be regular, or we are left with the highly improbable alternative that the irregular motions somehow cancel each other out. In that case, the great year would lose its significance as a sign of the rational ordering of the universe. If the great year recurs, and there is no suggestion in the *Timaeus* that it does not, then celestial motion must be regular and the solar system stable and free from any degeneration. The predictable recurrence of the great year is a sign of the cosmological stability of the *Timaeus*. Certainly one cannot, as Adam does, equate the great year of the *Timaeus* with the periods of God's help and God's abstinence.[17] The great year could be achieved in the reign of Cronus, (there is, however, no mention of the grand conjunction, and it is not the criterion by which that age comes to an end, cf. *Politicus* 269c and 272d), but could only be achieved by chance, if at all, in the reign of Zeus when the planets move increasingly chaotically, and contrary to *Timaeus* 39c ff., the time between conjunctions would be unpredictable.

One might also note the way in which in the *Timaeus* we can make precise calculations about the visible heavenly bodies in distinction to the position of the *Republic* 530a ff. The general idea that the visible heavens are amenable to calculation proliferates throughout the *Timaeus*,[18] as does the idea that events occur in the physical world *kat' arithmon*, 'according to number'.[19] Finally, we need to consider *Timaeus* 47a ff. where we are told:

> God devised and gave to us vision in order that we might observe the rational revolutions of the heavens and use them against the revolutions of thought that are in us, which are like them, though those are clear and ours confused, and by learning thoroughly and partaking in calculations correct according to nature (*logismon kata phusin orthotetos*), by imitation of the entirely unwandering (*pantos aplaneis*) revolutions of God we might stabilise the wandering (*peplanemenas*) revolutions in ourselves.

We can compare this with *Timaeus* 40b, where *planes* and *aplanes* are used of the planets and the fixed stars respectively. Either Plato has made an oversight, or *planes* is being used in two different senses. Outside of its specific use in astronomy of distinguishing planets from fixed stars, *planes* again takes on its sense of disorderliness, but here we are told that the motions of the heavens are not disorderly.[20] It is important to recognise that 47a ff. is not an isolated comment. As I shall discuss in more detail in Chapter 9, the analogy between the world soul, the visible manifestation of which is the heavens and the human soul is a central theme of the *Timaeus* and one that is worked out with some care. *Timaeus* 37a ff. insists that all the judgements of the world soul are true and that the circles of the same and the different move regularly.

IV. Celestial motion in later Plato

One objection might ask how is it that as the physical world is supposed to be an imperfect copy, there can be perfectly regular celestial motion. Here we need to consider the difference for Plato between ideal and uniform motion. If we were to describe the usual Newtonian scenario to Plato, of a body moving in a straight line at a constant velocity unless acted upon by external forces, would he consider such motion to be perfect? Surely not. *Laws* 893b ff. develops a set of criteria for the best type of motion. Some of these criteria are indeed to do with regularity, but others assert the primacy of self-generating circular motion in one place. It is clear from this is that we can remove some of the qualifications for perfectly rational motion without losing regularity or uniformity, as they are necessary but not sufficient conditions.

This appears to be the course that the *Timaeus* takes.[21] The motion of the whole universe is one of uniform rotation entirely free from the other six motions.[22] Consider how the motion of the fixed stars relate to this. At 40a *Timaeus* tells us that while they have translational motion due to obeying the motion of the same:

> With regard to the other five motions, they are motionless and still, in order that each might attain the greatest possible perfection.

One can show the regularity of this translational motion in two ways. First, it is equated with the motion of the same (the motion of the same is equivalent of the motion of the universe as a whole, and from *Timaeus* 34a that must be a regular motion), and secondly, we are told that it consists of only one of the six motions to the exclusion of the other five.[23] Such motion is less perfect than that of the universe as a whole in the hierarchy of motions, but this is achieved by complexity rather than irregularity.[24] So too with the planets. Their motion may be less perfect than that of the fixed stars, but this is because their paths are more complex, being subject to the motion of the different as well as that of the same. One might compare here time, which is a moving image of eternity, yet moves in a regular manner.

The matter of the highly controversial 'contrary power' (*enantian … dunamin*) at *Timaeus* 38d4, which (arguably) applies to Mercury and Venus and (arguably) deals with why these planets are always seen near the sun will be discussed at length in the next chapter. There are some complex matters of interpretive strategy which come out of this issue, the result of which in my view is that regular circular motion emerges unscathed.

The *Timaeus* does not, unlike the *Laws*, specifically condemn the description of the planets as *planeta*, because of the ambiguous use of this term. One point of interest concerning the condemnation in the *Laws* is that it reads like a strongly worded clarification of a matter on which Plato may have felt that he was misunderstood. At *Laws* 821a the Athenian Stranger says to Cleinias:

> My good fellow, we speak falsely nowadays when all the Greeks tell of those great deities, the sun and the moon … We assert that they never follow the same path, along with some other stars, those we have named 'wanderers'.

At *Laws* 822a the Athenian states categorically:[25]

> The usual opinion concerning the sun, moon, and other planets, that they at some time wander (*planatai*), is not the case; precisely the opposite is true. For each of these bodies always travel on one path, and not many, although this may not seem so.

This is entirely in accord with *Laws* 898ab, where the motion that the Athenian Stranger will argue is given to the cosmos by soul is:

> regular and uniform and in the same place and around the same things and in relation to the same things and according to one reason and one order.

One might compare the attitude of the *Laws* on accuracy and calculation with the similar comments in the *Timaeus* and the different attitude taken by the *Republic*. At *Laws* 967a the Athenian Stranger tells us:[26]

> Even at that time a sense of wonderment emerged concerning the heavenly bodies, and those who studied them accurately suspected what is now indeed believed, that if they were soulless and lacking in intelligence, they would not be amenable to wonderful calculations (*logismois*) of such accuracy.

So too the *Philebus* assumes the regularity of celestial motion. That the universe has a soul, and that the years, seasons and months are arranged by intelligence are emphasised at *Philebus* 30a and 30c respectively. One might argue that *Philebus* 30c (and indeed *Epinomis* 983c) are deliberately evocative of *Republic* 530ab and set out to correct the earlier position. *Philebus* 30d tells us that a kingly soul is behind this intelligence, and that there are others which carry out similar functions. That is very much in line with the *Timaeus* position, where the demiurge creates the order of the world, and then has lesser demons, whose embodiment are the heavenly bodies, mark out time.[27] This is important because we are told that pain and pleasure are to be equated with deterioration and restoration,[28] and, in line with the general tenor of the Laws, that it would be quite unfitting for the Gods to experience pleasure or pain. At *Philebus* 31d it is emphasised that pain occurs when the harmony of the living being is disrupted. One might think here of the way in which both the *Philebus* and the *Timaeus* describe the universe as a living being, and the way that the *Timaeus* takes great care to emphasis the harmony that constitutes its body, especially at *Timaeus* 35a-37d. If so then the heavenly bodies cannot suffer any deterioration and hence, like the *Timaeus* but unlike the Republic, their motions ought to be entirely regular.[29] One might also compare the general tenor of the remarks at *Philebus* 28d-33a with those of the *Laws* 967a ff. There is the same emphasis on the order of the heavens and the supposition that the stars have souls, and the same denigration of previous thinkers who thought otherwise. There is also no question that the *Epinomis* advocates regular celestial motion.[30]

V. The Politicus myth and the *Timaeus*

One might also consider some more specific points of difference between the *Timaeus* and the *Politicus*. The *Timaeus* is adamant that the cosmos, once created, can function in a stable manner without any attention from its creator and can only be dissolved by an act of his will.[31] In the *Politicus* though either God must perpetually guide the cosmos or it degenerates of its own inherent nature, and is only saved from sinking into 'an endless sea of unlikeness' by the active intervention of God.[32] While the *Politicus* tells us that the cosmos is mounted on a pivot, the *Timaeus* is adamant that it is not.[33] While in the *Timaeus* the cosmos is now *ageron kai anosos*, 'ageless and unailing', in the *Politicus* God has to make the world *athanaton kai ageron*, 'undying and ageless' in recreating the reign of Cronus after the current cycle.[34] As we have seen, the cosmos of the *Politicus* is a rational animal, but it is not a god and so not immortal, where the cosmos of the *Timaeus* is a god. A further consideration here is the nature of time. If the *Politicus* is operating with the conception of time found in the *Timaeus* then time is going to become increasingly chaotic along with the motions of the planets. As we are living in the degenerating cycle it is more likely that Plato is still operating with the *Republic*'s conception of time.

It would seem then that on cosmological grounds we can separate two groups of works, the *Timaeus*, *Philebus*, *Laws* and *Epinomis* as advocating regular and stable celestial motion, the *Phaedo*, *Republic* and *Politicus* as not. If we assume that Plato made a single move on this question, we might attempt to employ this as one criterion for the order of his works. While this would give us two nice groups, I would reiterate here my reservations about producing absolute orders for Plato's works. There are a couple of objections relating to the *Politicus* myth that we might reject though. One might be concerned that certainly in the *Phaedo* and the *Politicus*, and with the myth of Er we are dealing with material that is specifically introduced as myth. So too one might argue the main thrust of the *Politicus* myth is political and not cosmological,[35] and we do not, for instance, take every detail of the geography of the *Phaedo* myth or the cosmology of the myth of Er entirely seriously or ascribe these views directly to Plato. However, I do not require that we accept all the details of the *Politicus* myth (or those of the *Phaedo* or *Republic*), merely that we take some of the underlying principles which are used to construct it as being indicative of Plato's current views, as we would do with Plato's other myths.[36] The *Politicus* myth certainly shares some assumptions concerning cosmology and time with the *Republic* and the *Phaedo*. It is highly significant that when Plato re-uses the myth of Cronus to make some political points at *Laws* 713b ff., it is stripped of its cosmology, and in its new form fits with the regular and stable celestial motion position of the *Laws*. That we might take in conjunction with *Laws* 822a where Plato might be thought to clarify fully his position on the motions of the heavens, along with other trenchant passages in the *Laws*, as an indication that Plato is making it utterly clear what his new position on cosmology is.[37] The *Laws* draws the sort of political morals one might expect from the myth, which are in line with the Eleatic Stranger's comments at *Politicus* 275b. If the cosmology of the *Politicus* myth is not essential to its political interpretation, then why do we get this cosmology, especially if it is something that Plato does not hold with? It is also important to note the manner in which the myth is introduced. At *Politicus* 268de the Stranger says that the myth he is about to introduce 'may offer some amusement (*paidian*)' and asks young Socrates to:

> Give your full attention to this story, just as children (*paides*) do; for you have not entirely outgrown such games (*paidias*).

We are perhaps not meant to take the myth literally, but there are important points to be gleaned from looking at the structure of what is said, and that, even if the main thrust of the myth is political, must include the considerable amount of cosmology we are given in the myth as well.[38] As McCabe comments,[39] it is not clear why such a detailed cosmology is required to make the political points which some out of the myth, so we might suppose the myth to have come further purpose concerning cosmology or divine teleology.

One final concern might be whether any astronomer could accept the degenerating cosmology of the *Politicus* myth. That might depend on how much deviation is already supposed to have occurred and how quickly the rot is supposed to set in. Ancient Greek astronomy still had to resolve many anomalies concerning the motions of the heavens, which could be equated with the deterioration that had already occurred. The deterioration might also be very slow. Here it is important to emphasise that when at *Politicus* 270a8 we are given a comment on the length of the degenerating period (*pollas periodon muriados*), this is not 'thousands of revolutions' but something rather stronger, literally 'many ten thousands of revolutions', and its most likely meaning is 'countlessly many'.[40] If we make the reasonable assumption that one rotation of the universe is one day (that is the likely implication of *Politicus* 269a coupled with 269bc ff.),[41] then one thousand revolutions will be approximately three years, and ten thousand approximately thirty (or around 'one generation of men'). *Politicus* 269b, 272de, and 274de each make it clear that we are dealing with a timescale that encompasses many generations. Whatever the outcome of any dating considerations here, the *Phaedo*, *Republic* and *Politicus* share the position that the motions of the visible heavens will be irregular, unstable and not amenable to exact mathematical analysis, and that is contrary to the view of the *Timaeus*, *Philebus*, *Laws* and *Epinomis*.

VI. Reason and necessity

One objection here to the notion that the heavens move in a regular manner in the *Timaeus* is the question of the relation of reason and necessity. At *Timaeus* 48a we are told:

> The generation of this universe was the bringing together of a combination of necessity and intelligence. Intelligence ruled necessity by persuading her to lead the greatest part of the things that were generated towards the best. It was in this way, through necessity submitting to sensible[42] persuasion, that this universe was in the beginning brought together. If this was the manner of its generation, then really speaking one must bring in the way in which the wandering *aitia* acts.

What does Plato mean by necessity though, and how is it that necessity can be 'persuaded' by anything? Plato speaks of necessity accomplishing things of chance and disorder.[43] As reason only scores a partial victory over necessity, there is then some residual chance and disorder.[44] Now it seems strange that necessity should be associated with these things. However, one might take both chance and disorder in two separate senses, depending on what they are contrasted with. An event might be said to occur by chance because there is no causal chain that leads to its occurrence, contrasting chance with causal determinism.[45] However, one might also contrast chance with human design. If we were to blindly throw paint at a canvas, it would be mere chance if anything good were the result, though no causal chain need be broken if such an event were to occur.[46] Similarly perhaps, if the necessity that *Timaeus* speaks of produces anything of worth, this is chance in the absence of any teleological design rather than in the sense of any breach of causal connections.[47]

We can take the same sort of approach with disorder. The order that it could be contrasted with might be that of physical law, or that of a teleological arrangement of the phenomena. The ordinary emission of light, for instance, might be lawlike but disorderly (no order to wavelength or direction), relative to the stimulated emission of light and its ordering into a laser beam (ordered in wavelength and direction).[48] An ancient analogue here might be that the principles of military strategy apply to all groups of men, but some groups are well ordered formations while others are disordered rabbles. This is important as Plato often uses *taxis* and its cognates, with its connotations of drawing up into a good military order in these contexts, a good example being *Timaeus* 30a:[49]

> For god desired that all things should be good, and nothing paltry (*phlauron*) as far as was possible, he took over all that was visible, which was not at rest but in discordant and disorderly motion (*alla kinoumenon plemmelos kai ataktos*), and led it into order out of disorder, judging the former to be entirely better than the latter.

If we then take it that the partial victory of reason over necessity is the partial instantiation of teleology, then there is not a problem with perfectly regular behaviour by physical bodies. It may be that the attempt to instantiate completely the good produces a set of conflicting demands which cannot be jointly met. A good example of this might be the question of the human skull, discussed at *Timaeus* 75bc. In order for us to have acute perception, the skull ought to be as thin as possible. In order for it to protect our brains and ensure a long life, it should be as thick as possible. Similarly at 75e there must be some flesh around the skull for the purpose of temperature control, but as little as possible so as not to obstruct perception. If we think of necessity in this manner, then reason can only persuade it as far as logical possibility will allow.

Cornford questions how regular behaviour can be persuaded in any manner. The important thing to recognise here is that the demiurge performs a once and for all action when he orders the world from its primordial chaos. There is then no question of a ghostly hand interfering with causal sequences after the initial ordering.[50] The persuasion of necessity lies in this initial ordering. If the demiurge does not bother to order the matter of the universe, then anything good that occurs in the subsequent development of the cosmos will be purely due to chance.[51] As we have seen, the best that could be hoped for in the pre-cosmic state is that the like-to-like principle will have brought together clumps of like entities, most probably the four elements, and that will not constitute a well ordered cosmos. However, if matter is arranged such that good will result from the law-like development of the initial conditions, necessity is perhaps 'persuaded' by reason to produce something good. So the earth, with its mixture of earth, water, air and fire is created where necessity would have produced merely a clump of earth entirely surrounded by water. *Laws* 889b is the critical cosmogonical passage here, as we saw in Chapter 1§I. We cannot expect a like-to-like principle to bring together unlike things except by the merest chance. So too the elements exist due to *techne*. If the demiurge not only arranges the positions of the matter in the universe but also determines the shape of the fundamental particles, then further constraints enter the picture. If the shape of these particles must be the best possible, as the *Timaeus* insists, and the shape of the particles contributes to the properties that they possess, then the demiurge is saddled with some more constraints on what a reasoned arrangement of the particles may be able to achieve.[52] As Cornford points out, it is important that we do not make the demiurge omnipotent.[53] On my account, the demiurge is limited by causal necessity, which he can persuade only by altering the initial conditions of the system, and logical possibility which prevents him from instantiating conflicting demands of the good.

VII. Implications for causation

A further consequence of this analysis of the *Timaeus* is a strengthening of the connections between events. Sorabji has argued that for Plato physical bodies are incapable of perfectly regular behaviour, and so there can be no question of causal necessities. In certain works, such as the *Republic* and the *Politicus*, this would indeed appear to be Plato's position.[54] Sorabji further argues that Plato distinguished physical from logical necessity more sharply than a modern philosopher would as he associated necessity with the 'wandering cause' in the *Timaeus*, and Sorabji associates this wandering with passages from the *Republic* and the *Politicus* which imply that bodies behave irregularly:

> When Plato speaks of necessity here, he cannot be associating it with perfect invariability. For he states emphatically in the *Republic* (530b) and the

Politicus (269de) that what is bodily cannot always behave in the same way without alteration.[55]

Now, if the arguments of the previous sections have been correct, we can say two important things about regular behaviour in the *Timaeus*. First, contra the *Republic* position, the heavens do indeed move in an entirely regular manner. If we ask why that is so, then we must look to the fact that the heavens are no longer mounted on any sort of pivot and the heavenly bodies are now immortal, intelligent self-movers. Being intelligent they will act for the best and being immortal they will continue to do so, thus giving a solid basis for the ongoing order of the heavens. The gods, as the *Laws* and *Epinomis* emphasise, do not change their minds. I shall return to consider merits of such an approach in Chapter 6. All I wish to establish here is that the heavens move in an entirely regular manner in the *Timaeus*.

The second critical difference with the *Republic* and *Politicus* can be put like this. If we associate necessity in the *Timaeus* with an invariable like-to-like principle, then that principle may, in a regular, necessary and law-like manner, produce chance and disorder, where we contrast chance and disorder with a teleological ordering rather than regular, law-like and invariable behaviour. Thus necessity wanders from teleology but not from regularity. So unlike the *Republic* and *Politicus*, matter is capable of regular behaviour, and indeed persistent regular behaviour even if it is incapable of intelligent behaviour.

Clearly there are some important consequences for Plato's philosophy of science. In Chapter 2 in relation to the *Republic* on the 'deviations' of the heavens, we saw Sorabji remarking that 'This is contrary to the spirit of those who believe that nature obeys exceptionless laws'.[56] That is incorrect for the *Timaeus* both for the heavens, where the fact that the stars have souls means that they will behave intelligently, and for matter in general. In Chapter 7 I will further support this claim about matter by arguing that the basic entities supposed by the *Timaeus*, the two half-equilateral and the half-square triangles, can be seen as having stable properties.

Plato has been accused of reducing physical necessity to logical necessity in the *Phaedo*.[57] Whatever the case in the *Phaedo*, this is not the case for the Timaeus. In relation to one of the key cases for the *Phaedo*, the relation of fire and heat, it is interesting that at 61d the *Timaeus* says:

> Let us consider why we say that fire is hot. We may investigate this by noticing the dividing and cutting effect that it has on our bodies. That it has the property of sharpness is something we nearly all perceive. As for the thinness of its sides and the sharpness of its angles and the smallness of its particles and the speed of its motion, by virtue of all of which it is energetic and divisive, always cutting keenly whatever it encounters, these must be accounted for by recalling the origin of its shape, and it is rather this and no

> other property that divides up our bodies into small pieces and which naturally gives it the quality we call heat and supplies its name.

So is the relation between fire and heat here one of logical necessity? The *Timaeus* postulates different sizes of fire particles. Consider the possibility that our bodies are made of smaller fire particles than the ones we meet. The fire particles we meet will no longer have this effect on our bodies, and so cannot be considered to be hot. If that seems a somewhat artificial example (though the stars have bodies of fire, and the effect of fire on them might be interesting), consider the following. The *Timaeus* gives an analysis of colour.[58] Colour is due to interaction of the outgoing fire particles of our visual stream and incoming fire particles. When the incoming particles are larger then those of our visual stream, the stream is contracted and we see black. When they are smaller, the stream is dilated and we see white. So is there any relation of logical necessity here? Surely not, as there is no logical necessity that our visual stream is comprised of particles of a certain size and no logical necessity that contraction of the visual stream makes us see black.

One might argue that these examples are too strong, as it appears that causal as well as logical necessities are breached if fire is not always hot. That, it would seem, is achieved by making qualities relative to the perceiver. As will be important later in this book, there are two senses to this. There is the sense of the section of the *Theaetetus* which deals with the idea that knowledge is perception, whereby all our perceptual judgements are taken to be correct. There is also though the *Timaeus* sense, whereby qualities may be relative to us, but there is a physical basis both for the qualities and our perception of them, and we may be in error about our perceptions. So (in normal circumstances, and perceiving correctly) we feel hot when our bodies are being dissolved by fire particles. If our mental revolutions are disturbed though we may judge ourselves not to feel hot even when our bodies are being dissolved by fire. A less radical example may be that I may judge (erroneously) that this drink is sweet when in fact it is bitter, in the sense that it is carrying out the physical processes on my tongue associated with bitterness.[59] I take this to be a different case than to the following. If I am ill and my tongue is affected (say it is swollen) and a drink has a different effect on my tongue to usual (what would have the effect of contraction on a normal tongue and so would taste bitter, returns my swollen tongue to normal and so tastes sweet to me). In this case the effect is different but I judge correctly, in the other case the effect is the same but I judge incorrectly.

We need to go a stage deeper for the *Timaeus*' causal necessities, which hold between the entities of geometrical atomism. There is no breach of causal necessities here. The fire particles always behave in the same manner even if they would not be hot to one of the celestial deities. Indeed the *Timaeus* gives us several statements on how the elements interact, and on how large and small amounts of elements interact.[60] These would seem quite straightforwardly to be causal but not logical necessities. So too we might consider the like-to-like principle to be a causal rather than a logical necessity. One has to be slightly cautious here, as the like-to-like principle may describe an effect of the way in which particles in motion interact, rather than be a causal principle in itself.[61] The interactions of the particles would still constitute causal necessities, however.

It would appear that not only is Plato interested in exploring the possibilities of a cosmology which postulates regular behaviour, but is also concerned with explaining why there should be such behaviour. At *Timaeus* 61d above and in the *Timaeus* generally, Plato relies on the geometrical properties of matter to explain regular behaviour, and more strongly, the (causally) necessary connection between events. It is important that for the most basic entities those properties do not change, as will be argued in Chapter 7, and while water may decompose and turn into fire, the properties of the icosahedra of water and the tetrahedra of fire do not change either.

I will postpone discussion of the question of the nature and basis of the intelligent behaviour of the heavens until we have seen a little more of Plato's astronomy and cosmology. Here too Plato seeks for an explanation of regularities, but this time has too explain their apparently intelligent behaviour as well, and so attributes souls to the heavenly bodies. If those souls always act in the best manner they will act in an entirely regular manner, and here Plato invokes a different basis for understanding natural laws.

VIII. Political, social and moral decline

In addition to these changes concerning cosmology and time, it may also be possible to discern a change in Plato's attitude to the inevitability of political, social and moral (PSM) degeneration. At 273c in the *Politicus* myth, the Eleatic Stranger tells us:

> From the one who brought it together, the universe has received only good, but from its previous state it retains in itself and engenders in the living creatures all the harshness and injustice that has its origin in the heavens. When the world nurtures within itself living things under the guidance of the helmsman, it produces little evil and much good. However, when it becomes separated from him, it fares best during the time immediately after the release, but as time proceeds and it grows forgetful, the old condition of disorder gains sway more and more, and towards the conclusion of time little good and much of its opposite flourishes.

That seems to be as clear a statement as possible of an inevitable PSM decline. We might compare this with *Republic* 546a ff. where Plato has Socrates describe the inevitable descent from the ideal state through timocracy, oligarchy and democracy to tyranny. One problem with this description is that it is difficult to judge Plato's attitude here as he deploys the *paidia / spoude* (play/seriousness) contrast. At *Republic* 545d Plato has Socrates say:

> Shall we, like Homer, pray to the Muses to tell us how a faction first began the trouble, and believe that in their tragic style they speak as if in earnest (*hos de spoude*) employing high flown language, though they tease and play with us as if we were children (*hos pros paidas hemas paizousin*)?

One reading of this piece of 'serious play' might be that we should not take the details of each society and the passage from one to another too literally, but should be concerned with the general character types of each society and the tendency for society to degenerate.[62]

We should perhaps distinguish between a deterministic decline, where each stage is of necessity worse than the previous, and one where while there are factors which unchecked will cause a society to decline, conscious human action may, at least temporarily, counteract them. Those factors, as in the *Politicus*, may become stronger with time. Taylor has quite reasonably likened this passage to a medical textbook description of a terminal illness.[63] That describes the typical stages an unchecked ailment will progress through, while in an actual case there may be many complicating factors. Here conscious action by a doctor may slow the decline, temporarily check it or even temporarily improve the patient's condition, without removing the underlying cause of his or her ultimate decline and demise. Plato I suggest is committed to this latter possibility,[64] such that when we look at other works we will have to look for an absence of corrupting factors rather than mere evidence that society can make some temporary progress. Let us now consider the position of the *Laws*. At 769d ff., the Stranger tells us that legislation ought to be written down and duly amended if found wanting in practice, 'If the ordering of the polis is always to improve and to in no way decline.' The implication here would seem to be that not only can there be progress, but that the fate of the city lies in its own hands, rather than being subject to any degenerating forces. One of the key political questions asked by the *Laws* is outlined by the Stranger at 676b ff., where he says:

> Sometimes larger states have come to be from smaller ones, and sometimes smaller from larger, and worse states have come to be from better and better from worse ... We must attempt to understand, if it lies within our power, the reason for such change.

The rest of *Laws* Book III is then devoted to answering this question. Various factors are analysed here; at 688c ignorance and vice are discussed,[65] at 691a, financial acquisitiveness, at 697c a lack of communal spirit, at 704d geographical factors.[66] Yet at no stage in Book III, or anywhere else in the *Laws*, is any explanation in terms of an inevitable degeneration cited or alluded to. Indeed, the point of this investigation seems to be to argue that all of the causes of political degeneration are either avoidable or treatable.

After this discussion the Stranger introduces a cyclical theory of history where after periods of increasing social sophistication mankind is reduced to a primitive and rural existence by severe flooding. There is no sense of an inevitable decline here though, or of a decline alternating with a more positive phase. It is highly significant that in this context that when the myth of Cronus makes a reappearance at 713a ff., it is shorn of any commitment to inevitable degeneration. In the *Laws* then, we do not find any inevitable PSM degeneration of the kind we find in the myth of the *Politicus*. How much of a change that is from the Republic depends on how seriously we take the passage at 545d ff. Even if the *Republic* is not committed to an inevitable decline, there is an important change of tone in the *Laws* which cannot be fully accounted for by reference to the different subject matter.[67]

As for the *Timaeus*, all the passages which concern politics are to be found in the opening myth, and as I have argued, these have various literary and philosophical functions to serve which mean that we should not take them to be indicative of Plato's views. The *Timaeus* does have some interesting points to make about moral decline, however. Our souls, analogously to the world soul, are made up of two revolutions.[68] Our intellectual welfare is related to how uniformly our intellectual revolutions move, which is dependent on whether we take the trouble to master our incoming perceptions, or are mastered by them. Similarly, the key moral injunction for humans concerns the relation of the emotions and the intellect. At 42b *Timaeus* tells us concerning men and their emotions that:

> If they have mastery over these, then they will live justly, but if they are mastered by them, unjustly.[69]

While humans who do not take the trouble to master what impinges on them may become corrupt, there is no sense of any inevitability here, or of any increasing disorder in the world that leads to the corruption of mortals. In contrast to the *Politicus*, where the world inevitably becomes a more corrupting place, and where humans can do little about this, the *Timaeus* appears to provide a level playing field for mortals to determine their own fate. One might usefully compare *Timaeus* 86d ad fin. with the *Politicus* in the way that it analyses moral degeneration as something that is the sole responsibility of the individual soul, and it is notable that at *Timaeus* 87b ff. that treatable moral shortcomings are related to political and social difficulties. Especially important here is *Timaeus* 90d ff., where we are told that the goal in life is to rectify the disordered revolutions in our heads, because by doing so we will attain the best sort of life. I can find no indication that the *Philebus* entertains the idea of any PSM degeneration. It is interesting that the *Politicus* ties together the two concepts of cosmological and PSM stability or degeneration, and that in the later dialogues where we find cosmological stability we find PSM stability too. If the driving force behind such degeneration is cosmological or metaphysical, changes here might explain the move to PSM stability. It is interesting to note that according to Empedocles we are living in the increasing strife/decreasing love part of the cosmic cycle, and that the *Purifications* talks of the moral degeneration which is brought about by this increase in strife.[70]

IX. The implications of a stable cosmology

What I hope to have established in this chapter is that the *Timaeus*, like the *Philebus*, *Laws* and *Epinomis* but unlike the *Phaedo*, *Republic* and *Politicus* advocates a stable cosmology and motion for the visible heavens which is regular, circular and amenable to exact mathematical analysis. Along with that go some important cosmological changes. The cosmos is no longer thought of as turning on a pivot, contra the *Republic* and the *Politicus*, and the cosmos now has an immortal soul, against being soulless in the myth of Er and having a mortal soul in the second phase of the *Politicus* myth. There is also some evidence that the later dialogues have a different attitude to the inevitability of PSM decay, which may be linked to the shift to a more stable cosmology. I do not think we have the means to determine which of these changes is driving the others, though I would point out that these appear to be co-ordinated changes.

Depending one's view of what generated the difficulties in the *Republic*, there may also be an important metaphysical change, of the analogue for how mathematics relates to the world. In the *Timaeus*, the visible heavens are amenable to precise mathematical analysis, where the *Republic* allows this only for that which is accessible to reason and thought and is not visible to the eye. So too the basic entities of geometrical atomism, the scalene and isosceles triangles, are precise in their structure (1, $\sqrt{3}$, $\sqrt{2}$, and 1, 1, $\sqrt{2}$), and it is critical for the way that they fit together that they are.

As we shall see when we look at geometrical atomism in more detail, there is never any suggestion that these are not perfect triangles.[71] So instead of the geometrical objects and their instantiations analogue for the relation of mathematics and geometry to the world, we have a new analogue which allows some real precision. On these bases we have a new argument for the relative dating of Plato's later dialogues, with the usual caveats about such dating.

If the motions of the heavenly bodies are regular and stable, then there are some important consequences both for Plato's relation to the development of Greek science and for interpreting the *Timaeus*. This book will now pursue two general lines of enquiry. First, there is a great deal more to be said about the development of Plato's astronomy and how that relates to the development of Greek astronomy. We need to examine the notorious 'contrary power' of *Timaeus* 38d in some detail, as some argue this runs contrary to the idea of regular circular motion, and this debate raises some important issues of interpretational strategy. We also need to look at the development of Plato's cosmology in more detail, relating it to the development of Greek cosmology, and examining the possible nomological reasons Plato may have for teleology. These will be the tasks of the next two chapters.

A further point here relates to the realism/instrumentalism debate about Plato. Hetherington has argued that as the visible world for Plato is imperfect,[72] citing *Republic* VII, and that 'reality exists not in the visible world but rather the world of ideas',[73] Plato cannot be considered a realist in the orthodox sense. The fact that Plato distinguishes forms and particulars cannot be decisive against his being a scientific realist in a reasonable sense. Many modern scientific realists are realists about universals, and would accept at least some of what Plato might be thought to say about the contrast between universals and particulars. As emphasised earlier, it is also important to treat Plato as a proto-realist rather than to try to assimilate him to modern realism. The important question for scientific realists though is not what is most real,[74] but whether theories about the physical world can be considered to be true or not. Even if that is problematic in the *Republic* because of a theory of the imperfection of the physical world, or some inherent irregular behaviour (and that would need some further argument), it is not a problem any longer in the *Timaeus*.

The second general line of enquiry will be to investigate the consequences that regular and stable celestial motion have for how we view the *Timaeus*. One might object that my position on celestial motion and the relation of reason and necessity attributes too much in the way of regularity and stability to the physical world of the *Timaeus*, and by allowing reason so much sway, renders too much of the physical world comprehensible. I do not argue that the view I have given in this chapter is the only possible interpretation of the relation of reason and necessity, merely that it is one that accords with regular circular motion. What I shall investigate in the final part of this book is whether we can align key passages in the *Timaeus* with the idea of regular and stable circular motion, and produce a viable overall reading. I believe this is possible, and the result is a reading which is rather more epistemologically optimistic than in many accounts.

There are various questions here that will need to be considered. What are the implications of there being exactitude in at least one aspect of the physical world? What is the epistemological status of these stable heavens? It would seem that in the first place we can know that the heavens are stable and in the second that we are able to make some precise calculations concerning the heavenly bodies.[75] Perhaps, then, the astronomy of the *Timaeus* shares some of the optimism of the *Theaetetus* and the *Philebus* towards knowledge of the sensible. The whole account of the world given by the *Timaeus* is presented as a myth or likely story. Is it not then the case that the theory that the heavens move regularly is itself only likely, such that our calculations can only be as likely as that theory? Whatever Plato's reasons for taking this approach in the *Timaeus*, one can certainly point to places in other dialogues where the regularity of the heavens is asserted without the 'likely story' qualifier, most notably *Laws* 822a ff., 893b ff. and 966e ff., and so too throughout the *Epinomis*. There is also, though, the question of flux. There might be two problems here. First, one might worry as the *Theaetetus* does about how things that are in radical flux can be known. Secondly, one might worry how it is that the heavens can have regular motion and be stable if the things that constitute them are in radical flux. If an alternative reading can be established that makes the idea of regular and stable celestial motion sit well in relation to these issues, we will then have to deal with the question of what Plato intended in the *Timaeus*. First though, let us look in some more detail at his astronomy and cosmology.

5

Plato and the Development of Greek Astronomy

In this chapter I argue for two conclusions which do not, at first sight, appear to sit well together. First, I argue that although it shows some important advances on its predecessors, the astronomy of the *Timaeus* is actually quite crude and poor and has generally been overrated by the commentators. Secondly, I argue that the astronomy and cosmology of the *Timaeus* was of paramount importance to the development of these disciplines in ancient Greece. It marks the coming together of two previously disparate traditions, those of observational astronomy and philosophical cosmology, and the beginning of an enormously fruitful tradition in astronomy. In what way then is the astronomy of the *Timaeus* poor? There are highly evident discrepancies between what the model of the *Timaeus* predicts and phenomena well known in fourth-century Greece. So much so that one has to wonder about Plato's competence if we take this to be a literal model of the heavens. Commentators have therefore sought to close the gap between what Plato was clearly aware of and what the model can account for, by arguing that the astronomy is more sophisticated than it first appears. Paradoxically, I believe this makes matters worse because in doing so the model becomes ad hoc and disorganised, losing the emphasis on regular, circular and ordered motion. The harder we try, worse it becomes. I argue that we ought to abandon this approach altogether. It is much better that we allow the discrepancies to stand (there is evidence that Plato was aware of them) and treat the model, as I believe Plato intended, as a prototype. It is a first attempt at accounting for the motions of the heavens in terms of combinations of regular circular motions, and as such it indicates the basic principles one might employ in cosmology and in solving the problems of the motions of the heavens. While the idea of combinations of regular circular motion, especially in conjunction with an offset ecliptic is a major advance in astronomy, it is by no means meant to be the last word. One might add that, if Plato thought that the model of the *Timaeus* did account for all the phenomena, then there has to be a question as to why he issued a challenge to 'all who are engaged on these studies' to find the regular, circular and ordered motions that account for the phenomena. So while the model of the *Timaeus* may be a relatively crude prototype, it marks the coming together of observational astronomy and cosmology and lays the basis for perhaps the most successful programme of research undertaken by the ancients in the science.

I. The *Republic* and the myth of Er

To see how Plato's astronomy advances, we need to compare the astronomy of the *Republic* with that of the *Timaeus*. The *Republic's* myth of Er gives us a picture of the heavens as a spindle (616c ff.), and provides the following information:[1]

Position	**Object**	**Appearance**	**Band width**	**Speed** (vs. stars)
Outermost	Fixed stars	Spangled	Broadest	N/A
2nd	Saturn	Yellow	8th	5th
3rd	Jupiter	White	7th	4th
4th	Mars	Ruddy	3rd	3rd
5th	Mercury	Yellow	6th	2nd =
6th	Venus	2nd whitest	2nd	2nd =
7th	Sun	Brightest	5th	2nd =
Innermost	Moon	Lit by 7	4th	Fastest

The width of the rims of each of the bowls in the spindle is most plausibly interpreted as a reference to the relative distances of the planets, and the speeds are the orbital speeds of the planets. Much of this is interesting and useful. We have an order for the distances of the planets related to their relative speeds. Some of the physiologoi (Anaxagoras, Hippocrates of Chios and Democritus) considered the planets to be 'left behind' (*hupoleipesthai*), such that the planet left behind most is the slowest.[2] Here we have the view, typical in Plato, that the planet 'left behind' most is in fact swiftest in its own motion.[3] We have the moon lit by the sun, and there is at least a recognition of a relation between the orbits of Mercury, Venus and the sun.[4]

Attempts to associate the widths of the bands with the amount of deviation a planet has from the ecliptic, or possible sizes of epicycles fail, and it is far more likely that this is some sort of representation of spacing between the planets, though why these relations are chosen is not at all clear.[5] One might say then that the model of the *Republic* is not much of an advance on that of Philolaus (see Figs 12 and 13).[6] The counter-earth has been abandoned, though so has the idea of a mobile earth. As I argue in the next chapter, however, the Greeks had good reasons for their geocentrism. There is the basis for a decent theory of the phases of the moon, as the moon is now lit by the sun and is orbiting the earth (and not the central fire). But the basic principle of the planets having simple circular motion and all lying in the plane of the earth's equator remains the same.[8]

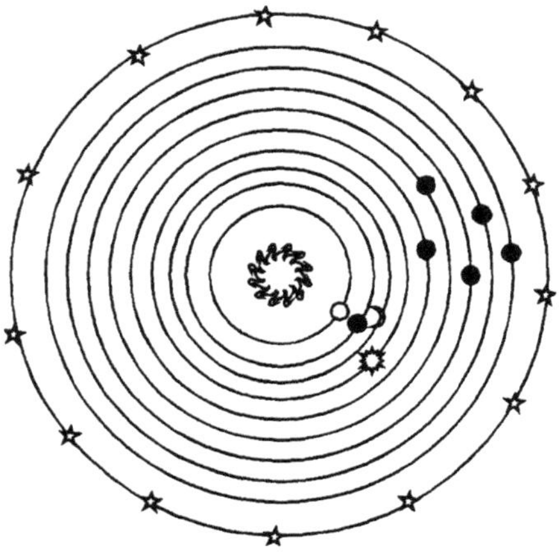

Fig. 12. Philolaus. All the heavenly bodies revolve around the central fire, which is hidden from the earth by the counter-earth. The bodies are (working outwards) the central fire, the counter-earth, earth, moon, sun, the five planets and the stars.

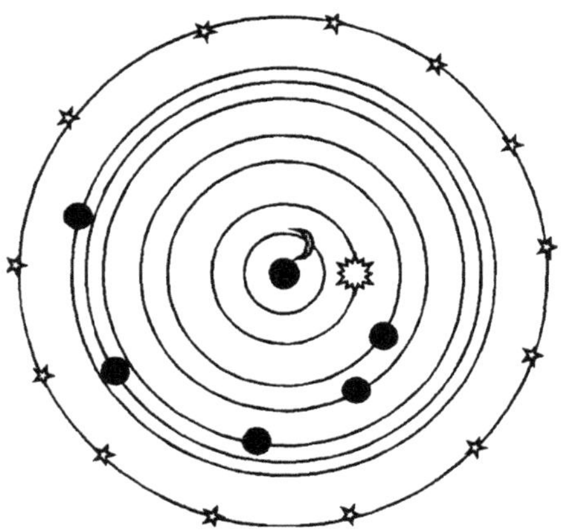

Fig. 13. *Republic*, myth of Er.[7] All the heavenly bodies revolve around the earth, at various constant speeds and distances, all motion again being in one plane.

Does the *Republic* refer to the retrogression of the planets?[9] It is possible that it does. The main question is the meaning of *epanakukloumenon* at *Republic* 617b2 in the description of the motion of Mars. The case for a reference to retrogression is that at *Timaeus* 40c5 *epanakukleseis* almost certainly does refer to retrograde motion.[10] Why though is Mars singled out when all the planets undergo retrogradation? Burnet suggests an emendation of the text, such that *Republic* 617b2-3 would then read *epanakukloumenon malista ton allon*, 'retrogrades more than the others', which is reasonable in astronomical terms as Mars has a greater arc of retrogression than the other planets. Burnet supports this with the fact that Theon of Smyrna has the addition when he quotes this passage, and that:

> In the first place he is apparently quoting from Dercyllides, who first established the text of Plato from which ours is derived. In the second place

> *malista ton allon* is exactly fifteen letters, the normal length of omissions from Plato's text.[11]

Dicks argues that Mars is singled out because it 'makes more than more than one circuit of the heavens between one loop and the next'[12] and so he takes *epanakukloumenon* to mean making an additional circle.[13] Knorr emphasises the phrase 'as it appeared to them' and suggests that this suggest a specific time at which the heavens were observed, when Mars was undergoing retrograde motion.[14] On the very limited account that the myth of Er gives us, it is probably not possible to tell whether this is a reference to retrogression or not. Whether or not the *Republic* refers to retrograde motion, it cannot account for it. The only possibility is the actions of the daughters of necessity. The key passage is *Republic* 617c5-d1:

> Clotho with her right hand grasped and turned the outer revolution of the spindle, leaving intervals of time between, while Atropos with her left hand did similarly with the inner. Lachesis grasped each in turn with each hand.

So Clotho moves the fixed stars and so gives the general motion of the heavens, while Atropos moves the sun, moon, planets slowly in the opposite direction. We might take it as significant that Clotho and Atropos create motion in opposite directions with opposite hands, especially as Plato refers to motion in the heavens as to the right or left.[15] So Lachesis, if we assume that she can touch all the appropriate rims using either hand, could either advance or retard each of them. There is a question though as to exactly what Lachesis grasps. Even if we assume that she only grasps the inner revolutions, so that the stars do not have sudden backwards and (extra) forwards movements, she is indiscriminate between the inner revolutions, that is between the planets, which do exhibit retrogression, and the sun and moon, which do not. I would suggest that the natural reading here is that Lachesis alternately takes inner and outer rings in both hands. Something more sophisticated and discriminating will be required, especially as there is no suggestion whatsoever that Lachesis could generate the variations in latitude observed during retrogression. The actions of the daughters of necessity look more oriented to keeping moving a model which rotates on a pivot and so slows down gradually (as a real spindle would), thus giving the deviations of *Republic* 530b than anything as complex astronomically as retrogression. The *Timaeus* then makes significant advances on the *Republic*. We not only have the ideas of regular and stable circular motion, but also the ideas that the motions of the planets are composites of such motions and that the plane of one of these motions is offset from the earth's equator.

II. The *Timaeus* model

The *Timaeus* makes some important advances on the model of the *Republic*. I shall begin by running through some of the basic phenomena of astronomy, to show some of the advantages of the *Timaeus* model over that of the *Republic*. If one tracks the motion of the sun against the background of the fixed stars during a year, one gets a plot like that shown in Fig. 14. This motion relative to the background of the fixed stars can be represented in various ways. The modern definition of ecliptic is simply the path of the sun relative to the fixed stars. By definition, the sun cannot deviate from the ecliptic. For us the earth-sun relationship is special, as we believe that the earth orbits the sun, and that the moon, but not the other planets, orbits the earth. For the Greeks, though, all these bodies orbit the earth, and so there is no reason to single out the sun for special treatment. So for the Greeks, the zodiac is the band within which all seven heavenly bodies move. For some of the Greeks the sun moves in a simple circle parallel to and in the middle of the zodiac, for others, notably Eudoxus and Callippus,

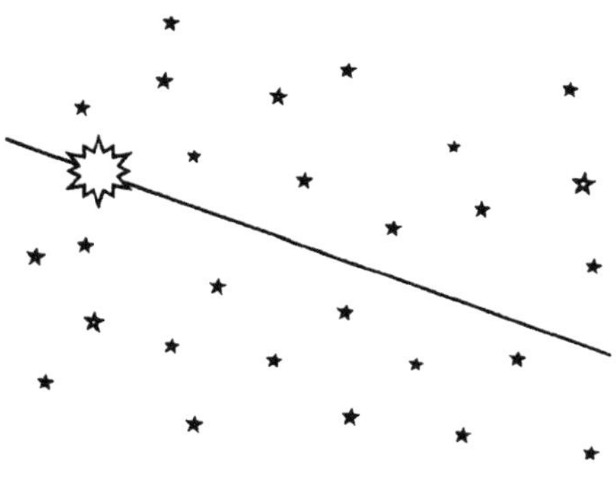

Fig. 14. Plot of the sun's motion. In order to make this sort of plot one observes the point on the horizon at which the sun sets (or rises) and then observes which of the fixed stars rises (or sets) at this point.

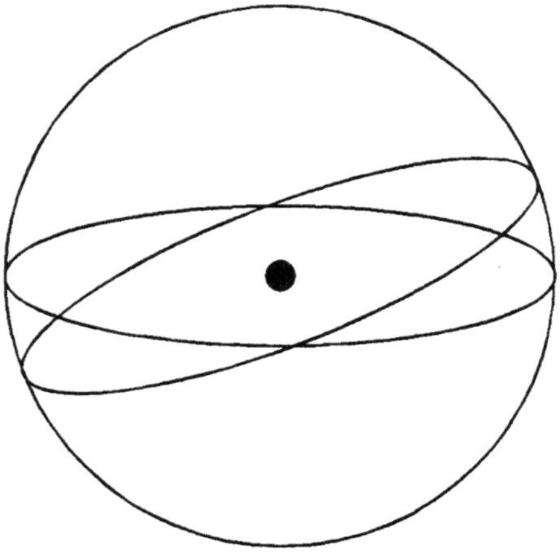

Fig. 15. The sun's motion. In a sort of pseudo-3D, we have the sphere of the fixed stars and the motion of the sun around the earth. The plane of the sun's motion, at approximately $23\frac{1}{2}^{\circ}$ to the earth's equator, is known as the plane of the ecliptic.[16] The standard ancient value was 24° (one fifteenth of a circle). What value Plato used, or whether he used a value, are unknown.

it deviates slightly from such a path. We can represent the motion of the sun and the position of the zodiac as in Fig. 16.

There is a further important phenomenon in relation to the sun which the ancient Greeks were well aware of. The sun does not set (or rise) on the same point on the horizon each day. The point at which it sets moves between a northerly and a southerly maximum. So from winter solstice this point moves north to a maximum of 23½° north of west (inclination of the ecliptic). We then have the longest day and shortest night, that is summer solstice (for the northern hemisphere). The point at which the sun sets then moves south. When it coincides with due west, we have a day of equal day and night, or equinox (autumn equinox). This point then moves to a southerly maximum, which produces the shortest day and longest night, or winter solstice. It then moves north again through a second (spring) equinox to the northerly maximum again. The points at which the

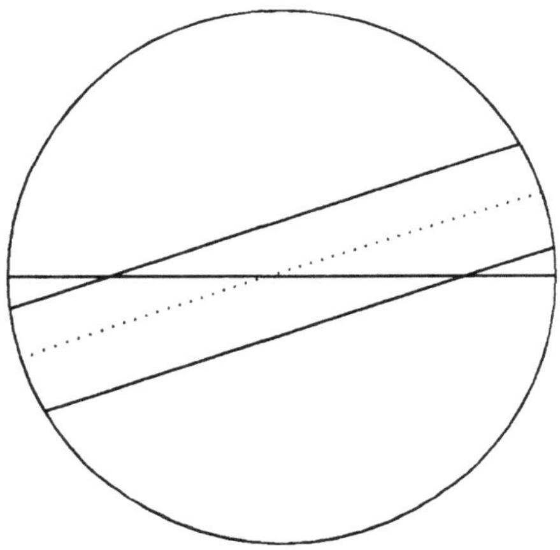

Fig. 16. Sun's motion and zodiac. The Greeks had varying estimates of the size of the zodiac though common at this time was 12°.[17] As will be familiar from astrology, the zodiac can be divided into twelve houses and then significance is attached to the positions of the sun, moon and planets relative to those divisions.

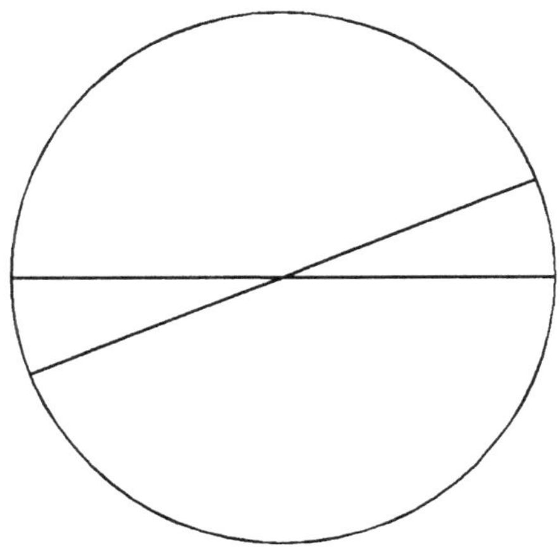

Fig. 17 Simplified motion of the sun. One can simplify Fig. 15 down to two dimensions as here.

sun reaches its maximum northerly or southerly position and then begins to head in the opposite direction were known to the Greeks as *tropai*, turning points.

These are some of the basic phenomena of astronomy, which were reasonably well known to the ancient Greeks. Let us now look at Plato's model in the *Timaeus*. It can be represented as shown in Fig. 18. The circle of the same rotates once every twenty-four hours, and generates the motions of the fixed stars. The circle of the different is set at an angle to the same, and is split into seven, for the sun, moon and five planets. The different generates motion relative to the fixed stars along the ecliptic.

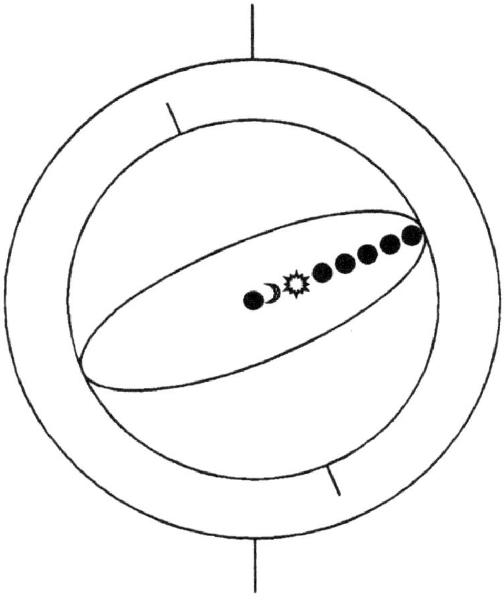

Fig. 18. Plato's two-sphere model. A pseudo-3D representation. Sun, moon and planets move at varying speeds and at varying distances from the earth but all their motions not due to the same are in one plane.

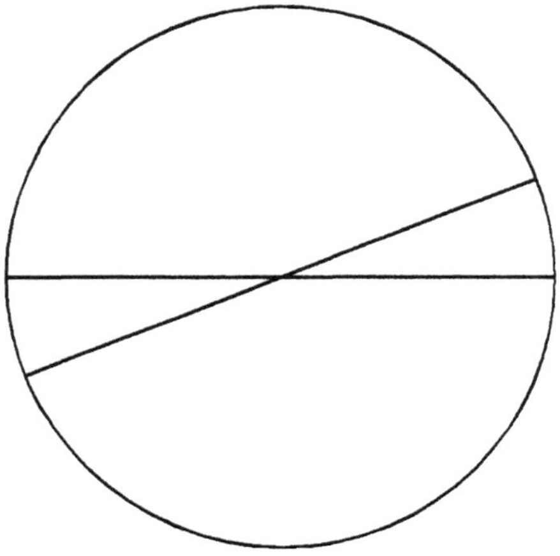

Fig. 19. Simplified representation of the *Timaeus* model. This emphasises that all planetary motion is in the plane of the different which is equivalent to the ecliptic.

Relative to previous models of the heavens, the major advance here is the idea that the motions of the heavens can be resolved into combinations of regular, circular motion offset at angles to each other. If one chooses appropriate orbital speeds, then so far the model looks quite good, and indeed it can give a very good, though by no means perfect representation of the movements of the sun. Clearly this is a major advance on the *Republic's* myth of Er. Although the sun, moon and planets have their own motions, there is nothing to indicate there that these are in any way inclined to the motions of the fixed stars. Indeed such an idea simply will not fit with the image of a spindle. While the Pythagoreans may well have known of the ecliptic, and indeed may well have divided up the earth into the relevant tropical and polar zones, they were not able to introduce an offset ecliptic into their cosmology.[18]

So the *Timaeus* model, like that of the *Republic*, can account (at least in a qualitative manner) for motions of the fixed stars and the periods of orbit of sun, moon and planets. In addition, the *Timaeus* model has the offset ecliptic and so can account for the motion of sun, moon and planets as offset to that of the stars. The model of the sun will also be able to give an account of the seasons, the varying height of the sun, its variations in setting points and the fact that it exhibits *tropai*. These are clearly significant improvements on the model of *Republic* and that of Philolaus. Let us now turn to the problems with the *Timaeus* model.

III. Retrogression and deviation in latitude

I will refer to the view of the *Timaeus* whereby the stars have one regular circular motion and the sun, moon and planets are assumed to have two regular circular motions (that of the stars and their own), with the earth central and static,[19] and no other motions whatsoever, as the 'bare model'. There are serious astronomical difficulties with this model. It cannot account for the retrograde motion of the planets, the deviation of the moon and the planets from the plane of the different, the relation of Venus and Mercury to the sun, the frequency of lunar and solar eclipses, the frequency of occultations of planets and stars, the inequality of the seasons or variations in the apparent size of heavenly bodies. As it would seem that Plato was aware of at least some of these phenomena, attempts have been made to find a more sophisticated astronomy in the *Timaeus* than is allowed by the bare model. An important question in what follows will be whether we should accept the principle that Plato believed that the model of the *Timaeus* could deal with all the astronomical phenomena he was aware of. A second critical matter is this. The bare model is very strong cosmologically, as it adheres strictly to the principle of regular circular motion. In terms of astronomy, however, it is weak in being unable to account for many phenomena that Plato was aware of, indeed more than is generally recognised. In order to give the *Timaeus* a reasonable astro-

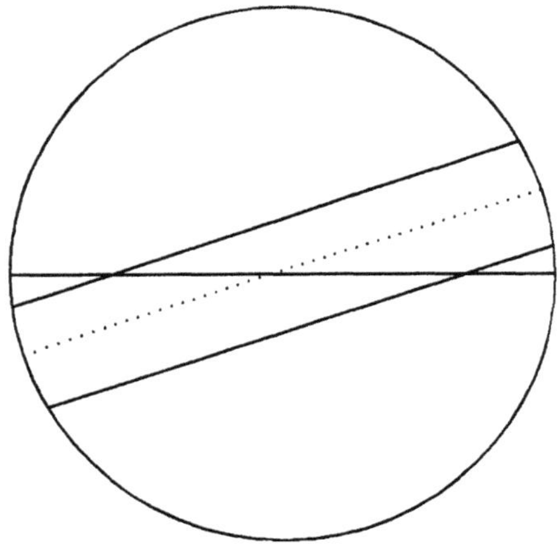

Fig. 20. The zodiac. Here the ecliptic (the path of the sun) is the dotted line, while the two plain lines denote the boundaries of the zodiac, the band within which the planets move.

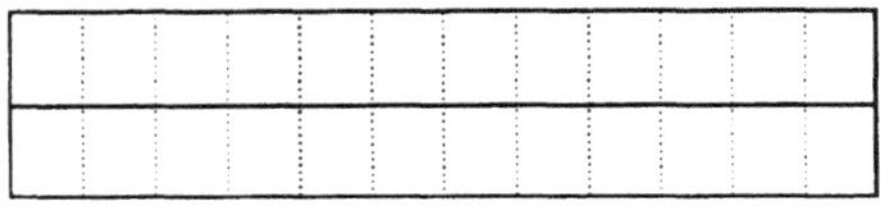

Fig. 21. Simplified zodiac. For simplicity, we can represent the ecliptic and the zodiac as here. We can then speak of longitude as parallel to the ecliptic/zodiac, and latitude as up or down deviation from the line of the ecliptic. The zodiac can be divided into twelve parts, or houses.

nomy in these terms, it is necessary to abandon the bare model and introduce a number of arbitrary motions which are either non-regular or non-circular or both. In what follows I argue that we need to do this to a greater extent than is generally recognised, and this not only compromises the principle of regular circular motion but threatens to wreck it completely. There is then a considerable tension between the astronomical and the cosmological merits of this model, and as we shall see later, this raises some important problems of interpretation both for Plato and for ancient astronomy and cosmology in general.

Let us begin with the questions of retrogression and deviation. We have seen the path that the sun follows throughout a year. The moon and the planets do not follow this path, but deviate to either side of it, and move relative to the fixed stars through a band around the path of the sun known as the zodiac. While the sun's motion is relatively simple, that of the five planets known to the ancient Greeks is not. Not only do they deviate in latitude from the ecliptic, they undergo a phenomenon known as retrogression. They progress relative to the fixed stars, but then become stationary, regress, become stationary and then move on once more.

Planetary motion, and retrograde motion in particular, shows up one of the key differences between geocentric Greek astronomy and modern heliocentric explanations. For us, the planets do not stop and then start

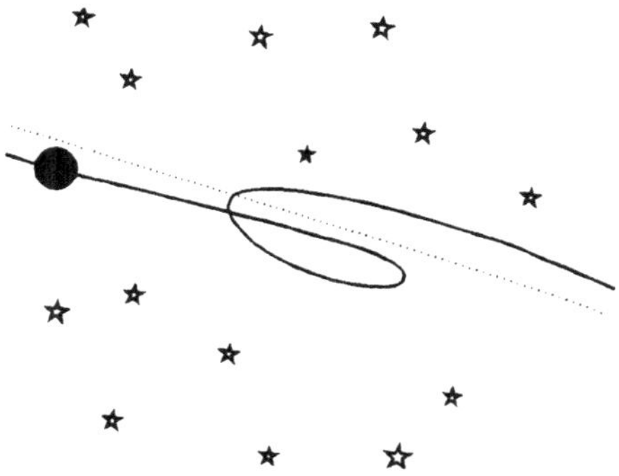

Fig. 22. Retrogression and deviation from the ecliptic. A schematic representation of planetary retrograde motion relative to the background of the fixed stars.

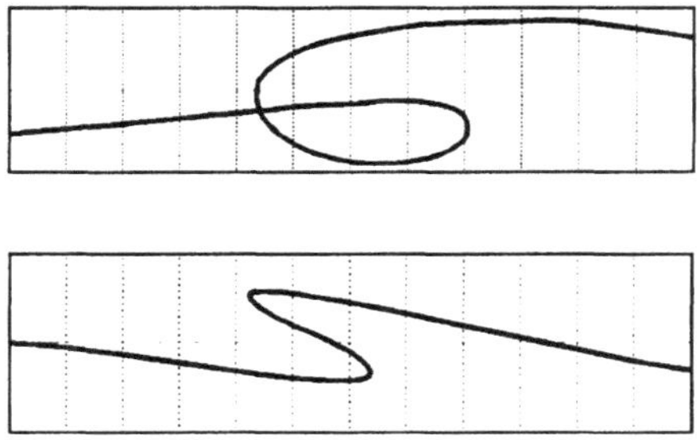

Fig. 23. Patterns of retrograde motion. We can represent this on a zodiac diagram. Note that the 'normal' path of the planet is not along the ecliptic and that retrograde motion involves deviation in latitude as well as regression in longitude. The second diagram gives another pattern of retrogression.[20]

moving again. They appear to do so against the background of the fixed stars because both the earth and the planet are in motion around the sun.[21] They have different sizes and speeds of orbit, and at certain times these factors combine to produce the appearance of retrograde motion for a viewer from the earth. To anyone who holds to a geocentric and geostatic view, however, retrograde motion has to be a real motion of the planets, and so complex schemes have to be devised to account for such motion. The reason that the planets deviate from the ecliptic is that the orbits of the planets around the sun are nearly in the same plane as that of the earth, but not quite.[22]

The bare model of the *Timaeus* cannot account for retrograde motion or deviation in latitude of the planets. The planets have only one motion relative to the fixed stars, moving around the circle of the different at a constant velocity, making retrogression impossible. The planets do not move out of the circle of the different, making deviation in latitude impossible. All the planets should follow the path of the sun according to the bare model, with no need for a zodiac as a band around the path of the sun.

The moon also deviates from the ecliptic, although it does not have any retrograde motion. The reason for this deviation is that the moon's orbit around the earth is not in the same plane as the earth's orbit around the sun. The inclination is quite small, as indeed are the inclinations of the

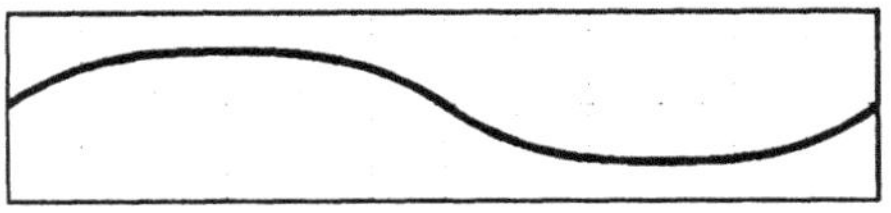

Fig. 24. The moon's motion through the zodiac. Greeks take the moon's deviation in latitude as 6° rather than the modern figure of 5°.[23]

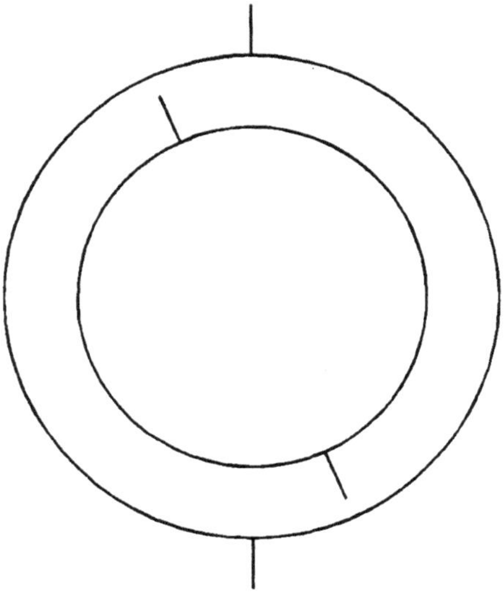

Fig. 25. Plato's two spheres. Here we have one sphere for the motion of the same and one sphere for the motion of the different, so one sphere for the daily rotation and one to generate motion along the ecliptic.

orbits of the planets to the orbit of the earth, which is why the zodiac is a relatively narrow band. It is important to note that the bare model cannot account for this deviation in latitude of the moon either, for the same reason as with the planets. This is of considerable significance as we shall see soon, as it plays an important role in the frequency of eclipses.

Before we move on, let us briefly look at how these matters are dealt with by Eudoxus. Plato's model for sun, moon and planets of the heavenly bodies can be represented as in Fig. 25. Eudoxus' model for the sun and the moon involves three spheres (Fig. 26). Eudoxus' model for any of the five planets is more complex (Fig. 27). As is well known, Eudoxus' model allows for both retrogression and deviation from the ecliptic for the planets. First, let us conceive Plato's model in mathematical terms. One sphere here generates the motion of the fixed stars, the other sphere, with its axis set at an angle to the first, generates motion along the ecliptic. Eudoxus' model for the planets has two extra spheres. Their function is to generate a figure known as a hippopede (Fig. 28). When the four motions are combined, with a suitable choice of axes and speeds, then the resulting motion gives an interesting representation of the motions of the planets. One gets both deviation from the ecliptic and retrograde motion (Fig. 29). While this is far from a perfect representation of the heavens, and there are indeed serious intrinsic problems, it is a major advance over the *Timaeus* model.

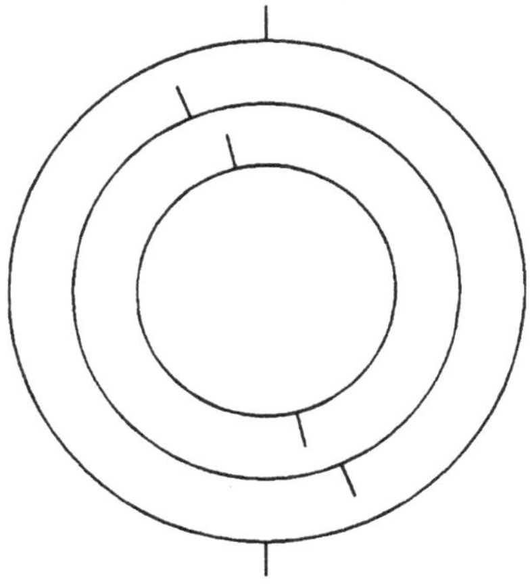

Fig. 26. Third sphere for the sun and the moon. Two spheres as ever generate daily motion and motion along the ecliptic. Another sphere generates some deviation from the ecliptic.[24] The angle at which it is set to the ecliptic gives the maximum deviation.[25]

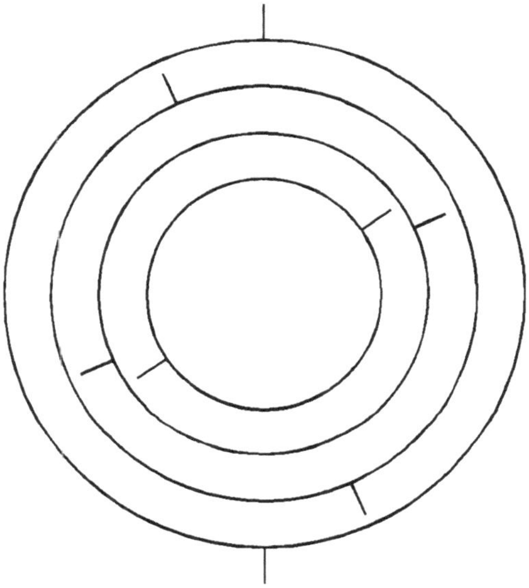

Fig. 27. Eudoxus' four spheres. Here we have the same first two spheres as with Plato, but this time we have two more to generate the hippopede.

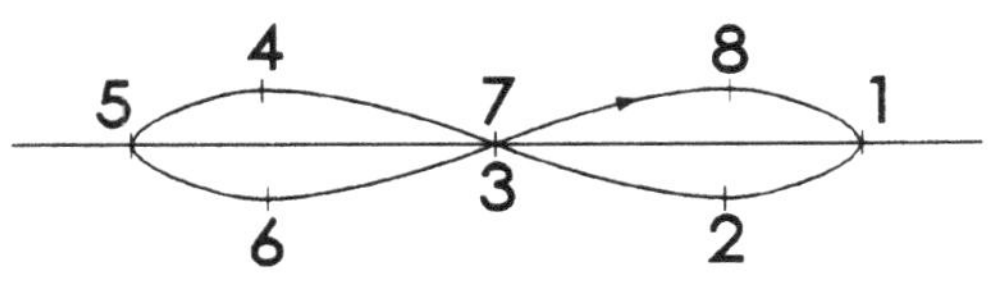

Fig. 28. Hippopede. The possible shape generated by the two spheres Eudoxus adds to Plato's model.[26]

Fig. 29 Resulting motion. The resulting motion of the four spheres then looks like this.

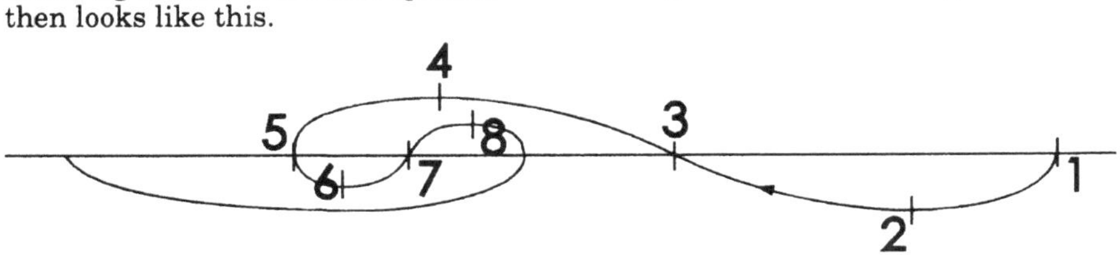

IV. Venus, Mercury and the contrary power

There is a further problem for the bare model which is that it cannot account for the relation of the sun, Mercury and Venus. Mercury and Venus are always to be seen relatively near to the sun in terms of angular separation, while the other three planets (Mars, Jupiter and Saturn) may be seen at any angle relative to the sun. The reason for this in modern terms is that Mercury and Venus are inferior planets, that is they orbit the sun inside the orbit of the earth, while Mars, Jupiter and Saturn are superior planets and orbit the sun outside the orbit of the earth (Fig. 30). The consequence of this is that Mercury and Venus must be seen as relatively close to the sun when observed from earth, while Mars, Jupiter and Saturn can be seen at any angle relative to the sun.

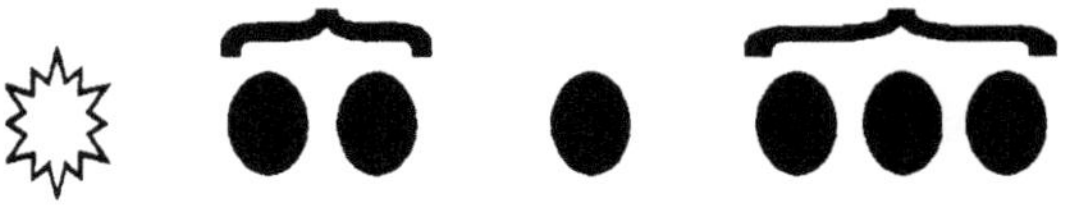

Fig. 30. Inferior and superior planets. Mercury and Venus have smaller orbits than the earth and are known as inferior planets; Mars, Jupiter and Saturn have larger orbits and are known as superior planets.

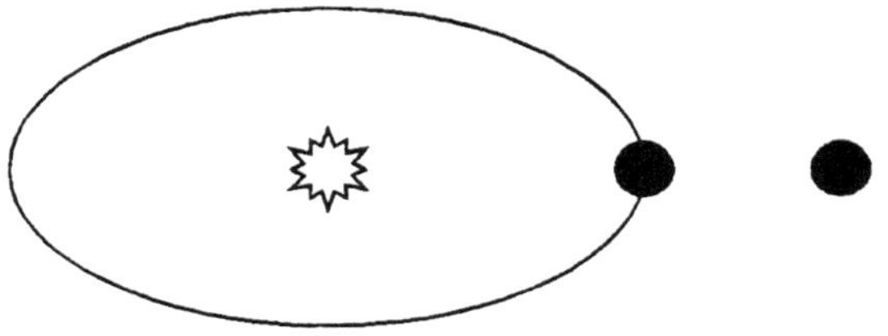

Fig. 31. Possibilities for inferior planets. Mercury and Venus, as they 'pass in front of' the earth, can be seen only at certain angles relative to the sun. Wherever the earth is in its orbit Venus and Mercury will be seen relatively close to the sun.

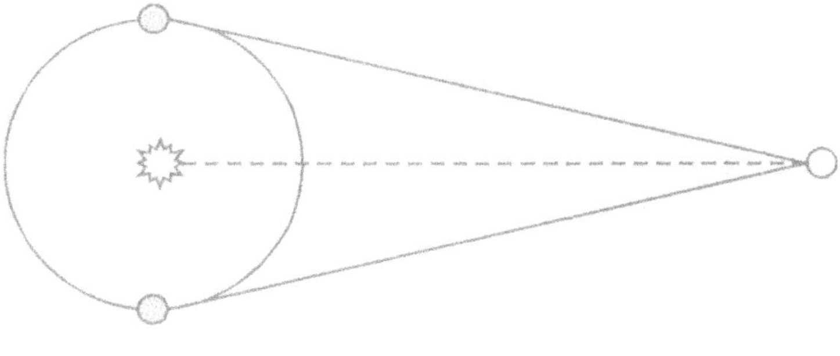

Fig. 32. Maximum angular separation of inferior planets. The maximum value for the angular separation of an inferior planet from the sun can be seen from this diagram. For Mercury it is around 23°, for Venus around 48°.

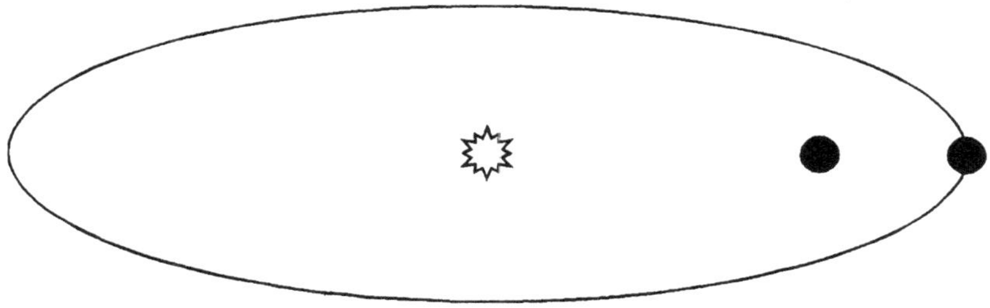

Fig. 33. Possibilities for superior planets. Mars, Jupiter and Saturn, as they 'pass behind' the earth, can be seen at all angles relative to the sun. The problem for any geocentric cosmology is to explain why this should be so, and indeed there was a considerable struggle to do so in a plausible manner (Copernicus was possibly the first to do so).

I do not wish to attribute to Plato a distinction between inferior and superior planets. He does recognise that there is something different about the orbits of Mercury and Venus, though, and that these planets ought to be grouped with the sun. Certainly this seems to be the case not only in the *Timaeus*, but in the *Republic*,[27] and the *Epinomis* as well, at 986e, and at 990b5 ff.:

> The fourth orbit and revolution, and also the fifth, have a speed nearly equal to that of the sun, and neither slower nor greater on the whole. It is necessary that of the three that with the mind most adequate to the task leads.
>
> The moon goes through her orbit most rapidly, bringing firstly month and full moon. Secondly it is necessary to understand the sun, the turnings he brings through the whole of his orbit, and the followers he brings with him.

Mercury and Venus are then always found in the vicinity of the sun. At some times they move more swiftly than the sun through the zodiac, at others more slowly, and can appear on either side of the sun, appearing to overtake it and then be overtaken.[28] As Plato puts it at *Timaeus* 38d:

> The morning star [Venus] and the star sacred to Hermes [Mercury] he placed in an orbit equal in speed to that of the sun, but possessing a contrary power (*enantian ... dunamin*) to it, whence it is that the sun, the star of Hermes and the morning star overtake and are overtaken by each other.

This raises the rather vexed issue of the nature and extent of the *enantia dunamis*, the 'contrary power'. The bare model cannot account for either the Mercury and Venus problem, or retrogression, though Plato appears to be aware of both.[29] As the only extra device we have available is the contrary power, there is a great temptation to seek an interpretation of the contrary power such that it can deal with both of these problems. Yet the contrary power is only explicitly applied to Mercury and Venus. According to Vlastos, this is because Plato is:

> explicitly denying us here an account of the 'others' (i.e. of Mars, Jupiter and Saturn), saying that this he might *perhaps* (*isos*) do adequately (*axios*) at some future time (38d6-e2): declining to go into the particulars of the motions and periods of the outer planets (to no one of which does he make so much as a single individual reference), he has no occasion to speak of their retrogradation.[30]

This I find unconvincing. Merely to say that Mercury and Venus have the contrary power is (notoriously!) not to give much of an account of them, and certainly is not an account of their retrogression (if this what the contrary power is) that cannot be equally tersely applied to the other planets. I have grave reservations about this extension of the contrary power. It is not clear that the contrary power is about retrogression at all. One could have an account of the relation of Mercury, Venus and the sun whereby Mercury and Venus at times progress more slowly through the zodiac than the sun, and so are overtaken by it, without Mercury and Venus ever being stationary or having retrograde motion relative to the fixed stars. That would cover all Plato says at *Timaeus* 38d, and it is notable that there is no reference to retrogression here. Indeed, the vast majority of the motion of Venus and Mercury is progressive. Against this, if there is a means by which a planet's progression can be slowed it is no great leap to create stationary points and retrogression. However, it is also quite possible that Plato treats the relation Venus and Mercury to the sun and retrograde motion as two distinct problems.[31] The fact that Plato uses the phrase 'contrary power' only for Mercury and Venus in relation to the sun and for no other celestial bodies while he has a serviceable phrase for retrogression is indicative that he recognised that there were two separate problems here.

As to the nature of the contrary power, Plato is quiet. One view of this is that Plato merely notes the behaviour of Mercury and Venus relative to the sun and each other but does not claim, on the bare model of the *Timaeus*, to be able to account for such behaviour. I shall return to this possibility later, but first let us look at two other interpretations of the nature of the contrary power.[32]

Knorr proposes that the contrary power refers to a 'periodic oscillating motion', which can account not only for the orbits of Mercury and Venus but for retrograde motion as well.[33] This he links to the roles of the daughters of necessity in the myth of Er, in particular Lachesis who occasionally grasps the rims of the hemispheres and turns them. He finds that:

> The passages we have surveyed in the *Timaeus* offer no explicit support for this proposal even if they are entirely consistent with it; but the *Republic* does.[34]

While this is a highly ingenious proposal, the *Timaeus* is not consistent with the myth of Er. In the myth of Er the cosmos turns on a spindle, and the actions of Clotho and Atropos are intermittent, which would fit well with the *Republic*'s position on celestial motion, but not with that of the *Timaeus*.

For the *Timaeus* model, Clotho and Atropos would be redundant. As for the role of Lachesis, as we have seen, one would have to be extraordinarily generous to get retrograde motion (and no other strange effects) out of Lachesis' actions, as there is no discrimination between which celestial objects she affects or how she affects them. That is a problem whether we consider the contrary power to deal with retrogression, or merely the orbits of Mercury and Venus.

A different approach is that of Cornford, who has the more general idea of self-motions for the moon and planets, the contrary power being one of these self-motions. According to Cornford, as the planets have souls, they can initiate self-motions. So all the planets have an axial rotation, the moon has the ability to go faster than the speed of the different while Mars, Jupiter and Saturn have the ability to go slower, and all of the planets, except the sun and the moon retrogress. In relation to retrogression he comments:

> This is the 'contrary tendency' (*enantia dunamis*) explicitly ascribed to Venus and Mercury, but also shared by Mars, Jupiter and Saturn. It involves variations in the speed of each planet, and intermittent counter-revolution accelerated to the point of bringing the main motion to a standstill and temporarily reversing its sense. (None of these self-motions distorts in any way the circular track of he planet's proper motion. So the planets do not 'stray' from one path to another, *Laws* 821, *Epin.* 982c.)[35]

Cornford constructs what at first sight looks a plausible *via media*. The *Timaeus* model can account for retrograde motion and for Mercury and Venus, and we get a minimal deviation from regular circular motion, as planetary motion remains strictly circular even if it is no longer regular. So the *Timaeus* model can account for all the phenomena Plato is aware of, and the tension between the cosmological and astronomical merits is to some extent resolved. Can this compromise hold? In the next section I argue that there are important phenomena that Plato was aware of which are not accounted for by Cornford's model.

V. Eclipses and occlusions

There is a difficulty with the bare model of the *Timaeus*, and indeed Cornford's model as well, which is not often considered but is critical to a proper evaluation of the *Timaeus*' astronomy. Quite simply, these models will produce far too many eclipses of the sun and the moon, and Plato cannot have failed to be aware that solar and lunar eclipses do not occur each month as these models imply. For these purposes, we can reduce the

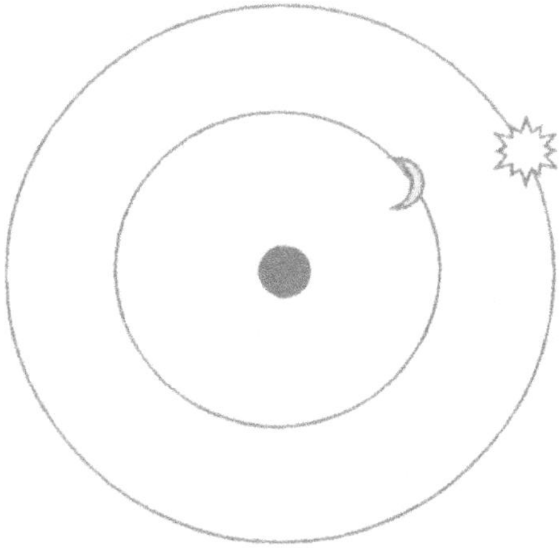

Fig. 34. Earth, moon and sun in one plane.

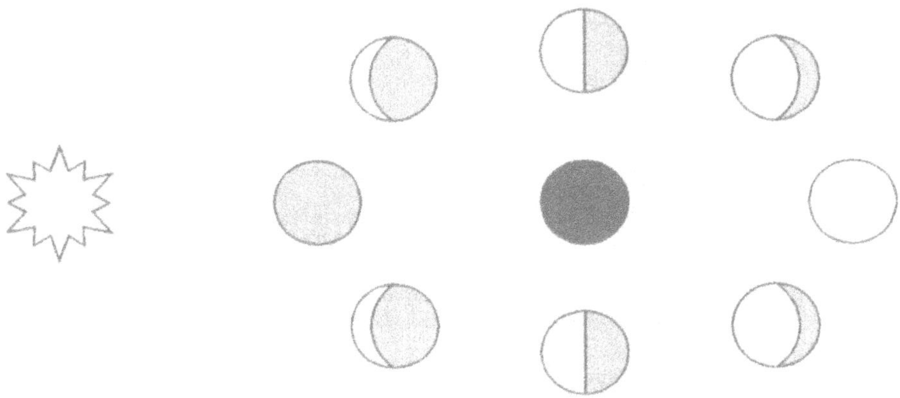

Fig. 35. Phases of the moon.

bare model to one plane, as we only need to consider the relations of the earth, sun and moon (Fig. 34). From this it will be evident that a great number of eclipses and occlusions will occur as sun, moon and all the planets are in the same plane. To understand the reasons for the regularity of eclipses of sun and moon, we need to understand how they are related to the phases of the moon (Fig. 35). As the moon is lit by the sun, how much of the moon we can see is dependent on the relative positions of earth, moon and sun. It should be clear that lunar eclipses can occur only close to full moon, and solar eclipses only close to new moon. They do not occur every new and full moon due to the inclination of the moon's orbit to the ecliptic. The points at which the moon passes through the plane of the ecliptic (the only time when eclipses can occur) are subject to complex change. Eclipses occur when new or full moon coincides with passing through the plane of the ecliptic.

From an understanding of the phases of the moon, it is evident that a solar eclipse can only occur at new moon and a lunar eclipse can only occur at full moon. If the moon's orbit of the earth is in exactly the same plane as the earth's orbit of the sun, then there will be an eclipse of the sun every new moon and an eclipse of the moon every full moon. The basic problem here is that Plato does indeed have the sun and the moon in the same plane, as they both move on the circle of the different. It was known at Plato's time that eclipses occur only at new or full moon, though not why. Empedocles may well have known the causes of eclipses in the sense of the alignment of sun, moon and earth, but nothing suggests that he could explain why they occurred at specific times.[36] In fact the moon's orbit of the earth is rather complex. The plane of its orbit is slightly inclined to the plane of the earth's orbit of the sun, and the points at which the moon passes through the plane of the earth's orbit are subject to change. Only when these points occur on, or very near to the line between the earth and the sun do total, or partial eclipses occur.

This is a rather neglected problem with the astronomy of the *Timaeus*, and none of the proposed more sophisticated models deal with it.[37] Yet the frequency of eclipses was something that was of great importance both for the Babylonians and for early Greek astronomy. One might well argue that eclipses are a more evident and more important phenomenon than planetary retrogression or the Mercury and Venus problem. There is not an easy solution to this problem for the more sophisticated models. Those of Knorr and Cornford have no deviation from the plane of the different, and even if the contrary power could deal with deviation, it cannot be extended to the sun, as it is the sun that other objects have this power contrary to, and there is no textual evidence for extending it to the moon. One is left with supposing some self-motion for the moon to generate deviation.

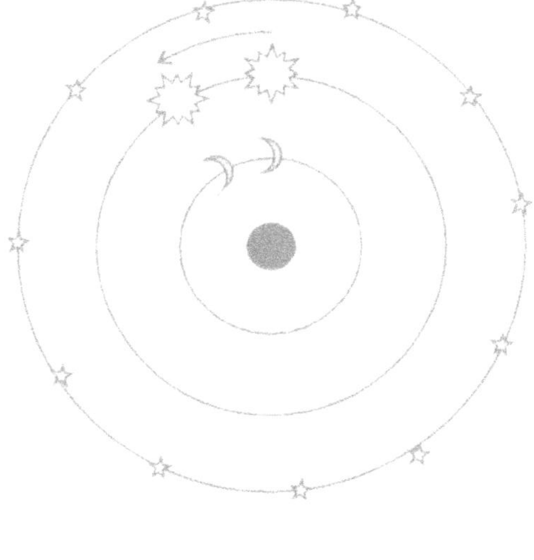

Fig. 36. Moon, sun and month. Month can be defined as either the time taken by the moon to return to the same position relative to the fixed stars or as the time taken to complete a set of phases, i.e. return to the same position relative to the sun. As the sun moves during this time, the latter is slightly longer.

Plato's model of the moon, however, does have its plus points. We do get a correct theory of the nature of total eclipses out of it, even if there is a problem about their frequency. Although Plato does not explicitly give us a theory of the phases of the moon, a solid basis for the correct theory has been laid, and *Timaeus* comments at 39c3-4 that 'Month occurs whenever the moon completes her own circuit and overtakes the sun.' This would seem to indicate that Plato at least recognises that the phase cycle of the moon is related to the relative positions of sun, moon and earth. Certainly that seems to be the case at *Epinomis* 978e.

It is certainly the case that Eudoxus made an attempt to rectify some of the problems here. As we have seen, he added two extra spheres to each of the planets in an attempt to generate retrograde motion. Eudoxus also added one sphere each for the sun and the moon in order to generate some deviation from the ecliptic. Two spheres as ever generate daily motion and motion along the ecliptic. Another sphere generates some deviation from the ecliptic.[38] The angle at which it is set to the ecliptic gives the maximum deviation.[39] Callippus adds two further spheres for sun and moon to account for variations in orbital velocity.[40] If you do not make some addition such as Eudoxus' third sphere, and as with epicycles and hippopedes there is no evidence for this in Plato, either you cannot account for deviation in latitude in general and the frequency of eclipses, or one must suppose some form of non-regular and non-circular motion.

With the bare model, in addition to the problem with eclipses, the moon will produce occultations of the other planets rather too often.[41] So too the planets will produce occultations of each other in sequences, and all the planets will produce occultations of the fixed stars on the same plane as that of the different, but no others. Again, we can simplify this down to one plane (Fig. 37). That such phenomena had been studied is attested by

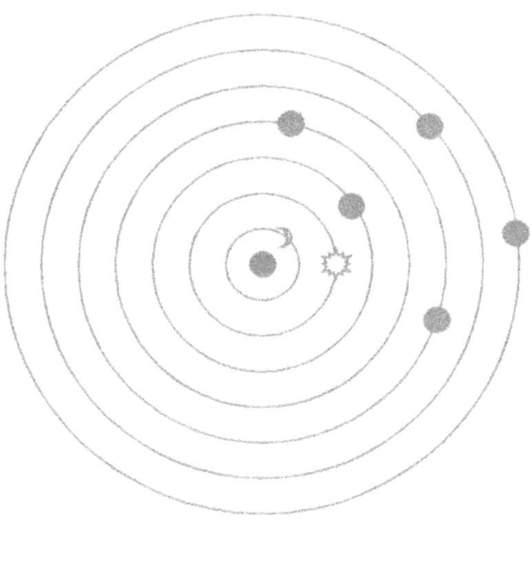

Fig. 37. Bodies on the plane of the different. Each inner body will occlude each outer body far too often, i.e. the moon will occlude all the other bodies roughly once a month. Only those stars which fall on the plane of the different will be occluded.

Aristotle at *Meteorologica* 343b8 ff. and b28 ff. where he refers to the Egyptians on these problems in relation to his theory of comets, as well as to Greek observations. He also refers to an occlusion of Mars by the moon at *De Caelo* 292a3 ff., and mentions similar accounts of other stars which are given by the Egyptians and Babylonians. There is a similar problem with the planets and the fixed stars, and from *Timaeus* 40c it is evident that Plato is quite aware of the fact that the planets at times obscure each other. When planets come to be in the same section of the zodiac, they are said to be in conjunction. There are several possibilities for what happens when planets pass each other. There may be a direct eclipse, there may be a partial eclipse, they may pass close enough that to the naked eye the two planets appear as one brighter, larger planet or they may be sufficiently separated that they merely pass each other. It is clear that, as with the sun and the moon, in the *Timaeus* model there will be far too many times when the planets are absolutely in line with each other and no times when they have the same longitude (measured along the ecliptic) but different latitudes (again relative to the ecliptic) which are necessary for the other types of phenomena. It would be surprising if Plato were not aware of this, and *Timaeus* 40c6 ff. may indicate a fair amount of knowledge of these phenomena:

> The dances of these stars and their juxtapositions (*parabolas*) with one another, the reversals and advances of their own cycles, which of the gods come into line (*sunapsesin*) with each other and which into opposition, which cover each other relative to us, and for what periods they each disappear and again re-appear (*katakaluptontai kai palin anaphainomena*) ...

If we take this passage to reiterate the same point, it is simply about eclipses and occlusions. However, if there is a contrast here between *parabolas* and *sunapsesin*, then we might take *parabolas* to indicate a close passing and *sunapsesin* to indicate being absolutely in line.[42] If so, Plato is aware of the difference between at least some types of phenomena associated with planets passing each other, but it is clear that his model of the *Timaeus* is not able to account for it. More radically, instead of taking *katakaluptontai kai palin anaphainomena* as 'disappear and again re-appear', again a reiteration, one might try something along the lines of 'veil themselves and again become bright'. If so, there are some interesting possibilities. This might be a reference to the fact that the planets have varying apparent size and brightness, though 'veil themselves' would suggest they disappear from view, and 'appear more brightly' would not be an attested usage of *anaphainesthai*. It may be a more specific reference to what happens when two planets are very close to being absolutely in line with the earth (same longitude, slightly different latitude), when they appear as one brighter object, though *parabolas* may have already referred to this phenomenon. More intriguingly, it might instead be a reference to the variations in apparent brightness shown by Mercury and particularly

Venus as they move in close proximity to the sun.[43] As they overtake or are overtaken by the sun they disappear from view (are lost in the brightness of the sun) only to reappear on the other side of the sun. If any of these possibilities are correct, then Plato seems to know a fair amount about these phenomena, and certainly rather more than his model of the *Timaeus* could account for. Note also that if the third possibility is correct Plato keeps separate his references to retrogression (40c5) and to the Mercury, Venus and sun problem (Fig. 38), and gives us a good catalogue of the problems for astronomy in the *Timaeus*.

Cornford's model cannot account for the frequency of eclipses, a phenomenon Plato could not fail to have been aware of.[44] So at least for the moon, Cornford's *via media* fails. Depending on Plato's knowledge, it may well fail for the planets as well. That the planets deviate in latitude may be inferred from the fact that when they pass each other, at times there is sufficient difference in latitude for them not to produce an occultation or merge into one bright entity, but to pass with clear space between them.

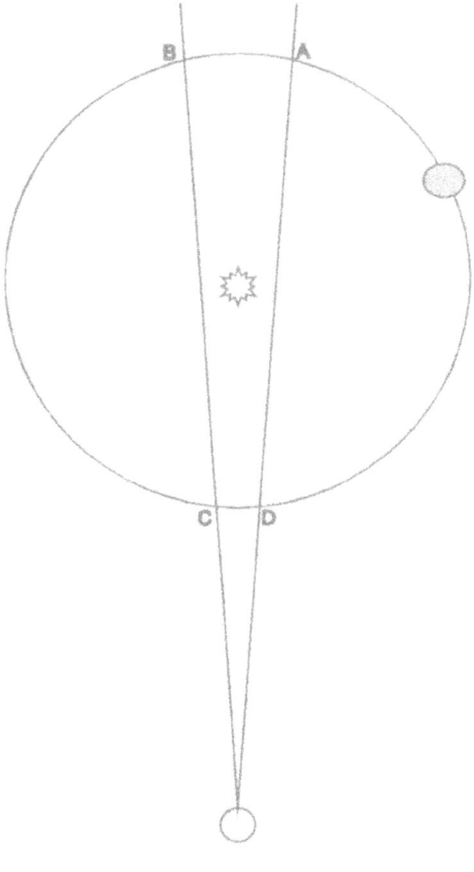

Fig. 38. Venus, Mercury and the sun. Both of the inferior planets (here Venus) are seen relatively close to the sun, and as Plato is aware, overtake and are overtaken by the sun. When Venus gets sufficiently close to the sun, as at A, it will no longer be visible, and will re-appear on the opposite side of the sun at B. Similarly it will vanish again at C and re-appear at D. It may be this phenomenon which Plato refers to at *Timaeus* 40c8-9.

So too it is evident from simply tracking the path of the planets against the background of the fixed stars. This need not be very sophisticated, as all one has too realise is that the planets do not produce occultations of the same stars on successive passes through the zodiac. One might make a case on the grounds of *Timaeus* 40c that Plato knew of either or both of these phenomena, though such a case would not be conclusive.

A further consideration here is the nature of retrograde motion. As emphasised in an earlier section, retrogression does not mean that a planet simply reverses its path, and often involves a significant deviation in latitude to produce the loop and zig-zag shapes. Plato's term for retrogression, if that is what it is, may be significant here.[45] At *Republic* 617b2 he uses *epanakukloumenon*, and at Timaeus 40c5 he uses *epanakuklesis*. This means 'circles back on itself'. It is quite possible that this is a reference to the loops produced in retrogression, and if so Plato must have realised that there is deviation in latitude by the planets. It is of course difficult to produce a watertight case here, and I do not pretend to do so. It is quite possible that Plato was aware of deviation in latitude by the planets, and on balance I would say it is likely that he was aware. If so, there are further problems for Cornford's model which cannot account for any deviation in latitude.

VI. Variations in planetary velocity and distance

As mentioned earlier, Plato's model for the motions of the sun is good but by no means perfect, even in relation to ancient knowledge of the heavens (Fig. 39). The bare model cannot account for the unequal length of the seasons, something that was known of in Plato's time. For the definition of the seasons, there are four critical days in the year. Winter solstice, the day of longest night, spring equinox, a day of equal day and night, summer solstice, the day of longest day, and vernal equinox, the other day of equal day and night. The seasons are the periods from solstice to equinox or equinox to solstice, but they are not of equal length. The Babylonians did not recognise this, but Euctemon and Meton had by the fourth century, and they had measured values for the inequality of the seasons.[46] By around 330 BC, Callippus had a remarkably good estimation of the inequality of the seasons. The following information is from the Eudoxus papyrus. This does not give the values for the final quarter of the year. Those given are calculated on the basis of a 365-day year. Nor are there figures for the first quarter for Democritus or Eudoxus, those given here being Neugebauer's conjectures. Neugebauer makes a tentative assumption that Eudoxus did not know of the inequality of the seasons, as he made no effort to account for this.[47] Equally, we might say he did know but could not account for this, and conjecture say 90, 92, 91, 92.

	Democritus	Euctemon and Meton	Eudoxus	Callippus[48]
Summer solstice to autumn equinox	(91)	90	(91)	92
Autumn equinox to winter solstice	91	90	92	89
Winter solstice to spring equinox	91	92	91	90
Spring equinox to summer solstice	(92)	93	(91)	94

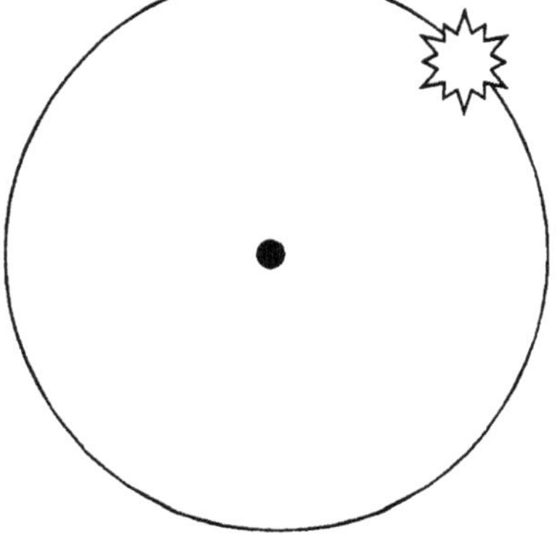

Fig. 39. Plato and the equality of the seasons. Plato's model is far too simple to account for the inequality of the seasons. All we need to consider here is the relation between sun and earth. If the sun moves in a circle at a steady speed there will be perfect equality of the seasons.

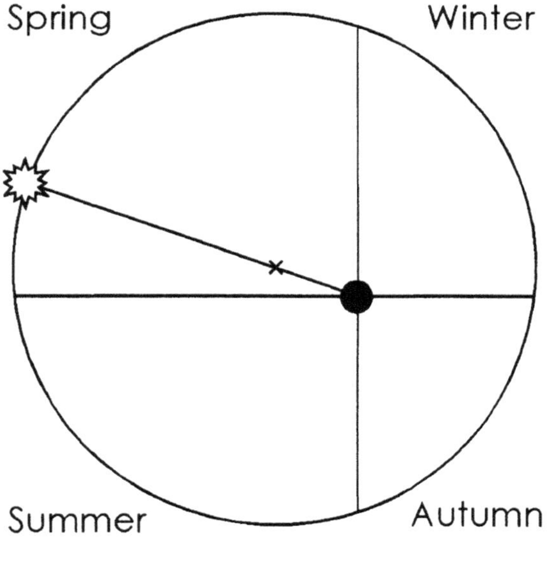

Fig. 40. Hipparchus' model. An example of what one must do to retain simple circular motion and account for the inequality of the seasons. Hipparchus' model retained regular circular motion but offset the earth from the centre of the sun's orbit to generate the inequality of the seasons.

Again we can simplify matters down to a plot of the motion of the sun on the different. In order to generate any inequality in the length of the seasons here the orbit of the sun would have to be made out of round (Fig. 39) or there would have to be some variation in the orbital velocity of the sun. The contrary power is not a possibility here as it is a power contrary to that of the sun, so we would need to assume some form of self-motion varying the speed of the sun.

Callippus, perhaps on the basis of the observation of the inequality of the seasons by Euctemon and Meton, goes beyond Eudoxus and adds another two spheres for the sun and two more for the moon to generate variations in velocity for the sun and moon.[49] Schiaperelli conjectures that the extra pairs were hippopedes, which with appropriate values would give reasonable estimates of the variations in velocity.[50]

The bare model of the *Timaeus* cannot account for variations in the apparent size of the celestial bodies. Did Plato know of such variations? According to Simplicius *in De Caelo* 504.17 ff.:

> These [unrolling spheres] of Eudoxus' school do not save the phenomena, not only those that were found later, but also those known before (*ta proteron gnothenta*) and recognised by them.[51]

The phenomena which Simplicius goes on to refer to are that Venus and Mars appear at times much brighter than at others, that there is variation in the size of moon, and there is also the matter of annular eclipses. The moon can produce solar eclipses while at varying distances from the earth. When it is near it covers the whole sun, when it is far away, a small ring of the sun is visible around the moon at total eclipse, the phenomenon of annular eclipse. Whether these phenomena were known to Plato is a matter of conjecture. The closest thing to direct evidence we have is *Epinomis* 978d7, where the moon is said to appear larger and smaller (*e tote men meizon, tote d'elatton*), which would not appear to be a reference to shape in relation to phase, especially as the next line appears to give us such a reference.[52] Again, one could suppose a self-motion for this, though that would be another breach of regular circular motion and a new departure for self-motions (which so far have been for longitude, then latitude, and now distance). If we accept that Plato knew of variations in apparent size or brightness, and that is quite possible, then there would need to be variations in the distances of the heavenly bodies. If so, and we put this in Plato's terms, we have now utterly wrecked regular circular motion as we have attributed deviations from it in all six motions which Plato recognises; left and right (around the plane of the different), up and down (deviation in latitude), forward and back (deviation in distance).

1. At *Timaeus* 38d, we have the problem of why Mercury and Venus and the sun overtake and are overtaken by each other. As argued above, I

believe we should take this to be a separate problem for Plato from that of planetary retrogression.

2. At *Timaeus* 39cd, we have the problem of the times of orbit of Mars, Jupiter and Saturn. This may be slightly more complex than is generally recognised, as immediately prior to this Plato has Timaeus say that month is the time the moon takes to come back to the same place in relation to the sun, thus giving a synodic month. Plato may then want the synodic periods of the planets (possibly as well as the sidereal periods if the great year is to be calculated, as that is done relative to the same). Note in relation to this that the *Epinomis* (990a) criticises Hesiod as one who is not a true astronomer, being someone who has only 'observed settings and risings' rather than studied orbits.

3. At *Timaeus* 40cd, the nature of the problems is more debatable, but I would argue that we have the problems of planetary retrogression, of juxtaposition and eclipses, of the times of eclipses and perhaps the disappearance and reappearance of Mercury and Venus as they pass the sun.

4. At *Epinomis* 978d7 we have a possible reference to variation in the size of the moon.

5. At *Timaeus* 47a5 we are told that: 'Sight of day and night, of months and the revolving years, of equinoxes and solstices (*isemeriai kai tropai*),[53] have caused invention of number.'

We can thus reasonably attribute at least a recognition of the phenomena of solstice and equinox to Plato. Plato also refers to solstices in the *Laws* (767cd and 945e) as a means of marking out the political year and refers to the turnings of the sun at *Epinomis* 990b. So too at *Phaedo* 98a Socrates says that he was determined to find out about:

> the sun, the moon and the other stars, their speeds relative to each other, their turnings (*tropon*) and what else they underwent.

Although he does not directly say so, it is beyond credence that Plato was not aware that lunar and solar eclipses do not occur every month. There are many phenomena which were known of by those interested in astronomy at the time of Plato or just after, so it is quite possible that he was aware of the problem of the inequality of the seasons, other evidence on variations in size and brightness, or the deviation of the moon and planets from the path of the sun.

VII. Prototype or finished article?

The consequence of a close examination of the astronomy of the *Timaeus* is that we appear to have an awkward dilemma. The evidence in favour of regular celestial motion in the *Timaeus* is very strong, as we saw in the previous chapter. This though leads to a poor astronomy, and surely Plato cannot have been so ignorant of astronomy to think this a good model. A good astronomy though would not only compromise regular circular motion, but would wreck it. Cornford's attempt to construct a via media of a reasonable astronomy with minimal deviation from regular circular motion also fails as it cannot account for the frequency of eclipses, something that Plato was certainly aware of, and several other phenomena that he may well have been aware of. The choice between a good astronomy and a good cosmology then appears to be a stark one.

Attempts to make the astronomy of the *Timaeus* more sophisticated by reading into it later developments in Greek astronomy are also highly suspect. Van der Waerden's suggestion that Plato held an early form of the epicycle theory lacks evidence.[54] Plato never mentions epicycles when he could describe this critical device for circular motion astronomy in a few lines, and there are more plausible interpretation of the contrary power. If Plato did use epicycles, then why do we find no trace of them in Eudoxus, Callippus and Aristotle when they could solve some of the problem of concentric sphere astronomy? We might, on the same grounds attribute hippopedes to Plato. As epicycles can solve some quite sophisticated problems, what degree of knowledge, and on what grounds, do we attribute to Plato? Zeyl has commented that when the circle of the different is divided:

> The planes of the seven smaller circles are not necessarily parallel to the ecliptic or to each other, but vary within the outer limits of the Zodiac.[55]

This would solve the problem of general deviation in latitude (but not retrogression), but I do not see any evidence that Plato held this highly important idea of having the motions of the planets offset to one another,[56] which was only introduced much later, nor do Eudoxus, Callippus or Aristotle make use of the idea when it would considerably simplify and improve their systems.

So we might try a different approach. Ought we to accept the principle that the *Timaeus* model was adequate for all the phenomena known to Plato? It is by no means clear that we should. If we consider this as a more general principle for ancient astronomy (everyone believed their own model could, etc.), then it will certainly not do. The most pertinent case would be that of Eudoxus. Most commentators would agree he almost certainly would have known of the inequality of the seasons, but could not account for it, and we have Simplicius' evidence on Venus and Mars, annular eclipses and the size of the moon, all of which were taken to imply variation in distance but could not be accounted for. He may well have known that his system could produce reasonable results for Mercury, Jupiter and Saturn but ran into serious difficulties with Venus and failed to produce any retrogression for Mars.[57] It is hard to believe that Eudoxus did not realise at least some of the shortcomings. One might argue something similar for Callippus, taking due account of his proposed improvements, and indeed Simplicius tells us that no one up to Autolycus could account for the phenomena he lists in relation to variations in distance. I would suggest that the great majority of ancient astronomers were aware of discrepancies between their models (whether merely mathematical or physical) and the motions of the heavens. This of course would be one of the major factors driving them forward. We need then some evidence or argument to support this principle in Plato's case, and that does not appear to be forthcoming.

There is a further interpretational issue here. Cornford's argument for irregular motion begins by citing Taylor on the contrary power. Taylor comments:

> The *Timaeus* does not tell us why the two planets and the sun in turns gain on one another. No explanation could be offered by a man who assumed all three to be revolving with uniform velocities in the same sense and with the same period in concentric circular orbits.[58]

Cornford replies:

> This seems to me to be a reason for concluding that *Timaeus* does not make all these assumptions, which would render the phenomena not merely inexplicable but impossible.[59]

There are phenomena which Eudoxus and Callippus would have known about which are impossible on their models. Do we then conclude that they did not adhere to regular circular motion? Or are we better off understanding their models as improvements on previous regular circular motion models, with Eudoxus and Callippus aware of some of their deficiencies?

It seems to me that Cornford's view falls between two stools and illustrates some of the difficulties of interpretation involved here. On the one hand, we have the bare model which cannot account for retrogression, proximity and the frequency of eclipses. Alternatively, one can produce a good correspondence between the *Timaeus* model and observations known at the time by a general application of the contrary power but lose the idea of regular circular motion. Indeed one can make the accuracy of such a self-motion model arbitrarily good. By invoking self-motion for linear retrogression Cornford saves observation but loses regular circular motion, but then for deviation from the plane of the different he saves regular circular motion but loses correspondence with observation (and indeed makes the phenomena impossible!). At this point he needs an argument as to why one should change strategy here, some criterion for limiting the application of self-motion. That, though, is highly problematic once we go beyond the confines of *Timaeus* 38d, as there is no textual reason for any such limitation and self-motion could account for anything.[60]

Plato could very easily have gone along the route of citing a generalised self-motion wherever the bare model failed to meet up with observations. This could have been done with a minimum of effort. All he has to do is add two or three words about the contrary power, or self-motion when he mentions retrogression and eclipses. Plato, however, does not make this explicit when he easily have done so and wisely leaves options for further investigation open. I would emphasise that all *Timaeus* 38d1-6 says is that Mercury and Venus have some power in addition to that which the sun has, and that may well be only a means of marking out the question of the proximity of Mercury and Venus to the sun from other astronomical problems.[61]

If we were to accept that Plato thought the *Timaeus* model to be adequate for the phenomena he was aware of, then it would seem that one, or some combination of the following would be the case:[62]

1. Plato blundered in believing his model was adequate for the phenomena of which he was aware.
2. Plato was ignorant of some basic and evident observational astronomy, which was well known to those around him.
3. Plato believed in some (non-regular, non-circular) self-motions and either did not realise or was not concerned by the extent to which this would sabotage the emphasis on regular circular motion in the *Timaeus* and Laws.

Yet there is nothing which compels us to accept this principle, which seems implausible in relation to many other ancient thinkers. What do we do if we abandon this principle? A far better option is simply to assume that Plato intended the model of the *Timaeus* to be a prototype which sets out some basic principles of cosmology. It would be a prototype of models which employ combinations of regular circular motion and motions for the sun, moon and planets offset to those of the fixed stars. As such, it is strong cosmologically though weak in terms of astronomy. Plato may well have been aware that this simple model cannot account for some major phenomena, but we have to take into account that this is a very early stage in model building. Although the model of the *Timaeus* is weak in terms of astronomy, it is a significant advance on the model which Plato gives in the *Republic*, which does not offset the motion of the planets. It would also be an advance on the model offered by Philolaus.[63]

Some points in favour of this idea. The general evidence for regular circular motion in the *Timaeus* would seem very strong, as we saw in the last chapter. Plato's concern, in general, is more with teleological cosmology. Another worry here is that, as Vlastos has pointed out, if Plato did employ self-motions, this really does not speak well of him. As Vlastos comments: 'The explanatory value of such an ancillary hypothesis would have been bogus.'[64] Indeed, any use of the contrary power/self-motion looks entirely ad hoc, and has no explanatory or predictive power. I will return to the question of the relation of Meno's paradox to astronomy and cosmology in a little while, but here I will just point out that any position created by a generalisation of the contrary power/self-motion would be an exceedingly unhealthy hypothesis.

If the model of the *Timaeus* is a prototype, then the role of the Platonic challenge reported by Simplicius is clear. There are problems in astronomy which the bare model of the *Timaeus* is not able to solve, which should be solved in terms of regular circular motion. Plato wisely leaves those difficulties open for further investigation, but how one should set about this challenge is clear. If Plato believed the model of the *Timaeus* to be a final one, or did not believe in regular circular motion, then the challenge makes little sense.

Why might Plato be optimistic that such phenomena as retrogression and the relation of Mercury and Venus to the sun might be solved by regular circular motion? Perhaps because Plato's model for the sun does something similar. If we consider the setting points of the sun on the horizon during a year, they progress to a northerly maximum, halt, move to a southerly maximum, halt and progress again. That problem is solved by imagining two regular circular motions for the sun in combination. One might describe the retrograde motion of the planets and the motions of Mercury and Venus relative to the sun in a similar manner, and hope that further regular circular motions will solve these problems.

The model of same and different, being effectively that of the mind of a visible god, has a great many other functions to perform in terms of epistemology, ethics and psychology as we shall see later. It is not solely an astronomical model and so we should not be overly surprised that it is strong on cosmological principle but weaker on astronomical detail.

Cornford believes that he can bring his view of the *Timaeus* in line with the *Laws* and *Epinomis*. On the question of circular motion, that is indeed so. However, Cornford believes that each of the planets has an intermittent self-motion which creates retrograde motion. As we shall see in the next chapter, this most certainly cannot be brought into line with the view of the *Laws* and the *Epinomis*, which are most insistent that the celestial gods always have acted and always will act in an entirely consistent way, and simply do not act in anything like an intermittent manner.

Finally, we get a good transition into the work of Eudoxus. With Plato we have regular circular motion and a two-sphere model for sun, moon and planets. Eudoxus can be seen as picking up the Platonic challenge with three-sphere models for sun and moon and four-sphere models for the planets. It is also possible to argue that Eudoxus attempts to remedy the major defects in Plato's prototype. Thus he can (at least qualitatively) account for retrogression, for deviation from the ecliptic, for the relation of Mercury, Venus and the sun, and, as he has three-sphere models for both sun and moon, he has the foundations for a much improved theory of the frequency of eclipses.[65]

If so, are we not better off considering Plato's model as a cosmologically strong prototype where he is aware of at least some of the astronomical deficiencies? It would be a prototype of models which employ combinations of regular circular motion and employ such motions for the sun, moon and planets offset to those of the fixed stars. Plato may well have been aware that this simple model cannot account for retrogression, proximity and eclipses, and here we have to take into account that this is a very early stage in model building. Although the model of the *Timaeus* is weak, it is an advance on the model which Plato gives in the *Republic* and on the model offered by Philolaus. It is also important to note that if the model of the *Timaeus* was a prototype, Simplicius' report of the challenge Plato set the astronomers makes sense, which it does not if Plato believed himself to have solved all the astronomical problems.

VIII. The astronomical and cosmological traditions

One might argue that in the Presocratic period there were two quite separate traditions of the investigation of the heavens. There was a tradition of observational astronomy, and of the meteorological calendars, the parapegma.[66] There was also a tradition of philosophical cosmology, perhaps beginning with Thales and certainly counting many of the most eminent Presocratic philosophers among its number. While one might point to Anaximander's rings and Pythagorean speculation concerning the harmony of the heavens, perhaps culminating in the cosmology of Philolaus, there is no really effective union of cosmology and astronomy. Cosmological models are not so far advanced as to take into account any but the most elementary observations. The heavens go around the earth, and there is a distinction between the planets and the fixed stars, but beyond that there is no attempt to make cosmological models fit with the results of observational astronomy in any detailed manner.

It is perhaps with Plato and Eudoxus that we find the first wedding of these traditions, something that is clearly hugely important for the history of astronomy and cosmology. One might object to giving such a role to Plato, arguing that Eudoxus really is the critical character. A standard objection here, that on the evidence of *Republic* VII Plato rejected or denigrated empirical astronomy, and so could not have brought about a synthesis of the philosophical and observational traditions, we have already dealt with in Chapter 2. The *Timaeus* shows awareness of many of the empirical problems of astronomy. A weakened form of this objection would be that Plato did no empirical work himself. This might be met in two ways. First, one might accept that Plato was not primarily an astronomer, but argue that he made a significant philosophical and theoretical contribution. Secondly, one might argue that during his own career Plato does make some significant advances in astronomy. Clearly Plato is not an observational astronomer, but the *Timaeus* would seem to indicate that he was at least conversant with the major astronomical phenomena.

There is of course a vexed question here of any influence between Plato and Eudoxus. If there was any influence, who influenced whom, on what subjects, and to what extent are all open questions.[67] There is also an issue for the dating of the *Timaeus* associated with Eudoxus' dates. Owen argues that there is no influence from Eudoxus on the *Timaeus* and that this implies an early dating.[68] Citing Apollodorus,[69] Owen places Eudoxus' death at 356-353, and states that he must have produced his system well before this, as he left Athens for Cnidus some years before his death.[70] The effect of all this is to push back the date of composition of the *Timaeus*, in Owen's opinion to perhaps before 369 and the composition of the Theaetetus. I reject the dating of Eudoxus based on the notoriously unreliable testimony of Apollodorus which would have his death in 356-353.[71] Gisinger[72] places Eudoxus' death in 342, and both de Santillana[73] and Laserre[74] give 337, after Plato's death in 348. There is also the problem of accommodating the evidence of Pliny,[75] who talks of Eudoxus dating Zoroaster as living 6,000 years before the death of Plato, implying that Eudoxus outlived him. While it may be 'commonly agreed' that Eudoxus came to the Academy in 368,[76] the alternative datings allow for Eudoxus to have formulated his full theory after Plato composed the astronomical passages in the *Laws* or even after his death in 348.[77] It may well be that Eudoxus came to the Academy in 368, but when Eudoxus began his work on concentric sphere astronomy and when he finished it are both quite open questions, and we need to recognise that it is quite possible that Plato only knew of it in a developmental stage. There is nothing that compels us to suppose any influence was in one direction only, and it may well be that Eudoxus did not complete his development of his homocentric sphere system until after Plato's death. It is possible that Eudoxus, armed with his homocentric spheres, convinced Plato that the motions of the heavens could be resolved into regular circular motions and this precipitated Plato's move between the *Republic* and the *Timaeus*. Alternatively, Plato's prototype of the *Timaeus* may have been the spark which led to Eudoxus' work on homocentric spheres. In the absence of any further evidence, we have to go with the existing texts. So Plato gives us the first extant account of a path for the planets offset from the daily motion of the fixed stars, and that is a significant step forward in astronomy. Owen argues that Plato's use of the term 'contrary power' in relation to Mercury and Venus means that he was not conversant with Eudoxus' system at the time of writing the *Timaeus*. Even if this is so, Eudoxus has plenty of time yet to formulate his system.[78] Owen is inclined to find some Eudoxan influence in a move from irregular to regular celestial motion between the *Timaeus* and the *Laws*.[79] However, I have argued at length that this change occurs between the *Republic* and the *Timaeus*. From the evidence that we have, it is not clear whether or how Eudoxus influenced Plato, or indeed whether or how

Plato influenced Eudoxus. If his influence is as Owen suggests, then it was certainly pre-*Timaeus*. Even if he did not influence the composition of the *Timaeus*, the doubts concerning his dates mean that we cannot draw any firm conclusions about the date of this work.

IX. Plato and Eudoxus

One way of attempting to break any tight link between Plato and Eudoxus is to argue that while Plato was a realist, or engaged in metaphysical speculation about the nature of the heavens, Eudoxus was an instrumentalist, merely interested in finding a mathematical scheme that would fit the motions of the heavens. Thus one can find Dreyer commenting:

> As Eudoxus made no attempt to connect the movements of the various groups of spheres with each other, it seems probable that he only regarded them as geometrical constructions suitable for computing the apparent paths of the planets.[80]

So too Heath remarks:

> It would appear that he did not give his spheres any substance or mechanical connection; the whole system was a purely geometrical hypothesis, or set of theoretical constructions calculated to represent the apparent paths of the planets and enable them to be computed.[81]

As Wright has argued,[82] typically Eudoxus is thought to have been engaged in 'geometry' rather than 'physics' because he gives no account of the substance of the homocentric spheres and no account of how they might be interconnected. Wright suggests that, relative to other considerations which I shall look at in a moment, this evidence is relatively weak. I would go further and say that this evidence does not discriminate between instrumentalist and possible realist positions at all. One does not have to be an Aristotelian realist, with spheres of aether, to be a realist on this matter. If one considers the planets to be intelligent entities, then there is no need to give anything more than the appropriate mathematics for each planet. As intelligent Platonic entities, they will of course be following the best sort of motion so naturally they will be executing (combinations of) regular circular motions.

Here it is important to recognise the nature of how the mathematics are presented. There are sufficient parameters to determine fully the motions of all the planets. This is considerably beyond what an instrumentalist needs to give. If the sole criterion were that of calculational simplicity, then Eudoxus could have the earth revolve and so remove one sphere from each of the planets and that of the fixed stars, resulting in a considerable simplification of the system. Or again for simplicity, the motion of the fixed stars could be considered to be common to all the planets, thus saving seven spheres.[83] This is not the path that Eudoxus takes, however. He gives a full description of the motions for each of the planets. If Eudoxus is a Platonic realist on this matter, each planetary intelligence will need to be cognisant of all the motions that it needs to make, though this is not required simply for the calculation of the motions of the heavens. If each planetary intelligence has enough information to govern its own motions, then the question of the interaction of the spheres becomes redundant. There are two senses to this, how the spheres for each planet interact to produce that planet's movement, and how the systems for each planet might interact with one another. There is no need for the interaction of the spheres for one planet. The planet knows what motion it ought to execute, as a combination of circular motions. There is no interaction between the systems, and none is needed. Each planet knows what it is doing, and its motion is independent of all the other planets. If Eudoxus was a Platonic realist, then he has no need to specify the substance of the homocentric spheres, nor has he any need to specify how the spheres interact.

So the fact that he does neither is equally compatible with Platonic realism as it is with an instrumentalist approach. This, of course, is not to say that on these grounds Eudoxus was a realist in the Platonic manner. The instrumentalist interpretation here though, as Lloyd has remarked, is a questionable argument from silence, and here I hope to have shown that that silence can just as well be taken in another way. There is no other reason to suppose that Eudoxus was an instrumentalist. As Wright argues, if Eudoxus were only interested in the representation of the phenomena and in prediction then the accuracy of the system would have had paramount importance, perhaps with the simplicity of calculation a secondary consideration. If one was looking merely for an instrumental solution in astronomy, I would suggest that one would not be in the least bit interested in the models produced by either Plato or Eudoxus. This is because they are simply not as good at representing the phenomena as other methods open to the Greeks at the time, or even as good as the Babylonian methods of extrapolation from the data. Eudoxus' system, as we have seen, had significant drawbacks, and it is hard to believe that he was not aware of at least some of these. As Wright puts it:

> From a purely predictive point of view there is no reason at all to avoid openly ad hoc modifications of that model, la Ptolemy, for the sake of bringing predictions closer to observations. For instance, standard moves within the project should have included trying non-uniform or eccentric motion, and centres other than the earth. But there is no trace of this sort of thing in the records of the enterprise.[84]

It is difficult then to see Eudoxus as an instrumentalist, and so he may have been influenced by Plato.

X. Meno's paradox and astronomy

Let us return to the question of Meno's paradox and its resolution. In previous chapters we have seen the use of a solution to Meno's paradox in relation to the shape of the ultimate particles. What is the problem that Plato faces in astronomy? There is certainly another underdetermination problem here. We could adopt any number of hypotheses which save the phenomena of the motions of the heavens. Clearly one can formulate infinitely many such hypotheses if there are no constraints on these hypotheses other than fitting the data. So where does one start in this indefinite multitude? In line with Plato's solution to underdetermination, we ought to start with a healthy hypothesis. So neither the Platonic challenge nor the *Timaeus* (nor indeed the *Laws* and *Epinomis*) will allow any hypothesis which does not postulate regular, circular and ordered motion. Effectively Plato and Eudoxus do 'some straight' line drawing, only this is now done in three dimensions and for moving objects, and move directly to the simplest, most elegant mathematical generation. As we saw in the previous section, in fact neither Plato nor Eudoxus can 'save the phenomena' as well as other calculational devices that were available to the Greeks. On this matter, Wright makes a comment with which I would wholeheartedly agree: 'If prediction was his primary aim, Eudoxus should have given up on his spheres early in his search.'[85] One strong suggestion as to why he did not must be that he was not an instrumentalist and considered homocentric spheres to be a very healthy hypothesis in the Platonic manner.

There is another aspect of Plato's astronomy which is perhaps under-appreciated – the drive for a unified account of the heavens. Previous views had been either been too simplistic in their unity and so had failed to provide a scheme which could hope to represent the celestial motions properly (so Philolaus, Anaximander), or, while saving the phenomena, had been decidedly fragmentary and ad hoc and so had not offered a unified conception of the heavens (so the parapegmata tradition, the Babylonians). Plato is the first to recognise the complexity of the heavens and offer (at least the prospect of) a unified theory of the motions of the heavens. While Plato's prototype model of the *Timaeus* falls short of this aim, an empirically adequate and unified picture of the heavens is an important aim of the following astronomical tradition. This brings out another aspect of the solution to Meno's paradox and to the Platonic challenge. Part of being a healthy hypothesis is having the ability or likelihood to be able to cohere with other hypotheses. In relation to this I would stress again that the Platonic challenge asks for motions which are regular, circular and *ordered*.[86]

In the history of science, some ideas become so commonplace we are apt to forget their significance, and so their provenance. It appears that those ideas have always been there, because their contraries are so strange to us. Such is the case, I suggest, with the regularity and order of celestial motion, and a unified picture of the heavens. Since Plato these have been the basic assumptions of astronomy. We should not forget that was not necessarily the case before Plato, not only with the Babylonians who referred to the planets as 'sheep' (*bibbu*) because of their wandering courses, and had fragmentary methods of depicting the motions of the heavens, but so too with some Presocratics, and indeed as I have argued with earlier Plato. As we shall see in the next chapter, while Plato considers the heavenly bodies to be gods, he steers very firmly away from the idea that these gods act in any intermittent manner or that they do anything other than what is good, such that they act more like laws of nature and less like the gods of classical mythology.

It is clear from the nature of Plato's solution of Meno's paradox, in combination with his views on the best sort of motion, what the healthiest hypothesis concerning the heavens would be. That Plato is answering Meno's paradox/underdetermination questions in the *Timaeus* is, I think, clear from his treatment of the questions of the number of cosmoi, the shape and size of atoms and the number of species. It would then be very surprising if Plato were to deviate from regular circular motions for the heavens without very good cause. This is a good reason for adopting an interpretation of the *Timaeus* which holds on strongly to regular, circular and orderly motion but allows Plato to acknowledge some of the astronomical difficulties associated with an analysis solely in terms of regular circular motion, and set a problem for solution in the Platonic imperative. So we should not take the *Timaeus* to be literal and finished cosmology, but a prototype attempting to instantiate a solution to Meno's paradox. This is why the contrary power/self-motion would be an unhealthy hypothesis, and so would seem to be a highly unlikely interpretation of what Plato means at *Timaeus* 38d4. Again, contra Vlastos, what inspires Plato's advocacy of regular, circular ordered motion is not a 'metaphysical fairy tale'. It answers to real epistemological problems of how to go about doing astronomy.

One final comment on Plato and astronomy. One of the general weaknesses of ancient science is that many of its practitioners presented their work as finished and dogmatic systems. As suggested in Chapter 1, it is a modern rather than an ancient response to say 'We do not know yet, there are problems, but we hope to find an answer along the following lines'. One very important exception to this may be Plato and the tradition he begins in astronomy.[87] The weaknesses of the astronomical model of the *Timaeus*, which are considerable and greater than is generally recognised, are offset by the methodological and cosmological insights, and the fact that the model is (and is intended to be) the prototype for a whole new tradition of astronomy and cosmology. It is overwhelmingly likely that Plato recognised at least some of the defects of the *Timaeus* model, just as Eudoxus and Callippus would have been aware of the problems with their own models.

6

Plato and the Development of Greek Cosmology

One of the most remarkable aspects of Presocratic thought is the rapid increase in the sophistication of its cosmological models. From Thales and the earth floating on water through to the first complex cosmology of the *Timaeus*, there is a steady but swift overcoming of conceptual obstacles. Why that should be so is beyond the scope of this book, though I would look to the Greek formulation of the idea of a cosmos as an ordered entity comprehensible to human reason, and the rejection of mythological and supernatural explanations in favour of philosophy and debate.[1] Though not as spectacular, there is also an advance in observational astronomy, and ultimately a union of this tradition with that of cosmological speculation. The task of this chapter will be to assess Plato's role in the development of Greek cosmology. First, I argue that he takes up the Presocratic concerns with the position and stability of the earth. Plato can then be seen to consider the application of these concerns about the earth to the cosmos as a whole, along with a new concern, the integrity of the earth and the cosmos. As Plato's thought develops, we can see some more sophisticated answers being produced on the nature of the cosmos. These developments mark an important advance in Greek cosmological thinking. One aspect of Plato's cosmology which has received severe criticism, however, is his description of the cosmos and the celestial bodies as animate, intelligent, ensouled deities. Pedersen comments:

> In many ways, the *Timaios* reflects an animistic and anthropomorphic philosophy foreign to the Ionian philosophers' fundamental attitude, although there are a few similar ideas.[2]

So too we can find Burkert saying:

> Plato thought it an inescapable conclusion that the orderly motion of the stars is due to their having souls; it is a voluntary, chosen order. Here sophisticated Greek science harks back to the pre-scientific way of thinking and comes to rest in it.[3]

At first glance Plato would indeed seem to be making a backward step here, returning to precisely the sort of animistic or supernatural explanations the early Greeks had done so well to reject. Matters are not quite so simple though, and in fact Plato is very careful about what he attributes to the cosmos and celestial bodies in an attempt to explain why their behaviour is regular. He has to do this with very few resources, and his attempt is interesting, ingenious, and, in fact, very helpful for the future progress of Greek astronomy and cosmology.

I. Plato and the Presocratics

If we accept Aristotle's report, Thales produced the first cosmology:

> Others say that the earth rests on water. For this is the most ancient account we have received, which is said to have been given by Thales of Miletus, that it stays in place through floating like a log or some other such thing (for none of these rest by their nature on air, but do on water), as though the same argument did not apply to the water under the earth as to the earth itself.[4]

Quite possibly in reply to the sort of difficulty Aristotle points out come a new generation of cosmologies which rely on air to support the earth.[5] Air does not fall and so is perhaps more suitable as a supporting medium than water. Aristotle also tells us:

> Anaximenes, Anaxagoras and Democritus say that it is the flatness (of the earth) that is the reason why it stays still, for it does not cut the air below but covers it as flat bodies are able to do.[6]

Thales and Anaximenes are prime examples of what Furley would call a parallel as opposed to a centrifocal conception of the cosmos.[7] In a parallel cosmology, heavy objects fall in one direction, and do so in parallel lines. Thus we have the problem of why the earth does not fall in a similar manner.[8] One can contrast this with the centrifocal cosmos. Aristotle is the classic example here as everything in natural motion moves towards (or moves away from) the centre of the cosmos. The cosmos here has no direction of fall but does have a central point. Plato's cosmology is undoubtedly a precursor to Aristotle's, though Plato lacks the strictly centri-focal dynamics which make the Aristotelian cosmos classically centri-focal.[9] There are of course precursors to Plato on this matter too. One might look to Parmenides, especially Fragment 8 42-5, on the well rounded sphere balanced in all directions, though how the meditation on being of Fragment 8 relates to the more explicit cosmology of Fragments 10, 12 and Aetius II, 7, 1 is not at all clear. Alternatively one might look to the Pythagorean cosmology of Philolaus with its concentric orbits around the central fire,[10] though again both lack centri-focal dynamics as far as we know.

There is some controversy over the nature of Anaximander's cosmology.

The question is whether he gives a full equilibrium argument for the stability of the earth, which would imply a central earth and no direction for the cosmos, thus giving at least a precursor of a centrifocal cosmos, or whether he only gives a partial equilibrium account. In the latter case the earth is supported by air, such that we are back with the parallel cosmos, but the flat top and bottom of the drum-shaped earth are kept at right angles to the line of fall of the cosmos by the equilibrium of the drum. The outcome of this debate is important for how we value the full equilibrium argument for the stability of the earth that we find in the *Phaedo*. Aristotle tells us:

> There are some, such as Anaximander among the ancients, who say that (the earth) rests on account of its likeness. It is fitting that what is established in the centre and has equal relations to the extremes should not move up, down or to the side. It is not possible for it to move in opposite directions at the same time, and so necessarily it remains still.[11]

This would appear to attribute a full equilibrium theory to Anaximander. The key piece of counter evidence though, as cited by Furley, comes from Simplicius:

> Anaximander thought that the earth remains in place *because of the air that supports it* and because of its equilibrium and uniformity.[12]

I side with Panchenko and Schofield against Furley here in giving greater weight to the position Aristotle describes,[13] especially as it is supported by Hippolytus.[14] At *Phaedo* 108e Plato has Socrates say:

> I am now persuaded that, firstly, if the earth is round and in the centre of the heavens, then it requires neither air to prevent it falling (*pesein*) nor any other necessitation of this sort, but the uniformity of the heaven itself in every way and the equipoise of the earth itself is sufficient to restrain it. For something which is equipoised and is placed in the middle of something homogenous cannot yield to being moved aside in any way (*oudamose klithenai*), but will remain steadfast.

There is no doubt that this is a full equilibrium argument, and Plato almost certainly has a spherical earth in mind here.[15] It is possible that there is another layer of sophistication here, one that is only required once one gives up the parallel cosmos. Plato uses *piptein* and *klinein* to describe what the earth will not do. *Piptein* has connotations similar to the English verb to fall. It can be used of a city, of a government, or of a soldier in battle, and so can carry a sense of downfall, demise or ruination.[16] So Plato's worries here may then go beyond why the earth does not drop, to why the earth doesn't fall apart, if nothing holds it together. With a parallel cosmos, the integrity of the earth, at least in one direction, is a simple matter. Pieces of earth would move downwards, but are supported either by water or air and so cohere together. Even if integrity is not a concern here (and admittedly this evidence is rather thin), it is in the *Republic*.

The *Phaedo* gives us the conception of the free floating and immobile earth. The myth of Er then makes a significant conceptual leap to begin to talk of the integrity of the cosmos and how the cosmos might be supported. Here the cosmos needs external bonds, *ta hupozomata* (616c2), like the brace that passes under or around the hull of a trireme, rather than being self-sufficient.[17] Or if *ta hupozomata* are internal to the hull, then the cosmos needs internal bracing to hold it together.[18] It is notable that here the cosmos is not referred to as a living entity and does need some bracing, while in the *Politicus*, *Timaeus* and later works the cosmos is a living entity (mortal in the *Politicus*, immortal in the *Timaeus*) and needs no such bracing.[19] Another important contrast we saw in Chapter 4. Where the cosmos turns on a pivot (*Republic*, *Politicus*) and does not have an immortal soul (no soul, *Republic*, mortal soul, *Politicus*) there is irregular celestial motion and degeneration. The planets too have no souls and require the daughters of necessity to move them.

The *Timaeus* then has some significant developments on these views. The cosmos is all there is, so it can have no external bracing nor can it rest on a pivot. While I have so far stressed a point relating to celestial motion here, the move from a supported cosmos to a free one which exhausts all there is looks to be a significant conceptual leap, analogous to the move from a supported to a free-standing earth in Presocratic thought. The integrity of the cosmos (and indeed of all the celestial bodies) is explained by their having souls, so no bracing is required, and as these souls are immortal, there is no degeneration. So too the motions of the cosmos and the celestial bodies are generated by their souls. While there is a sense in which we can easily refer to top and bottom of the cosmos in the myth of Er (as the cosmos is on the knees of necessity) *Timaeus* 62c ff. makes it very clear that such terms as above and below are not suitable. One must make due allowance for the highly metaphorical language of the myth of Er and the fact that cosmology is not Plato's primary concern there, but still there are important differences in the structure of the cosmos from the *Republic* to the *Timaeus*.

II. Plato and geocentrism

Plato gives us a cosmos with a central, stable and static earth. That the earth is static, at least in the usual sense, I take to be the upshot of the debate about the rotation of the earth in the *Timaeus*.[20] The key passage is at *Timaeus* 40b8-c1:

> Earth, our nurse, is packed around (*illomenen*) the axis which stretches through the universe, was contrived to be guardian and craftsman (*demiour-*

> *gon*) of night and day, first and eldest of the gods who have come into being with the heavens.[21]

The initial problem here is that *illomenen* (or *eillomenon*) can mean 'packed around' or 'revolving around'.[22] If Plato had a proper motion of the earth in mind, there are several words which he could have used to express himself more clearly.[23] What the earth is required to do to produce a coherent cosmology might be inferred from other passages. *Timaeus* 36e1-3 tells us that soul was:

> woven everywhere from the centre to the extremity of the heavens, and enveloped in a circle from the outside, revolving on itself.

One would expect then that the earth would revolve with this and so there would be no relative motion between the earth and the fixed stars, if no other factors were involved. However, *Timaeus* 39c1-2 quite reasonably says:

> So for these reasons night and day came to be, the period of the single and most intelligent revolution.

For that to be the case, there must be relative motion between the earth and the movement of the same. One should also note that the earth is described as the 'first and eldest of the gods which have come into existence with the heavens' (*Timaeus* 40c2-3). The other gods, i.e. the stars and planets, all have motions befitting their intelligences, so perhaps the earth ought to as well. If it does have motion, this ought to be a simple motion befitting the eldest. This counts heavily against the view of Burnet and Taylor that the earth moves up and down on the central axis of the cosmos.[24] As Taylor recognises, such a motion could be used to explain the frequency of eclipses.[25] However, this motion would be arbitrary and non-circular, and hardly fitting for the earth.[26] It could not account for deviation or occlusions in general, as moving the earth up or down would make the sun, moon and all the planets appear to deviate the same side of the ecliptic at once. I do not see that we can support the idea of a motion up and down an axis, nor would such a motion be consistent with views expressed in the rest of the *Timaeus*.

The standard resolution of this matter is that the earth is considered to have a motion equal and opposite to that of the same and so is effectively static.[27] This is a simple motion (regular circular motion in one place) and would produce the required contrast with the same to produce night and day. While the effect on the earth is clear enough, whether one calls this a motion of the earth is another matter. Aristotle comments:

> There are also those who hold that the earth is in the centre and is wound

> around, i.e. moved (*illesthai kai kineisthai*) about the axis through of the universe, as is written in the *Timaeus*.[28]

I take the report by Plutarch, who reports Theophrastus, that Plato in his old age regretted giving the earth the central place in the cosmos, and thought it ought to go to some worthier body, to be unreliable.[29] There is nothing in Plato's later writings to suggest this, nor is there any evidence from his contemporaries to support Plutarch.[30] Plato then does nothing to offset the geocentric trend of Greek cosmology, and little to offset the geostatic trend. How should we view that? First, we might consider what the alternatives here were. According to Aristotle,[31] all those who thought the cosmos to be unique and finite believed that the earth was at the centre of the cosmos. Of course the atomists also believe that our earth is at the centre of our cosmos, even if there are indefinitely many other cosmoi. The only dissenters here were some of the Pythagoreans.

Why did the Greeks accept geocentrism? If one is beginning cosmology, the most natural assumption to make is that the earth is static and central. There are generally perceptible consequences of being in motion, and as the earth shows none of these, it is reasonable to conclude that it is static and it is the heavens which move.[32] If we look at the heavens over the course of a night, the stars appear to circle around us and so it is reasonable to assume that the earth is central. So one might argue that while the initial impulse to geocentrism is understandable, the Greeks never developed anything more sophisticated. Perhaps this is because of a religious prejudice on their part, most explicit when the Stoic Cleanthes believed that Aristarchus should be charged with impiety for suggesting that the earth moved.[33] Perhaps this is because of an anthropocentric prejudice, that man should be at the centre of the cosmos, possibly derived from theology or mythology. It may be the case that any or all of these factors influenced the Greeks, but we should not allow this to obscure the fact that the Greeks had good arguments for geocentrism.[34] The basic fact that there are no perceptible effects of the motion of the earth is a very strong consideration, and remains so up until Galileo and the development of a physics which can deal with a mobile earth.[35] This is one reason why Copernicus' heliocentric hypothesis was not accepted for over sixty years.[36]

III. The Greeks and geocentrism

A further major argument in favour of geocentrism in the ancient world is the question of parallax. If the earth is in motion, then parallax effects, due to the change of position of the observer should be evident. The Greeks were aware of parallax effects, but could detect none due to the motion of the earth.[37] This remained a problem beyond the Copernican revolution, as stellar parallax was not detected until Bessell in 1838. To explain away the lack of parallax, heliocentrists had to argue for a much larger cosmos

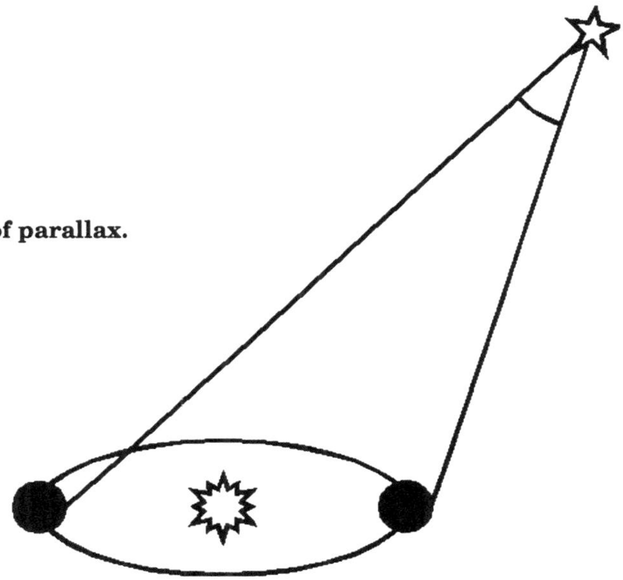

Fig. 41. The problem of parallax.

where parallax effects were too small to be measured by current instruments. Often such a cosmos would be implausibly large by the standards of the time.[38] Fig. 41 illustrates the problem. If the earth orbits the sun in one year, then there is a significant difference in its positions at six-month intervals. This ought to be detectable relative to the fixed stars, which should appear to be in slightly different positions. In fact this is difficult, as the amount the earth moves is very small compared to the distance of the stars, and parallax was not detected until 1838. A further difficulty here lies in choosing which star with which to make parallax measurements. Clearly it is important to choose one relatively near, such as Alpha Centauri, where one is talking in terms of single figures of light years, rather than one that is tens or hundreds of light years away, or even further. Prior to the establishment of a reliable cosmological distance ladder, this is not a straightforward task.[39] One consequence of this is that with each improvement in observational accuracy, advocates of a mobile earth had to increase their estimates of the size of the cosmos such that parallax would be too small to be detected.

There is the further difficulty that if the earth is in motion, why does the moon follow it? Gravity is not a possible answer here, and for Plato the like-to-like principle cannot be invoked as the moon is a star and so is made mainly of fire, and so will not be attracted to the earth. For Aristotle the moon following a mobile earth is inconceivable. Would changing to heliocentrism account for the phenomena better or allow for a simpler and more elegant cosmology for the ancients? The answer on both counts is no.

The phenomena can be met just as well with a geocentric cosmology. There is nothing, before the invention of the telescope, which demands heliocentrism.[40] The cosmological model of Philolaus, apart from its observational drawbacks, is not simpler than Plato's models, and is actually a complex model in its special requirements. If we look to later developments, then if Aristarchus' model is to be able to save the appearances, then the motion of the earth would have to be far more complex than simply a circular orbit around the sun and a daily rotation. The demands of the inequality of the seasons and the precession of the equinoxes would make the motion of the earth extremely complex if the phenomena were to be saved. It should also be noted that Copernicus' system was not significantly simpler than the Ptolemaic model,[41] and it was not until Kepler's work on elliptical orbits sixty years later that the Copernican hypothesis became viable in this respect.

Finally, there is a problem which in my view has been underrated in assessments of the ancients. This is the question of the integrity of the earth. The problem is that if the integrity of the earth is based on its position, then that integrity disappears as soon as the earth is put in motion, or otherwise moved from that position.[42] Aristotle of course has a very strong account of the integrity of the earth. The theory of natural place and natural motion means that earth and water will move to the centre of the cosmos unless hindered from doing so, and so the earth will be at the centre of the cosmos, and will have a great deal of integrity and coherence. However, if the earth were in motion it would no longer have this integrity, nor would heavy objects fall in the usual manner.[43] Aristotle's account of the integrity of the earth and phenomena which we would explain in terms of gravity leads inexorably to geocentrism. So too Plato's equilibrium account of the integrity of the earth in the *Phaedo* would require a central and static position for the earth, as indeed would Anaximander's. If one is going to relate integrity to position, and in the absence of a conception of gravity this has to be an option, then a central position appears the most plausible. This in turn leads almost inexorably to a central and static earth. This is one reason why I argue that the lack of anything like a modern conception of gravity runs very deep in Greek cosmology. One of Copernicus' major difficulties in making heliocentrism plausible lay in accounting for the integrity of a mobile earth. He has to give up the Aristotelian account that went with Ptolemaic cosmology, and falls back on a rather implausible version of a like-to-like principle.[44]

If we look to the *Timaeus*, then Plato's cosmogony is strongly geocentric due to the like-to-like principle. The subsequent dynamics which are generated by the like-to-like principle are not as strongly centri-focal as Aristotle's in principle, as if the earth were moved the dynamical centre point would move too, but as there seems to be no reason for the earth to move in practice Plato's later view is nearly as strongly centri-focal as Aristotle. If the earth were moved for Plato, I suggest it would return to the centre again, under the influence of the like-to-like principle and the constrictive principle generated by the motion of the same, though that might be quite a slow process, unlike in Aristotle. But then we return to the *Phaedo* principle – something equipoised in something uniform has no reason to move. That principle in alliance with the cosmogony means the earth will always be central for Plato. Perhaps this also encapsulates the difference between modern and ancient attitudes on this matter. We ask why the earth should be central and stable, the Greeks ask why it would not be. The Greeks, Plato included, had good reasons for their geocentrism.

IV. The animate heavens

Plato considers the heavenly bodies to be gods, to be animate, to be intelligent and to have souls, and so too the cosmos as a whole. Coming after the materialism of the atomists, is this not a regressive move in Greek cosmology? After all, Plato is well aware of the views of the atomists and Anaxagoras' theory that the celestial bodies are hot stones.[45] So why does Plato adopt this position? As we saw in Chapter 1, Plato has a very low opinion of what matter can do alone and unaided. Even if matter is properly formed (has intrinsic order), then physical principles (such as the like-to-like principle of *Timaeus* 62c-e) are not sufficient to explain how the cosmos acquires or sustains order. For Plato, regular and orderly behaviour requires further explanation, and so the motions of the heavens require explanation. *Laws* 967b puts it like this:

> Those who engaged in these matters accurately would not have been able to use such wonderfully accurate calculations if these entities did not have souls.

It is worth looking closely at what sort of entities Plato postulates here, what they do and what they are supposed to explain. *Timaeus* 30b describes the cosmos in these terms:

> Thus in accordance with the likely account it is necessary to say that this living, ensouled intelligent cosmos has in truth come about through the foresight of god.

The *Timaeus* states that the cosmos does not have many of the things we would usually associate with an animal (or even, within either a Pagan or a Christian tradition, with a god). From *Timaeus* 33b onwards we are told that this 'animal' is perfectly spherical, has neither eyes nor ears as there is nothing external to see or hear, there is no air for it to breathe, no need of any organ to receive food or to excrete the remains, as it is entirely self-sufficient and nothing comes in or goes out. As it needs neither hands to defend itself nor feet to stand on, and has no need of legs or feet to propel itself, it has no limbs, and at 34b we are told it is a god. To say the least, this is a somewhat strange animal, and certainly could not be considered to be an anthropomorphic god, even if it does have intelligence and soul.[46] There may well be assorted criticisms of some Presocratics implicit here too, notably perhaps the Pythagoreans. Quoting Aristotle's lost *On the Pythagoreans*, Stobaeus tells us:

> The universe [according to the Pythagoreans] is unique, and from the infinite it draws in time, breath and void, which distinguishes the places of separate things.[47]

So too the idea of a unique, self-sufficient and exhaustive cosmos cuts against the views of the atomists. There are advantages that come out of this idea of the cosmos though. If the cosmos is like an animal, then we have an analogue for the integrity and the internal organisation of the cosmos, and the ongoing order of the cosmos. A typical ancient contrast here is between the body with life and mind and what happens to the body after death.[48] What does this god do though? All it does is revolve uniformly in one place. At *Timaeus* 34a we are told that the cosmos was given:

> A motion proper to its body, that of the seven motions which is best suited to reason and intelligence. Therefore he made it move in a circle, revolving of itself uniformly and in the same place, and he took from it all trace of the other six motions and kept it free from their wanderings.[49]

Here I would emphasise that this is entirely regular circular motion with no deviations. This is so for the fixed stars and the planets as well. The stars are spherical and intelligent like the cosmos (40a) and are divine and living creatures (40b). As with the cosmos, they have motions befitting their intelligence, and so we are told at *Timaeus* 40a:

> To each of these he gave two motions, one being uniform and in the same place, always thinking the same thoughts concerning the same things, the other being a forward motion obeying the revolution of the same and similar. With regard to the other five motions, they were motionless and still, in order that each might attain the greatest possible perfection.

The planets are brought into being in order to generate time. We are then told at *Timaeus* 38e:

> When each of the bodies required to help in the production of time had been bought into its proper motion, as generated living creatures having their bodies bound with living bonds and having learnt their assigned duties, they began to move according to the motion of the different.

As I have argued, we have no reason to suppose that the planets deviate in any way. They follow the same and the different, where the fixed stars follow the same and the cosmos rotates in one place. The essential point here then is that the celestial bodies have no freedom of action (or no desire to deviate from regular circular motion). They have the intelligence to carry out their assigned duties, and not to do anything else. It is their intelligence which explains their regular and orderly behaviour. That they move, and do so without any external compulsion, is explained by their having souls. For Plato, the soul is a principle of motion. So at *Laws* 895c we find:

> *Athenian Stranger*: If we should see that this motion (self-generating motion) had come about in something composed of earth, water or fire, whether separate or in combination, what should we say was the state of that thing?
> *Cleinias*: You are asking me is whether we should speak of this thing as being alive?
> *AS*: Yes.
> *Cleinias*: Emphatically so.

So too at *Laws* 897b:

> If the whole course and motion of the heavens and everything that are in them are similar in nature to the motion and revolution and calculations of intelligence (*nous*), and work in a related manner, then clearly we must say that the best soul has charge of the whole cosmos and leads it along these selfsame courses.[51]

Although the cosmos and the heavenly bodies have life, soul, divinity and intelligence, they have these in a highly circumscribed and attenuated manner, from a modern point of view.[52] If this is describable as vitalism, it is a highly depersonalised one where the attributes of animate beings that are required for the description of the cosmos have been carefully sorted from those that are not.[53] This allows us to distinguish quite sharply between Plato and the mythological and magical traditions. The key matter here is that these souls/gods are not capricious. They always act for the best so will always act in the same manner. There is in principle nothing unpredictable or irregular about their behaviour.[54] This is something that the *Epinomis* is very clear on at 982d ff. The stars have been acting in this manner for a very long time,[55] and do not ever decide to do anything different. In terms of theology, it is significant that Plato's god always acts for the best, in sharp distinction with many previous Greek conceptions of the gods.

Why does Plato not simply express this in terms of physical laws, or just leave the behaviour of the heavens as a set of regularities? Plato's proto-realism will not let him rest with the idea of bare regularities. The idea of physical law though is something that has been hard won and was not something that was simply available for use by the Greeks. Why that should be so and why Plato answers in the way he does I shall discuss in detail in the next section. In the meantime it is worth noting that a typical seventeenth-, and indeed eighteenth-century view is that in some manner god underwrites the laws of nature and is responsible for the regular behaviour of matter.[56] Being a Christian god, he always acts in a good manner, and so we see a regular and well arranged world. Whether we can know the laws which god has in mind or are limited to knowledge of regularities depended on the nature of god (the voluntarism debate) and our relation to him. Thus Berkeley urged an early form of instrumentalism, while Newton and Boyle were more optimistic. This was important because the mechanical philosophy, in limiting matter to extension and motion, had removed all the scholastic and occult qualities which might be used to explain why bodies behave in an orderly manner. Even those who dissented from the mechanical philosophy, such as the Cambridge Platonists, were keen to emphasise regular behaviour.[57] Thus we find More saying:

> I affirm with Descartes, that nothing affects our sense but such variations of matter as are made by differences of motion, figure, position of parts, etc., but I dissent from him in this, that I hold it is not mere and pure mechanical motion that causes all the sensible modifications of matter, but that many times the immediate director therof is the spirit of nature, one and the same everywhere, and acting always alike upon the like occasions, as a clear minded man, and of a solid judgement, gives always the same verdict in the same circumstances.

Plato is not alone then in the history of scientific though in using divine entities to explain regular behaviour.[58] The key question is whether these entities are law abiding or not, and for Plato they most certainly are. Lennox argues that Plato's teleology is unnatural.[59] In the sense that teleology is imposed on the world by the demiurge, and so is not a natural, original feature of the world, that is so. It is also important to recognise that Plato's demiurge is a god subject to natural law and regular behaviour, as are his demi-gods, and so there is a sharp contrast with the caprice of the gods of Greek myth.[60]

V. Nomological realism

Plato's views on the explanation of regularities are worth thinking of in terms of the cave allegory. The prisoners in the cave are chained facing the wall, and see only the shadows of the objects being carried along the road behind them. The state of the prisoners might be described as Humean.[61] They may associate the voices they hear with the passing shadows which they see, as at *Republic* 515b:

> If the prison had an echo from the opposite wall, then whenever a passer by spoke, do you think they would they suppose anything other than that the voice belonged to the shadow?

They cannot though make any further investigation. So too they can associate one object or a sequence of objects with what objects will follow, but again can investigate no further than this. At *Republic* 516cd Socrates says that perhaps there was:

> Some honour and praise among them and prizes for seeing most sharply the passing things, and being best able to remember which of them came before, which after and which together, and so had the greatest power in predicting which of these was to come.

As emphasised in Chapter 2, however, for Plato we both can escape and ought to escape from the cave. Clearly there are many other epistemological issues which Plato wishes to focus on here, but one point might be put like this. If the world were such that there were only regularities, or we only had access to regularities, then we would not be able to escape the cave. Important here is Plato's distinction between *empeiria*, experience or practice, and *techne*, art or skill. As we saw in Chapter 2, often when Plato contrasts the two, *empeiria* is coupled with *tribe*, literally rubbing or threshing, though in these contexts it carries a sense of learning by routine or perhaps trial and error, and may well carry certain pejorative connotations.[62] Plato supposes that there is always a *techne*. However, if there were not something more than regularities, or as Plato puts it 'what usually happens', then there would be nothing to base the *techne* on. So we find at *Gorgias* 500e4:

> Cookery seems to me to be not a skill (*techne*) but an empirical knack (*empeiria*), while medicine investigates the nature of those it treats and the reason for what it does, and has some account to give of each of these things, does medicine. But the other, concerned with pleasure, at which its whole treatment is aimed, works towards this in an entirely unskilful (*atechnos*) manner, not having investigated the nature of nor the reason for pleasure, unreasoning (*alogos*) in every way, and so, one might say, making no distinctions, and by routine and knack (*tribe kai empeiria*) it preserves only a memory of what usually happens, and that is how it is able to provide pleasures.

So too at *Phaedrus* 270b, this time the contrast being medicine and rhetoric:

> In both cases it is necessary to analyse a nature, of the body on the one hand, and of the soul on the other, if you intend to have not only practice and experience (*me tribe monon kai empeiria*), but skill (*techne*), when attempting to bring about health and bodily strength by means of medicine and diet, or to instil will-power and virtue by means of discourses and rules for the conduct of life.

I do not wish to attribute to Plato an understanding of the modern distinction between regularities and laws, nor an understanding of the modern conception of physical law. He does though require something beyond regularities, and as I have argued for other issues, this narrows the gap between Plato and some modern positions. Vlastos may well believe that the laws of nature are 'radical contingencies ... for which no further reason can be given',[63] and so he sees a need to excuse Plato by commenting that there was 'no Hume in Plato's past'.[64] There are many modern philosophers of science though who would disagree with Hume, and object to the dissolution of the 'causal into the casual', as, I suspect, would Plato.[65] Vlastos admits there are problems of supporting counter-factuals on a regularity theory,[66] but there are also other serious difficulties.[67]

1. There is the problem of vacuous laws. If laws are simply true universal generalisations, any universal generalisation (however vacuous or silly) counts as a law if it has no counter instances.

2. There is the problem of non-instantial laws. Newton's first law ('All bodies on which no net external force is acting either remain at rest or move in uniform velocity in a straight line') is non-vacuous but has no instances. If one amends the regularity theory to 'universal generalisation which have instances', in order to exclude vacuous laws, Newton's first law is ruled out.

3. There is the problem of missing values. Laws are quite general (F=ma ranges over all values of F, m and a) and so range over infinite instances, only some of which can have been actualised. Regularity theory can only support laws for those instances which have been actualised.

4. There is the problem of accidental generalisations. 'All dogs with three legs and one ear are called lucky', if true, is as much of a law as anything offered by physics.

5. There is a problem for some 'epistemic' regularity theories have the implication that there are no unknown laws. Whether something is a law depends upon whether scientists treat generalisations as laws. Scientists cannot treat what they have not discovered, so there are no unknown laws.

This is not to say that the realist position is problem free, the latest theories running into problems of identification, that is giving an account of the relation of necessity that makes law rather than regularity, and inference, that is if we hold that necessity, however construed, holds between universals, what is it that allows us to say anything about particulars.[68] Interestingly for what follows, Dretske offers a legal analogy to explicate the inference relation.[69] Vlastos writing in 1969 held that the regularity theorists were more populous, and while that has probably changed in the last thirty years, the important point is that the realist position is still a viable one.[70]

There is another important aspect to Plato's distinction between *empeiria* and *techne*, to do with the development of science. One might argue that firstly humankind developed technology, the ability to manipulate and predict phenomena without being able to produce a proper explanation of those phenomena. The first ages of mankind can be defined in relation to the dominant technology, thus stone age, bronze age, iron age, etc. Only after this do proper scientific explanations develop. Exactly when and where is too large an issue here, though personally I would place this with the Milesians and the creation of the idea of the cosmos and the rejection of mythological and supernatural explanations. One distinguishing feature of a society that might be said to possess science then would be a consciousness of the distinction between science and technology. That, I suggest, is first clearly formulated by Plato, and that is an important point in the development of Greek science. While his distinction between *empeiria* and *techne* is clearly not an exact match for the modern distinction between science and technology, it does show a recognition of something beyond manipulation and prediction based on 'experience and grinding away' (*empeiria kai tribe*), which we cannot give a proper account of.[71] We might also differ from Plato in what we believe to be on the *techne* side of this distinction, but the key phrase (*Gorgias* 500e) would seem to be that medicine at least 'investigates the nature of those it treats and the reason for what it does, and has some account to give of each of these things'.

VI. Civil law and physical law

We can now examine why, in order to explain regular and orderly behaviour, Plato attributes soul and intelligence to the heavenly bodies. He does not have at his disposal a conception of gravity or action at a distance, or indeed a notion of physical law in the modern sense at all. The like-to-like principle can explain the regular behaviour of matter, but as we saw in Chapter 1 it is insufficient to explain the order and the continuing apparently intelligent behaviour of the heavens. In order to produce and sustain good order intelligence is required. So where the proto-physical law of the like-to-like principle fails we need to turn to something else. The Greeks also had the idea of *nomos*, law (in the civil sense) or convention, which was often contrasted with *phusis*, nature. Thus the idea of a law of nature in this sense would be something that might well look rather strange to the Greeks, and indeed where this phrase occurs in Plato and Aristotle it looks rather paradoxical.[72] When other ides fail, though, one might try to apply the idea of civil law to nature. However, there is an immediate problem. Why should nature obey something like civil law? It is a matter of choice, of intelligence and moral character for humans to obey such laws. Thus Plato gives the cosmos and the celestial bodies some semblance of intelligence and soul in order that they may be able to obey laws. In order to get perfect regularity, those entities must not only be able to make the choice to obey the laws of celestial motion, but also have to be perfectly rational, that is gods, in order that they do so all the time. *Timaeus* 47bc is of considerable importance here, as it contrasts the entirely unwandering (*pantos aplaneis*) revolutions of god with the wandering (*peplanemenas*) revolutions in ourselves. Something else must be added here, which is that the principles of celestial motion must be the best sort of principles otherwise perfectly good souls would not freely choose to obey them. Here then we see another reason for Plato's teleology. These principles need to be good in order to be obeyed and the celestial bodies need to be good to obey them without any further compulsion.

Another important factor to come out of this is the ongoing work of intelligence in the cosmos. As we have seen, it is one thing for the cosmos to come into an organised state, quite another for it to continue in that state. Plato has criticisms of the atomists on this matter, but here we perhaps see an important reply to Anaxagoras, by showing just how intelligence may continue to operate in the cosmos and contribute to its organisation and stability. The latter is important as for Plato the physical world is in some sort of flux. Just how radical that flux is will be a topic for the next chapter, as will the relations of the heavens to that flux. The fact that the heavenly bodies have divine souls will be important in this context for the ongoing stability of the heavens. One might also note that it is the heavens where reason scores its greatest victory over necessity, as is marked by the transition from the works of reason to those of necessity at *Timaeus* 47e. That is not surprising perhaps as we have the continued influence of intelligence.

This is not of course Plato's only attempt to explain consistent behaviour with something more than a regularity theory. If we look to *Phaedo* 102b-105b, there is the 'bringing on' relationship for hot/fire and cold/snow,[73] odd/3 & 5 and even/2 & 4.[74] So too if we look to *Timaeus* 61d ff., we find an account of the properties and behaviour of atoms related to their shapes. It is interesting to note that in each of Plato's three accounts of why this should be so he begins to address the modern problems of identification (what is the necessity involved in these relationships?) and inference (if relations hold between universals, why do particulars follow them?). In the *Phaedo*, the nature of the necessity is the relation between forms (though it is not entirely clear in the *Phaedo* what that is), and inference is validated by the model of participation of particulars in forms.[75]

In the *Timaeus* we have a different sort of necessity based on geometrical atomism. Although the *Phaedo* does not use geometrical examples, it is not hard to see how it might. What is different about the *Timaeus* is that it connects fire with swiftness and ability to cut of necessity via the geometrical structure of the fire atom.[76] With the account of the heavens in the *Timaeus* and *Laws*, the nature of necessity is again an issue, as is why individual gods will act in a rational and regular manner.

One can of course raise objections against all of these accounts, but all I want to point out here is Plato's struggle towards nomological realism in the absence of an idea of physical law. To some extent I would agree with Vlastos, when he suggest that Plato's comment to Leucippus and Democritus might well have been:

> If you must have rational necessity in nature, you cannot get it from regularities which are matters of brute fact. The only kind of necessity known to me is that of mathematics and dialectic.[77]

In relation to this it is interesting to see how much of the fundamental structure of the physical world is given a mathematical, or more accurately a geometrical basis in the *Timaeus*. That in itself is a significant move in Greek cosmology, which I shall discuss in more detail in the context of geometrical atomism. While Plato never manages a proper conception of physical law, the geometric cosmology may well have been influential in its later formulation. The Renaissance becomes interested in the application of mathematics to nature, and we can find Galileo saying:

> Philosophy is written in that vast book which stands forever open before our eyes, I mean the universe; but it cannot be read until we have learnt the language and become familiar with the characters in which it is written. It is written in mathematical language, and the letters are triangles, circles and other geometrical figures, without which means it is humanly impossible to comprehend a single word.

One way in which the scientific revolution may have come to the idea of physical law is via the though that if nature is fundamentally geometrical, then perhaps it obeys laws something like of geometry. Plato's geometrical cosmology was a significant step forward, even if he was not able to exploit to the full all the possibilities that it offered.

VII. Analogues for physical processes

It is sometimes said that the Greeks had an organic picture of the world, while we have a mechanistic one.[78] Clearly that is an overgeneralisation, but there is a significant difference in style of explanation between the ancients who did hold a more organic world view and modern views. Since the scientific revolution of the seventeenth century, and the overthrow of scholasticism and the natural magic tradition, we have become accustomed to scientific explanations having a certain form. Rejecting the scholastic modes of explanation and occult qualities, we explain in terms of physical causation and mathematically framed laws of nature.[79] As we consider the world to be ultimately made up of small particles, ultimate physical explanation is in terms of those particles and their properties.[80] We use mechanical models as analogues for physical processes. We also use physical and mechanical analogues to describe biological processes. This is perhaps most clearly seen with the seventeenth-century view that animals were no more than very complicated pieces of clockwork, clockwork being the key metaphor for the mechanical philosophy of the seventeenth century.[81] Clearly this sort of explanation has become more sophisticated, but what I shall term the direction of explanation remains the same. We explain biological processes in physical and mechanical terms.

There are though other possibilities, and we might reverse this direction explanation, and use biological analogues and for physical processes, as Plato does with the cosmos and the celestial bodies. Aristotle is perhaps a good example of how biological analogues can intrude into physics in a more subtle manner, and this is something which is continued by the Stoics.[82] With modern science having moved away from the use of biological analogues, the essential question is the following. Is it evident, in the context of fourth-century Greece, that one should be using mechanical rather than biological analogues? The answer to that is quite clearly no. There are several reasons for this. If one wishes to explain order and regularity, mechanisms are not a good option for the Greeks. As Furley has argued, mechanism, and particularly clockwork, are paradigms of regularity and predictability to us.[83] The mechanisms available to the Greeks though, say the cart and the winch do not offer such an example.[84] It is not surprising then that even where the Greeks might be described as materialists, they are not mechanists. It is worth noting that even the atomists use biological analogues for physical processes as well. As we saw in Chapter 1, Sextus tells us that Democritus bases his argument for a like-to-like principle out of the vortex on both animate and inanimate things. Animals congregate with animals of the same and similarly in the case of inanimate things, such as seeds that are being riddled and pebbles on the sea-shore.[85]

Biological analogues were not the only Greek alternative to mechanical ones. As Lloyd has argued, the Greeks have quite a variety of analogues for the cosmos. These can be grouped under the ideas of the cosmos as an artefact, as a political entity or as a living entity, but even then there is a great deal of variety within the groups.[86] It is important to appreciate that the Greeks can get a good deal from these analogues. They feel the need to explain the unity of cosmos, the current and ongoing order of the cosmos and the nature of processes within the cosmos. Thus the idea of the cosmos as something analogous to an animal or a *polis* would seem quite reasonable to them.[87] It is easy to forget that these are significant questions for the Greeks. For the Christian tradition it is God who creates and sustains the unity of the cosmos, and gives it its current and ongoing order. For modern science, the cosmos is not a unified entity in the same way. This is due to many factors, not least a more sophisticated understanding of the nature of space and big bang cosmogony. However, the unity of the cosmos is a problem for those who believe in multiple worlds, as one has to give some sort of account as to what individuates the cosmoi. The ancient atomists would possibly answer that by referring to the membrane which is supposed to form around each cosmos.[88]

What then have we got in the way of objections to Plato's idea of the cosmos and the celestial bodies as animate and intelligent entities? Plato is careful to limit the characteristics of these entities in such a way that they always behave in a regular manner. His view, then, is not regressive relative to the Greek advances in terms of natural explanation versus mythology and supernatural explanation. We can hardly blame him for not employing ideas such as gravitation and physical law which were not available to him. Plato makes an interesting effort at explaining the regular behaviour of the heavens with what he has available, which are biological analogues and the notion of civil law. Out of that come the nomological reasons which Plato has for teleology. Again, in context biological analogues would have appeared at least as viable a possibility as mechanical analogues, even given that the Greeks had mechanical analogues in a reasonable form, and it is less than clear that they did. Plato gets a good deal of explanatory mileage out of these analogues on questions which are of considerable importance for Greek cosmology. We are right at first reading to be worried by Plato's account of the animate and intelligent heavens. However, if we place his ideas carefully in context, and assess the resources that were available to him, it is possible to see Plato making a very reasonable attempt to explain why the heavens move as they do.

At the beginning of the twenty-first century, we perhaps ought to be more open to alternatives to the mechanical conception of the world. Why so? Because the last century saw the gradual dissolution of the classical mechanical picture of the world. This begins with the rise of thermodynamics, and the idea of entropy and the irreversibility of many physical processes in the latter part of the nineteenth century.[89] Relativity broke down some mechanical conceptions of matter with the interconversion of matter and energy. Quantum mechanics dealt further blows to the mechanical picture with the idea of inherent probabilism and indeterminism, and the idea of wave/particle duality. The ultimate difficulty we have found is that the micro world is simply not like the macro world. While it was possible to use macroscopic analogues for the micro world (particles being like billiard balls being a typical example) it is no longer possible to do so. The physical micro world does not operate along mechanical lines.

I am not suggesting here that biological analogues for physical processes have made or will make a comeback. I merely wish to point out that the mechanical models for physical processes which were predominant from the seventeenth to the nineteenth century are not the be all and end all of science. Doubtless they were a viable and plausible possibility at their time, and they facilitated a good deal of progress, but science is now moving beyond them. In due course, we may come to understand why we found such mechanical analogues so alluring, at a specific stage in the development of science, relative to the analogues which succeed them. In the meantime, we should be as generous to the ancients. There was good reason why biological analogues for physical processes should have been attractive to them.

VIII. Plato and subsequent astronomy

Even if Plato's ideas about the nature of the cosmos and the celestial bodies were reasonable in context, one might still be worried about the effect that they had on subsequent astronomy and cosmology. In fact, with the context of Greek geocentrism, they were important in allowing some significant advances. What explanations of the motions of the heavens do the Greeks have? Effectively they have three sorts of answer, the Platonic, the Aristotelian and the vortex. Plato's answer is that the planets have the ability to move themselves, and regulate their own motions. The heavenly bodies are gods, are animate and have souls. As we have seen, it is the intelligence of these entities that guarantees their regular behaviour, and they are animate and ensouled to that extent and no more. Alternatively there is Aristotle's answer of nesting spheres of aether, with the full complement of driving spheres and counteracting spheres. This looks somewhat more mechanical in its orientation, and certainly the planets are more constrained by what is around them. They are though executing their natural motion and are moving due to their love of the prime mover, and are described by Aristotle as being divine,[90] so this view is not as distinct from Plato as that of the atomists and Anaxagoras, who hold that purely material celestial bodies with no active properties of their own are driven round by a vortex.[91]

One can see from this brief survey that the idea of a force such as gravity controlling the motions of the heavens is a long way off for the Greeks. Even if the Greeks had been in possession of a conception of gravity though, considerable difficulties lay in the way of their utilising such an idea to explain the motions of the heavens. Why so? For the Greeks there is a complexity of the motions of the heavens which cannot be explained by a single force emanating from a homogeneous spherical earth. That complexity is generated inherently by geocentrism allied to a static earth. In such cosmologies all the motions of the heavens have to be real motions. None of them can be apparent motions due to motion of the observer on the earth. Even in a model as simple as Plato's, how can one explain the different motions generated by the same and the different as due to a single force in a spherical and homogeneous earth? The position rapidly becomes more difficult as the complexity of the model grows. The addition of extra spheres for the sun and moon and the hippopede generating spheres for the other planets in the models of Eudoxus and Callippus makes it even more difficult to see all of these motions in terms of a single force emanating from the earth.

It should also be evident that like-to-like principles and other ideas which the Greeks use to account for what we would think of as gravitational phenomena will not be of any help here either. Aristotle's model is highly ingenious in that it gives a coherent picture of the cosmos tied to a general theory of motion and natural place (which does give an account of many gravitational phenomena), even if it does require a special account of celestial motion. However, there are problems with accommodating further astronomical sophistication within the model. At this stage there is already too much sophistication for the vortex theory, and that may be so even with Plato's simple model of the *Timaeus*. While a single vortex might give a first account of the motions of the heavens, it is hard to see how (or indeed why) two could combine to give the motions of the same and the different.[92] When one moves on to the Eudoxan system with 27 circular motions, the idea of vortices is no longer plausible. The later invention of more the sophisticated devices of epicycle, eccentric and equant pose even greater problems, and create difficulties for the Aristotelian answer as well.

The epicycle

The epicycle (Fig. 42) is the basic device of later Greek astronomy. By a suitable choice of speeds of rotation retrograde motion can be produced. One advantage over the hippopede is that different, and more accurate shapes of retrograde motion can be generated.

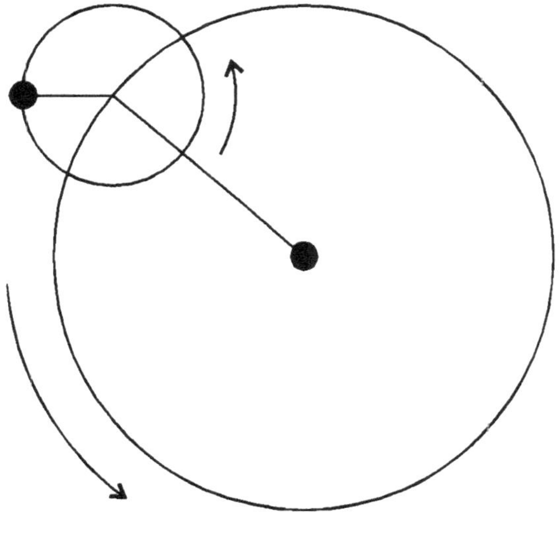

Fig. 42. Epicycle.

The problem here, for both the application of gravity and for Aristotle, is why does a planet revolve around an imaginary point (and why does that point itself revolve around the earth?). A further problem for Aristotle, and indeed one that is inherent in the concentric sphere scheme, is that epicycles generate differences of distance from the planet to the centre of the cosmos.

The eccentric circle

One of the difficulties with concentric sphere astronomy was that all the celestial bodies were held at a constant distance from the earth, but some show considerable variation in apparent size and brightness. With the eccentric circle, the earth is now moved from the centre such that there is variation in distance to the planet (Fig. 43). The eccentricity here is greatly exaggerated to demonstrate this effect. We can add in an epicycle again to generate retrograde motion (Fig. 44).

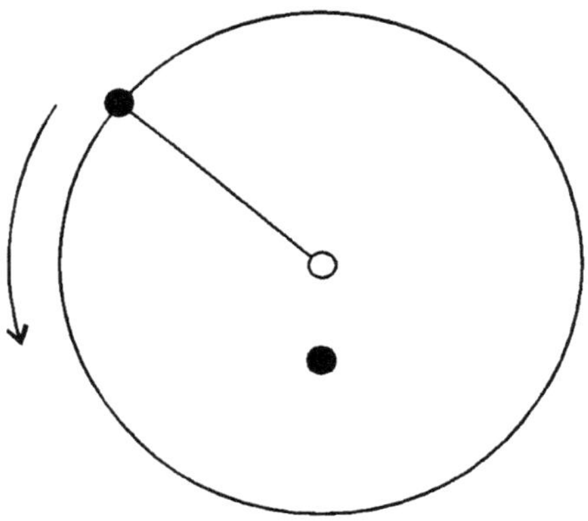

Fig. 43 Eccentric circle.

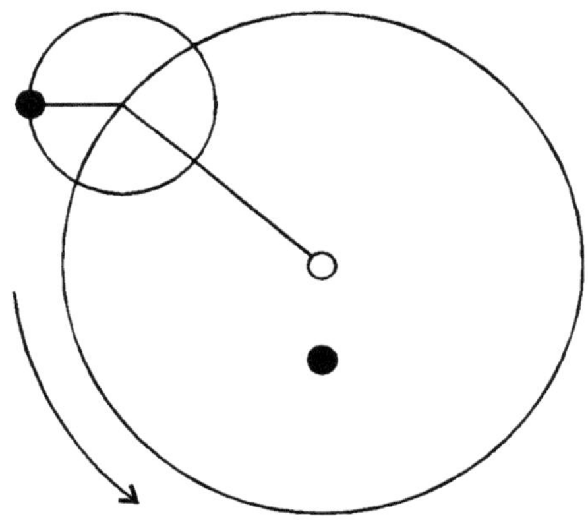

Fig. 44. Eccentric circle with epicycle.

The critical problem here is that we have a split between the physical centre of the cosmos, the earth, and the rotational centre of the cosmos, the eccentric point. It is then hard to see how a force emanating from the earth could control the motions of the planets, or why there should be a force emanating from the point in space which is the rotational centre.

The equant

The planets do not move with uniform motion, even when they are not retrogressing. So too there are variations in the speeds of the sun and moon. One way of generating variations in angular velocity from combinations of regular circular motion is by the equant device (Fig. 45). We have again an eccentric so the earth is offset from the centre of the cosmos, and an epicycle. We also have the equant point, about which there is constant angular velocity. The angular motion of the planet will then be constant viewed from the equant point, but not from the earth, nor from the centre of the deferent circle. Here the planet will travel from A to B in the same time as it travels from B to A. From the equant point that is a split of 180° for each section, but clearly that is not so viewed from the earth. Again the eccentricity here is greatly exaggerated to demonstrate the effect. Note also that the epicycle will both affect the shape of the orbit and vary the planet's angular velocity.

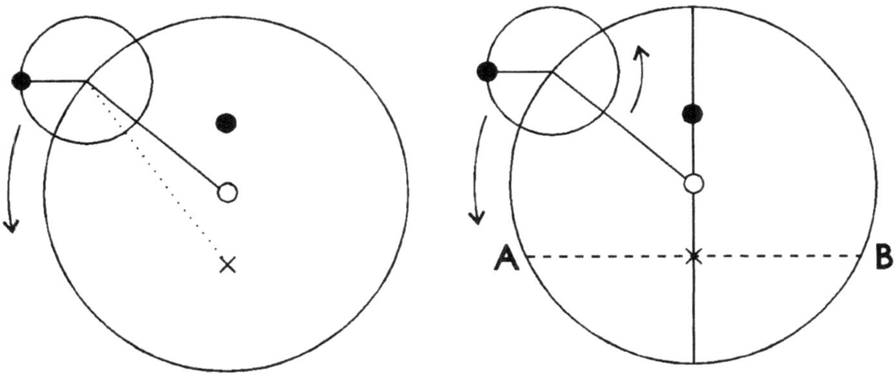

Fig. 45. Equant.

The equant poses similar problems to the eccentric. Now we have another point in space, the equant point, and it is hard to see how that could control the motion of the planets. Thus it is very difficult to conceive of a force emanating from the physical centre of the cosmos or the rotational centre of the cosmos which will control the orbits of the planets. This is a problem that cannot be solved until the orbits of the planets are simplified and the physical and rotational centres of the cosmos are in the same place again.

The position of Ptolemy is of interest here. While his cosmology is generally thought of as Aristotelian, there are also some significant Platonic elements in it, especially in relation to planetary motion.[93] In the *Suntaxis* and indeed also the *Tetrabiblos*, there is no mention of any celestial spheres. The later *Planetary Hypotheses* may present a slightly different picture. Ptolemy presents the 'nesting spheres' hypothesis in this work (which is absent from both the *Suntaxis* and the *Tetrabiblos*), whereby the spheres of each of the celestial bodies fit together perfectly with no overlap or void between them.[94] Ptolemy is adamant in the *Planetary Hypotheses* that there are no reacting spheres.[95] Indeed, in his new system there is no need for them, as he conceives of each of the heavenly bodies as having its own source of motion.[96] Each of the heavenly bodies moves itself without any assistance or interference from any other body.[97] This means that there is no transference of motion between spheres (Ptolemy specifically rules this out)[98] and so the motions of the outer heavenly bodies are not transferred to the inner ones.[99] It would seem then that far from being a hindrance to progress, Plato's ideas about the motions of the planets made up the only scheme within which Greek astronomy and cosmology could make progress. That would certainly seem to be so within geocentrism (and as we have seen, the Greeks have good reasons for their geocentrism), but would also be true for Aristarchus as well, for without the idea of elliptical orbits, it is not possible for the sun to control the orbits of the planets.

It is important to recognise that the move to gravity as the controlling factor in planetary astronomy was a long one that came in stages. While Copernicus hypothesised a sun centred cosmos, he retained the idea of regular circular motion. While he had ingenious means for replacing equants, he needed to retain epicycles and eccentrics in order to account for the non-circular orbits of the planets. So while Copernicus gave us heliostatism, that is a cosmology where the sun does not move, he did not give us heliocentrism, a cosmology where the sun is both the physical and rotational centre of the cosmos, as the eccentric device splits the physical and rotational centres. So too Ptolemy gave a geostatic but not a geocentric cosmology as he too used the eccentric. Kepler's proposal of elliptical orbits for the planets is perhaps the critical stage, as now the physical and the rotational centres of the cosmos are then reunited (Figs 46 and 47). It is only at this point that one can begin to think in terms of a force emanating from the sun which might control the orbits of the planets. This idea indeed occurs to Kepler, but he chooses magnetism as opposed to gravity as the controlling force. It is not until later in the seventeenth century that we get gravity as the controlling force, and later still that we get the critical $1/d^2$ formulation.

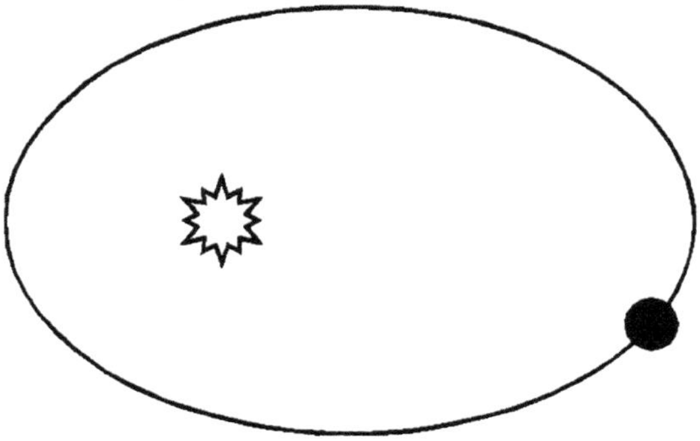

Fig. 46. Kepler's first law. Kepler's ellipses reunite the physical and rotational centres of the cosmos, which are both at one of the foci of the ellipse. This greatly simplifies cosmology. That one uses only one ellipse per planet, and not an ad hoc collections of epicycles, eccentrics and equants also simplifies astronomy.

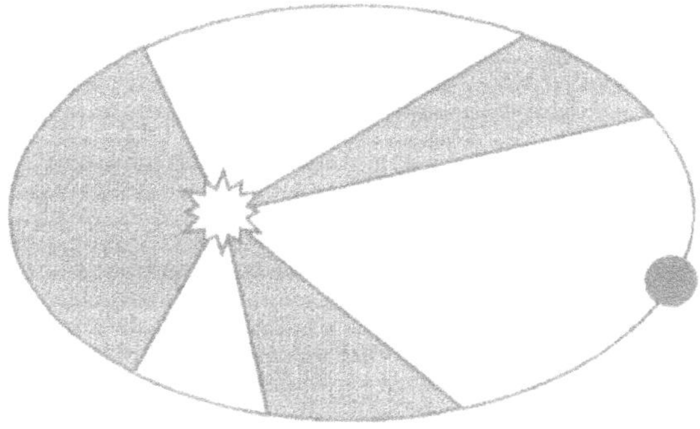

Fig. 47. Kepler's second law. Kepler's speed law generates the variations in a planet's orbital velocity as a natural consequence of the elliptical orbits. Equal areas are swept out in equal times by the line from the planet to the focus of the ellipse. Planets then move more swiftly as they approach the sun, less swiftly as they recede from it.

IX. Plato and subsequent cosmology

I shall conclude this chapter by looking briefly at Plato's influence on subsequent astronomy and cosmology. One reason for giving details of the work of Eudoxus and Callippus in the previous chapter has been to show just how immediately fruitful the challenge set by Plato was. They tackle precisely the problems that Plato sets in the *Timaeus* in precisely the way he would like. So too we have seen that the next phase of astronomy, involving epicycles, eccentrics and equants, also held to the Platonic challenge and so too was extremely fruitful.[100] There is no doubt about the importance of Eudoxus and Callippus in the history of astronomy, nor of those who replaced the concentric sphere scheme with epicyclic astronomy such as Hipparchus, Apollonius and Ptolemy.

Some, however, have been highly critical of Plato. By demanding circular motion (solely on the basis of metaphysical speculation), they argue, Plato set astronomy on the wrong track for 2,000 years, only for Kepler to reject Platonism and put things right again. This would be a harsh judgement for several reasons. First, it is the fate of virtually all scientific ideas eventually to become outdated and so in a sense become a bar to progress. That the Platonic challenge produced progress at the outset, and indeed for many years is indisputable. It is also the case in the history of science that some ideas seem so natural and evident that it is easy to forget that someone actually introduced them and at the time that was an important step. It is important in relation to this to remember that the Platonic challenge asks for regular and ordered motions as well as circular motions, and implicit in this is the idea of a unified theory of the heavens. Since Plato there has been no question that we should be looking for all of these, except circular motion. If the objection is merely that Plato got it wrong, so much is evident, and what is more important is whether circular motion was a plausible and viable possibility at the time.

Secondly, Ptolemaic astronomy and circular motion were indeed rejected in the Copernican Revolution. However, there is a considerable amount of analysis required in sifting which elements of the Platonic challenge were rejected and which elements may have proved inspirational in instigating and driving forward the Copernican revolution. There is no simple rejection of Platonic principles here. Copernicus straightforwardly accepts regular circular motion:[101]

> The movement of the celestial bodies is regular, circular and everlasting – or else compounded of circular movements.[102]

He applies this so strictly that he will not accept the equant. Thus he says of Ptolemy's system:

> These theories were not adequate unless certain equants were also conceived; it then appeared that a planet moved with uniform velocity neither on its deferent nor about the centre of its epicycle. Hence a system of this sort seemed neither sufficiently absolute nor sufficiently pleasing to the mind … I often considered whether there could be found a more reasonable arrangement of circles, from which every apparent inequality could be derived and in which everything would move uniformly about its proper centre, as the rule of absolute motion requires.[103]

One might also compare Copernicus' following criticism of his predecessors with *Laws* 898ab on celestial motion being part of one plan and one order:

> They do not employ the same principles, assumptions or demonstrations for the revolutions and the apparent movements … moreover they have not been able to discover or to infer the chief point of all, i.e. the form of the world and the certain commensurability of its parts. But they are in exactly the same fix as someone taking from different places hands, feet, head and other limbs – shaped very beautifully but not with reference to one body and without correspondence to one another – so that such parts make up a monster rather than a man.[104]

So too Copernicus makes reference to a 'best and most orderly demiourgos',[105] and to 'godlike circular motions'.[106] Copernicus' precise motivation and intentions are a matter of considerable debate. What is generally agreed, though, is that Copernicus' main concern was not with the empirical adequacy of the Ptolemaic system,[107] but rather with its structure, and that Copernicus was influenced to some extent by Neoplatonic thought.[108] It is clear that Copernicus adhered very strictly to regular circular motion,[109] and that this far from being a bar to progress may well have precipitated the chain of thought leading to the heliocentric hypothesis.

If we look to Kepler, his Neoplatonism is even more marked.[110] One can point in particular to his attempt to derive the ratios of the orbits of the planets from the five Platonic solids. Kepler though rejects circular motion, and is often held to have thus broken with Platonism. One question we might ask here is simply, What shape should Plato have used? If one is going to have more than a purely descriptive astronomy, one must take some view on the basic motions of the heavens, and that brings us back to Meno's paradox. What hypothesis should we propose? From an indefinite number of possible motions circular motion must have seemed by far the most plausible and viable possibility, especially from a geocentric viewpoint, and as we have seen the Greeks had very good reasons for their geocentrism. The stars appear to move in virtually perfect circles, so a good first guess at the motions of the planets has to be circular motions or a combination of circular motions. Elliptical orbits for the planets are in fact quite hard to conceive of even with heliocentrism. This is because the orbits are not actually perfect ellipses, and because the eccentricity of the ellipses varies considerably from planet to planet. The orbits are not elliptical because of perturbation (Figs 48 and 49).

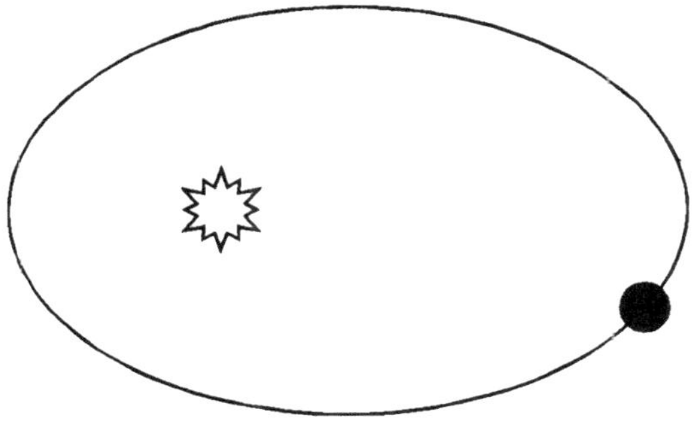

Fig. 48. Two bodies. If we have a two body system with no external influences then we get a straightforward elliptical orbit. The only forces operating are those between the orbiting and the orbited bodies.

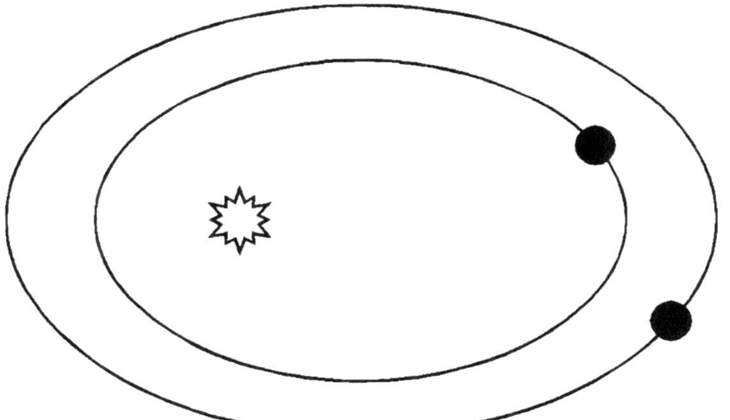

Fig. 49. Perturbations. When we add another orbiting body, if we consider only the forces between the bodies and the sun, then we have two perfect ellipses, as here. However, we also have to take into account forces operating between the two orbiting bodies, which will distort their elliptical orbits. These distortions are known as perturbations.

When Kepler eventually breaks with circular motion, it is important to recognise what he gains by this. As explained in the previous section, there is the important reunification of the physical and rotational centres of the cosmos, which allows the idea of a force emanating from the sun to control the motions of the planets. Allied to this is the rejection of the Ptolemaic devices of epicycle, eccentric and equant which had to be applied in a relatively ad hoc manner to each planet. What this achieves is an enormous simplification of the heavens, and a unification of the account of the motion of each of the celestial bodies, and again one might look to *Laws* 898ab and its demand that there be one plan and order to the cosmos. By breaking the letter of the Platonic challenge in rejecting circular motion Kepler is able to preserve its spirit or regular, ordered motion and one plan. Indeed that spirit is very much alive today, but has become so much a background assumption that we rather take it for granted. It is also interesting to consider Kepler in relation to *Timaeus* 54a, where Timaeus says that anyone who can produce more beautiful shapes for the basic triangles would be welcomed as a friend rather than a foe. So too here perhaps. I strongly suspect that Plato would welcome Kepler as a friend rather than a foe.

7

Geometrical Atomism – Flux and Language

Plato's geometrical atomism is in my view somewhat undervalued. Pyle's magisterial *Atomism and its Critics* gives little mention to it, and concentrates on the tradition of Leucippus, Democritus and Epicurus.[1] This is a shame, as geometrical atomism is an important and interesting theory in itself, as well as having important relationships to the earlier theories of Leucippus and Democritus and later theories such as those of Descartes. It is interesting that Plato presents geometrical atomism using a letters and syllables analogy, as Aristotle tells us that for the atomists the differences between bodies are caused by the atoms and are due to:

> shape, arrangement and position; being, they say, differs only in 'rhythm, touching and turning', of which 'rhythm' is shape, 'touching' is arrangement, and 'turning' is position; for A differs from N in shape, AN from NA in arrangement, and Z from N in position.[2]

We shall also see something very akin to Empedocles' four-element theory under attack, as well as the ideas of the atomists and the Milesians. As we shall see, there are interesting parallels between the use of the letters and syllables analogy in the *Timaeus* and other late works, notably the *Theaetetus* and the *Philebus*.

Examining geometrical atomism will also involve questions concerning the nature of the receptacle, the degree to which the world can be said to be in flux and what we can say about the world around us. So far, I have argued that in the *Timaeus* the heavens move in an entirely regular manner and are entirely stable.[3] One might feel that runs against the general tenor of some passages in the *Timaeus* and creates a significant tension, in particular with 27d where a radical disjunction is made between what always is and never becomes, and what always becomes and never is. There are several possible strategies for resolving this tension. First, we might look to *Timaeus* 29c, where after arguing that we can only expect a likely account of a likeness *Timaeus* says that if we cannot produce an entirely consistent account we should not be surprised but should gladly accept the most likely account and not seek any further. So we might shrug our shoulders and accept that such tension is inevitable.

With passages where several interpretations are possible, we might look to reduce this tension by interpreting them so as to produce the greatest overall harmony. We saw an example of this in Chapter 4 where I gave an interpretation of the relation between reason and necessity which would allow for regular and stable celestial motion. A second strategy would be to argue that there is sufficient difference between the nature of the heavens and the rest of the cosmos to allow for the stability of the heavens. Clearly the fact that all the celestial bodies are gods will be significant here, though there are some other important differences as well. If this fails to release the tension entirely, then we can fall back on the first strategy having at least dissolved some more of the tension. The third, and by far the most radical strategy, is to counter-attack. If one can harmonise a sufficient number of passages with the idea of regular and stable celestial motion, then one might become suspicious of the orthodox interpretation of any remaining recalcitrant passages. That is something I shall explore in Chapter 9. In this chapter and the following one I shall argue that it is possible to produce a reading of geometrical atomism, the receptacle and flux that underpins the stability of the heavens and allows us to talk in an informative manner about the physical cosmos.

As commented in the general introduction, I take a different line on the relation of the receptacle to geometrical atomism to most commentators. Commonly, the receptacle is treated on its own, or geometrical atomism is interpreted such that it fits conclusions drawn from an analysis of the receptacle. As will become clear in this chapter and the next, I believe there are serious problems with the receptacle which can be resolved by the introduction of geometrical atomism and a reformulation of the nature of the receptacle. So in effect I take geometrical atomism as the dominant theory to which we need to fit an interpretation of the receptacle, rather than vice versa.

I. The receptacle and language

There are many problems with the *Timaeus'* account of the relation of the basic constituents of the world to each other and to the receptacle, but here I wish to focus on three. There might be insufficient stability to underwrite the stability of the heavens, there might be problems with how we ought to refer to anything physical, and there is the matter of the relation of the account of the *Timaeus* on these matters to that given in the *Theaetetus*. At 49b Timaeus introduces the problem of the relation of language to things that are in flux, initially taken to be earth, air, fire and water by:

> Speaking of each of these, to say which ought really to be termed water rather than fire, and which by any name rather than each and all, so as to employ language which is sound and secure, is hard (*pisto kai bebaio chresasthai logoi, chalepon*).[4] We see, so we suppose, water congealing and

> becoming stones and earth, and this same thing when dissolved or dispersed becoming air, and air becoming fire by combustion ... and thus, it appears, they pass into each other in a cycle of birth.

The problem we face in attempting to construct a secure logos is that the elements that we refer to appear to change into one another.[5] Along with others, I take this to be a problem concerning identifying references.[6] If we identify some stuff as water, and it changes into air, should we now identify the same stuff as air? If we do, what happens to any distinction between the elements – what name, 'rather than one and all' ought to be applied to each? If so, this will then function as a critique of anyone holding any of these substances to be basic, which will certainly include Empedocles, anyone who holds that there are fire atoms and even the Milesians. The key passage is perhaps the following, which has been the subject of much debate concerning both its translation and interpretation. At 49d ff. Timaeus tells us:

> Since, then, each of these never appear the same, which of them can we assert with confidence and without shame to be this – whatever it is – and not something else? It is not possible, but much the safest way to speak of them is as follows. Whatever we perceive as always changing into something else, like fire, in each case we should not call fire 'this' but 'the suchlike' (*hos pur, me touto alla to toiouton hekastote prosageurein pur*). Nor should we call water 'this' but always 'suchlike', nor anything else as this if it had some stability, among the things we indicate by the expressions 'this' or 'that', and think we indicate something; for they slip away and do not wait to be described as 'this' or 'that' or any term which attributes permanence to them.[7]

The central debate concerns the phrase, *me touto alla to toiouton hekastote prosageurein pur*.[8] The traditional reading is that *touto* 'this' and *to toiouton* 'suchlike' are competing predicates for the subject *pur*, fire. The phrase then concerns ways in which we may talk of fire, one proper and one improper. So we ought not to call fire 'this', but we can call it 'suchlike'. The alternative reading takes *touto* and *toiouton* to be competing subjects for the predicate *pur*, the sense now being that words such as fire which we now apply to transient phenomena are better applied to more stable entities. So if we are to use 'fire' properly, we should only use it to refer to entities which are 'this', and not those which are 'suchlike'.

Both of these interpretations are acceptable renderings of the Greek.[9] Which we choose to accept, however tentatively, will be dependent on more general considerations.[10] As we shall see, this is not the only issue related to the question of flux where there is such a choice. As indicated in the introduction to this chapter, my concern is whether these passages can support the regular and stable motions of the heavens. So too as Gill comments, the decision which reading to adopt:

> reflects a decision about the status Plato grants to physical phenomena. On the traditional reading he legitimates talk about such objects. On the alternative he proscribes it.[11]

So on strategic grounds I opt to explore the first of these alternatives. This is not to say that the alternative reading is not a viable one, nor is to say there is not a great deal of interest to be said for the alternative reading, both in relation to the rest of the *Timaeus* and other works, notably the *Theaetetus*. I do have concerns about the alternative reading in that it seems in many versions to generate four types of entity for *Timaeus* 48e-53b where the text is adamant there are three.[12] I also have worries as to how well the alternative view can account for the relation of the entities of geometrical atomism to the receptacle. I shall return to these matters a little later, but let us see what we can make of the traditional reading first.

A major concern over the traditional reading was the apparent discord between the *Timaeus* and the *Theaetetus*. It has been argued that the *Theaetetus* post-dates and corrects the *Timaeus'* view on the relation of flux and language.[13] It is surely true, so Owen argued, that if everything is in radical flux, then we cannot successfully refer to anything at all, and this is a better position than the 'lame plea' of *Timaeus* 49d ff. that we can refer to the four elements as 'the suchlike'.[14] The alternative view was pioneered by Cherniss,[15] possibly with the aim of defusing the argument concerning flux for dating the *Timaeus* before the *Theaetetus*. At *Theaetetus* 182c ff., Socrates and Theodorus have the following exchange concerning the Heraclitean position and its relation to language:

> *Socrates*: Let us ask them, are all things changing and in flux? ... Have they both kinds of change that we distinguished, of place and of quality?
> *Theodorus*: Of course; they must if they are to be completely in change.
> *So.*: If they only underwent change of place, and not of quality, then we would be able to say what the flowing things that change position are. Could we speak in this manner?[16]
> *Theo.*: We could.
> *So.*: But since not even this stays still, what flows flowing white, but changing, so that there is a flux of this very thing also, the whiteness, and a change of colour, lest it be convicted of staying still, is it possible to name a colour with the result that it is correctly named?
> *Theo.*: But how could one contrive that, Socrates? Or indeed for anything else of this sort, if it always slips out from under us as we speak, being in flux?

The flux described in this passage is a very radical one indeed. Not only is there nothing which is free from qualitative change, but everything changes in every respect at every successive instant, 'lest it be convicted of standing still'. The passage that is supposed to create problems for a late dating of the *Timaeus* is the following at *Theaetetus* 183a:

> *So.*: If everything is changing, every answer to any question is equally correct, both to say 'thus' and 'not thus', or if you wish, 'becomes', if we are not to employ any expression that will bring them to a standstill.
> *Theo.*: You speak truly.
> *So.*: Except this, Theodorus, I said 'thus' and 'not thus'. It is necessary, though, not even to say 'thus'. For 'thus' would no longer be changing, and nor would 'not thus'. Indeed, there is no motion in 'this', either. Those who hold this theory must establish some other phrase with which to express it, as by their own hypothesis they now have no words, unless 'nohow' is allowed. That might be the most fitting expression for them, as it is indefinite.

Similarly at *Cratylus* 439d8-12 one cannot say that anything in flux is either *ekeino* or *toiouton*. So according to the *Theaetetus* we cannot refer at all to items which are in a radical flux. It is important to note that at *Theaetetus* 182c9 above, if something only underwent *phora* and not *alloiosis* we would be able to say what it was. What then allows us to refer to things in flux in the *Timaeus*? The leading candidate here is the receptacle. There is no question that the receptacle can be called *touto*, and that it us supposed to be free from change, and perhaps it is this which gives a sufficient basis for being able to call the elements *to toiouton*. At 50b ff. *Timaeus* introduces the gold example to explicate this:

> If someone were to mould all the shapes out of gold and without stopping remodel each of these into the rest, then should he point one of them out and ask what it is, by far the safest answer in truth would be that it is gold, but as for the triangle and the other figures that occur, it would be wrong to describe them as having being, as they change even as we state them, and we should rest content if they willingly receive the description of the suchlike with some safety.

So where we have flux, and here we would seem to have a radical flux,[17] we can refer to what is changing by *to toiouton*. What is it about the receptacle that allows this, and does the *Timaeus* give us a coherent account of the receptacle?

II. Some paradoxes

I take the primary task of the receptacle to be to provide a licence for *to toiouton* expressions. If so, it would seem that the receptacle cannot be merely a theory of space. If something is changing, and I am attempting to make an identifying reference, it does not seem to be particularly helpful to say that this change is occurring in space. Space may well be unchanging and so be a suitable referent of *touto*, but could hardly be said to constitute or lend any stability to the phenomena occurring in it. If a chameleon changes colour, I need to know something about the constitution of the chameleon, not about the tree that it is sitting in.

In order to license *to toiouton* references to the things in flux, then, the receptacle must be related in some constitutive way to them. Certainly the metaphors of the receptacle being like gold, or a moulding-stuff, or an odourless base for scent would seem to invite some form of material interpretation of the receptacle.[18] We might discuss here whether Plato has one theory of a material receptacle, and whether that is a coherent theory. It is quite possible that shapes relate to gold in a different way from how scents relate to bases or mothers to offspring etc., and that at least some of these are incompatible when we come to consider how the receptacle relates to forms. An alternative to keep in mind though is that Plato may give us several conceptions and wishes us to consider theories of a material receptacle in general. As we saw with pre-cosmic chaos in Chapter 1, he may provide something more flexible and more powerful than a straightforward statement of cosmogony. Whatever degree of pre-cosmic chaos we might suppose, a teleological ordering is required. Here, however we conceive of a material receptacle there may be difficulties.[19]

In order for the receptacle to remain a proper referent of *touto*, it must remain changeless. At *Timaeus* 50d4-51b2 great emphasis is laid on the characterlessness of the receptacle. This is so in order that it does not distort what is in it, but also the receptacle needs to be characterless in order that it be changeless. As Prior comments, 'Its nature, one is tempted to say, is that it has no nature.'[20] Now we might frame a slightly different problem. In order to be entirely free from change the receptacle must be characterless. If it is entirely characterless, it is difficult to see how it relates to the things in it. Gill argues that in the gold example, the gold is not constitutive of the shapes as *shapes*.[21] I would rather put this as a dilemma. If the gold is constitutive, then the receptacle is in grave danger of changing as the shapes change, and so its status as the referent of *touto* is endangered. If it is not constitutive, then the receptacle cannot licence *to toiouton* references.

These problems are perhaps best exemplified by the struggle Timaeus has at 50bc to produce a coherent account of the receptacle and its relation to the things in it. This passage follows directly from the gold example where we have been told that the triangles moulded out of gold can be referred to as *to toiouton*. *Timaeus* 50b5-c2 then tells us:

> The same account applies concerning the nature of that which receives all bodies. It must always be called the same, for it never departs at all from its own character (*dunameos*). For since it always receives all things, in no way whatsoever does it assume a shape (*morphen*)[22] similar to any of the things which enter it.[23]

That seems quite reasonable for the receptacle in order to retain its status as a referent of *touto*. However, immediately following this, at *Timaeus* 50c2-5 we are told:

> For in nature it is a moulding-stuff (*ekmageion*) for everything, changed and shaped (*kinoumenon te kai diaschematizomenon*) by the things which enter it, and because of this appears different at different times.

Timaeus cannot have this both ways and have a coherent account. This problem I shall refer to as the receptacle paradox. One might add that if the receptacle only appears to be changed (but is not changed), it is not really constitutive, and if it appears to be changed (and is changed) then it ceases to be a proper referent of *touto*. At *Timaeus* 50c4-6 we are told:

> The things entering and leaving the receptacle are copies of those ever existent, and are stamped (*tupothenta*) from them in some hard to explain and marvellous manner (*dusphraston kai thaumaston*), which we will shall follow up later (*hon eisauthis metimen*).

This Timaeus conspicuously fails to do in 48e-53b. All we are told is that the receptacle partakes in the intelligible in a most difficult and hard to grasp manner (*aporotata* ... *kai dusalotaton*) at 51a7, and at 51b and 52d that the receptacle is liquefied and ignified. Whether the geometrical atomism introduced at 53c ff. will help with these matters we shall see in a moment.

There is a further problem with considering the receptacle to be material on the basis of what we have seen of *Timaeus* 48e-53b. There would seem to be very little we can say about matter. Timaeus struggles to say anything positive about the receptacle, and of course there will be serious epistemological problems with anything so utterly characterless.[24] At *Timaeus* 49a3 it is something difficult and obscure (*chalepon kai amudron*), at 52b it is grasped without sensation by some bastard reasoning (*logismoi tini nothoi*) and is the subject of a dream.[25] Something that cannot be explained in terms of anything more basic and has no character seems in itself inexplicable. If it is not constitutive, it is hard to see how it helps to explain the phenomena, and anyway if it is characterless there would seem little that it can explain. It is difficult then to see matter, if we equate it with the receptacle, as anything which is explanatory or is itself subject to explanation. Perhaps *Timaeus*' dream, as McCabe comments, is as problematic as Socrates' dream in the *Theaetetus* (201e ff.), where the basic elements turn out to be unknowable.[26]

If we treat the receptacle as being in some way constitutive of the phenomena, we run into two major problems. It is not clear that the receptacle can play the role required of it in being so related to the changing phenomena that it can licence *to toiouton* expressions, and in being independent of them so as to preserve its own stability (the receptacle paradox). If the receptacle is both inexplicable and non-explanatory, it seems a poor theory of matter. Can the introduction of geometrical atomism at *Timaeus* 53c help with these difficulties?

III. Geometrical atomism

In my view it is of the utmost importance to recognise that there are significant differences between the four-element theory put forward at *Timaeus* 48e-53c and the geometrical atomism introduced at 53c. In the theory of 48e-53c, there is no mention of any sub-structure to the elements. The only mention of *stoicheia* prior to 53c ff. is at 48b9 which explicitly denies these elements (earth, water, air, fire) are *stoicheia*. Each of the elements of 48e-53c may transmute into any other element. In the later theory, what we suppose we see here is contradicted by the ban on the transmutation of earth and the other elements.

Of considerable importance is the fact that at 48e-53c, the entire discussion of the elements is conducted at the perceptual level. This is emphasised very strongly in the opening passages. We see, so we believe (*hos dokoumen*, 49b9 ...*horomen*, 49c1) the transmutation of earth, water, air and fire. They form a cycle 'so it appears' (*hos phainetai*, 49c9, cf. *phantazomenon*, 49d1), and we see things such as fire constantly changing (*kathoromen*, 49d4). This would seem to line up with the tripartite ontology of form, receptacle, and that which is perceptible, which is stressed at 48e3 ff., 50c7 ff. and 51a4 ff., and 51e6 ff. What is created by form and receptacle is visible (*horaton*, 49a1), sensible (*aistheton*, 52a5, cf. 52a7), and at 51a the mother and receptacle of the visible and sensible (*horatou kai pantos aisthetou*, 51a4) must not be called by any of the names of the elements. At 52b the receptacle is not accessible by perception (*anaisthesias*, 52b1). There is no discussion of anything which is below our perceptual threshold rather than in principle imperceptible.

Timaeus 56bc however is very specific. The atoms of the elements cannot be seen (*ouden horomenen*, 56c1-2) because of their smallness. Only when many of them are gathered together can a mass of them be seen (*horasthai*, 56c3). We can take it that the individual basic triangles and the complexes they form as the faces of atoms are also too small to be seen. These differences may have a significant bearing on Plato's intentions. He may well wish to show that the phenomena are in some sort of flux, requiring a reform of our references to them, without being committed to a theory where the ultimate constituents of the phenomena are in flux. The point being driven at by the first theory may then be a conditional one. If we operate only at the level of perception, and we equate changing phenomena, such as the four-elements, with the ultimate constituents of the world, then we will be unable to make any identifying references, or even any *to toiouton* references. This may be a criticism of current four element theories, and may be pushing a theoretical point about fundamental flux.

Empedocles may be a significant target here, though the point would apply to anyone who believes that earth, air, water and fire are basic. That such an attack is made on these elements for failing to be stable does not mean that the idea of a stable basis for phenomena is necessarily abandoned. If we suppose there to be some form of stable, amorphous substrate, such as the receptacle then we have the problems outlined in the last section. The later introduction of atomic sub-structure may be an attempt to rectify these difficulties by postulating stable physical entities at a level well below the threshold of human perception.

At 53d *Timaeus* postulates two sorts of triangle as the *archai* of the elements, but suggests that there are some higher *archai* than these. It is a reasonable assumption that these *archai* (or as Gill terms them, simples), will be stable. I doubt that anything that is not stable would qualify as an *arche* for Plato. We are also quite justified in using this assumption as an hypothesis to see what sort of account ensues. I would agree with Gill that: 'The important point is not *what* turns out to be basic but that there be ultimate simples.'[27] If there is something basic and simple, these simples can be called *to toiouton*, and this gives a sufficient basis for the elements to be called *to toiouton* as well. If so, then the problem of the relation to the *Theaetetus* is resolved. The objection there was to a flux where everything changed place and quality. In the *Timaeus* we have the foundation of some things which while they change place do not undergo *alloiosis*. While Gill is reticent about identifying the two fundamental types of triangles, which I shall call the stoicheic triangles, as these basic and simple entities, I am not.[28] I take one of the key points of the introduction of geometrical atomism to be precisely this, to introduce some basic physical constituents, and the stoicheic triangles are clearly strong natural contenders here. As I shall argue in the next few sections, there is a very strong case for considering the stoicheic triangles both as basic and as free from any change to their intrinsic order.

One might argue then that there is a case for the stoicheic triangles being called *touto*, if they are stable, the two standard objections being that they undergo *phora* or *alloiosis*. The very nature of the *stoicheia* (whether they be the stoicheic triangles or something else) precludes *alloiosis* being a problem. That they are in motion may not be a problem either, because at *Timaeus* 52de, 57c and 88de we are told that the receptacle moves, and is shaken like a winnowing-basket.[29] It is also interesting that at *Timaeus* 49e7 ff. we are told that only that which the four elements are created from or are dissolved into may be referred to as *touto*, and the elements ultimately are created from and dissolved into the stoicheic triangles. If the basic problem here is identifying references, again it would seem that the stoicheic triangles can quite safely be identified. To say that this is the position that the *Timaeus* adopts would probably be to push matters too hard, though it may be something that Plato is leading his readers to consider. However, it may be that the fact that the receptacle moves and can be called *touto* is an oversight on Plato's part, or one might argue that Plato gives us several images of the receptacle and we should not push any of them too hard. That what the elements can be dissolved into can be called *touto* may be restricted to the elements and receptacle model, or perhaps the dependence of the stoicheic triangles on the receptacle prevents them being called *touto*.[30] Gill argues that the simples, whatever they may be, will be called *to toiouton* because they are distinct from their relevant forms and move around.[31]

IV. Letters, syllables and triangles

Are the stoicheic triangles the ultimate constituents of the physical world in the *Timaeus*, and are they free from change? Let us begin with a distinction which the *Theaetetus* makes between two sorts of change. At *Theaetetus* 181c Socrates says:

> What I want to enquire is this; do they say that there is only one kind of change (*kinesis*),[32] or, as it seems to me, that there are two? ... Do you call it change when something exchanges one place for another, or spins around in the same place? ... Let this then be one form of change. But when something remains in the same place, but grows old,[33] or becomes black instead of white or hard instead of soft, or undergoes any other sort of qualitative change (*alloiosis*), isn't it right to say that this is another form of change? ... There are then two forms of change, change of quality and change of position (*phoran*).[34]

Consider this in relation to the basic entities of the world, the *stoicheia*. That they move around is not in question. What is more intriguing is whether they undergo any change in themselves. Flux theories here come in many shapes and sizes. These changes might involve one, many or all of the properties of a basic entity and those changes might take place at any rate up to the radical flux of change every instant. One might contrast such flux theories with the idea that the basic entities are entirely stable. To put this in terms of a distinction I used in Chapter 1 for Plato's *stoicheia*, perhaps the *stoicheia* undergo no change in their intrinsic order even if their extrinsic order changes. The ancient atomists held such a theory, and we can compare an elementary modern physics view of the world. According to this there are three types of simple, uncompounded particles, electrons, protons and neutrons, which are the exhaustive physical constituents of more complex entities such as atoms and molecules. These elementary particles change velocity and position, and enter into bonding relations with each other, but do not undergo any change in themselves. Complex entities which are ensembles of these basic particles on the other hand, such as atoms or molecules or clouds on a windy day may indeed be said to be in some form of flux, without compromising the stability of the basic particles.

In the above passage from the *Theaetetus* I translated *alloiosis* as 'qualitative change', as the *Theaetetus* is concerned here with perceptible qualities. One might though, as in the two examples above, have a theory where the *stoicheia* are not perceptible so we will have to talk of their characteristics. We can then distinguish between two quite different theories of how things change. I shall term these theories of fundamental flux and of fundamental stability, depending upon whether the *stoicheia* undergo *alloiosis*. The position ascribed to Heraclitus in the *Theaetetus* is clearly one of fundamental (and indeed radical) flux.[35] At *Theaetetus* 181d Socrates poses a hypothetical question for the Heracliteans:

> *So.*: Do you hold that everything changes in both ways, changing both in place and in quality (*pheromenon te kai alloioumenon*), or that some things partake of both, and some only one?
>
> *Thes.*: By Zeus, I don't know what to say; I suppose they would say both.
>
> *So.*: If they did not, my friend, then it would seem that things are both in motion and standing still, and it would be no more correct to say that all things move than all things stand still ... then since it is necessary for them to change, and since it is not possible for something not to change in every way, all things are always suffering all kinds of change.[36]

So too, one might argue, the flux described in the *Cratylus* is a fundamental one.[37] I have no disagreement with this, but I shall argue that it is quite possible that the flux of the *Timaeus* as one of fundamental stability. A second way of setting up the contrast here would be to employ a distinction that the *Sophist* uses between change relative to itself and change relative to other things. This would bring us close to Irwin's distinction between aspect (a-) change and self (s-) change. The following arguments are intended to establish that in the *Timaeus* there are some entities which undergo no change of character, no change in relation to themselves and no s-change as defined by Irwin.[38] Let me outline where I believe these basic entities can be located. At 48b the *Timaeus* says concerning earth, air, fire and water:

> No one has as yet revealed their generation, but we speak as if we know what fire and each of them are, postulating them as the basic principles (*archas*) of the physical universe, although it is not fitting for them to be in any likelihood compared to syllables, even by a man of little insight.

This might be read as a rejection of the letters and syllables analogy which commonly occurs in later Plato.[39] However, if the two most basic types of triangle that the *Timaeus* postulates are taken as the letters, and the atomic faces they go to make up are the syllables, then the atoms will indeed not be syllables, although the letters and syllables analogy may still apply. The *Timaeus* refers to the two most basic types of triangle using the substantive *to stoicheion*.[40] This is important as *stoicheion* carries the general sense of simplest component part, and a more specific sense in Plato of letter as opposed to *sullabe*, 'syllable'. I shall refer to these two basic types of triangles as the 'stoicheic triangles', and the compound triangles that are atomic faces as 'complex triangles' (see Figs 50 and 51).[41]

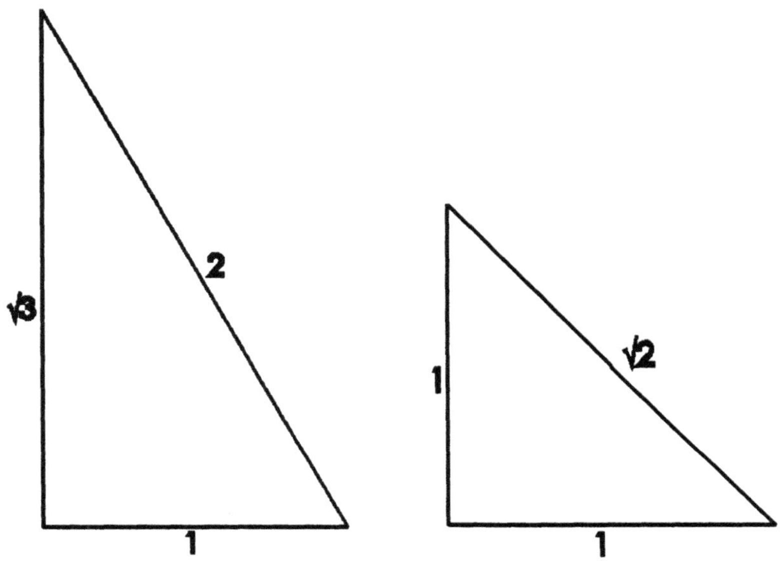

Fig. 50. Stoicheic triangles – the half-equilateral and the half-square. The *Timaeus* claims that earth, water, air and fire should not be called elements, not even syllables. The basic unit of the physical world would seem to be these two sorts of triangles, the half-equilateral and the isosceles, often referred to as *to stoicheion*. That the proportions chosen include 2 and 3 may have considerable significance for the history of cosmology and the question of how the world may be structured mathematically, as I discuss later.

The terminology I use for geometrical atomism is as follows:

I use elements to refer to earth, water, air, and fire.

I use atoms to refer to the individual cubes, octahedra, icosahedra and tetrahedra of the elements.

I use complexes or complex triangles to refer to the faces of the atoms.

I use stoicheic triangles to refer to the two basic triangles, the half-equilateral and the half-square.

Geometrical atomism has a further refinement, in that there are different types of each element. Friedlander, adding due caveats about the relation to modern atomism, has termed these 'isotopes'.[42] As Cornford has argued, this does not mean that there are different sizes of stoicheic triangles.[43] This would both strain the Greek of *Timaeus* 57d and would be extremely

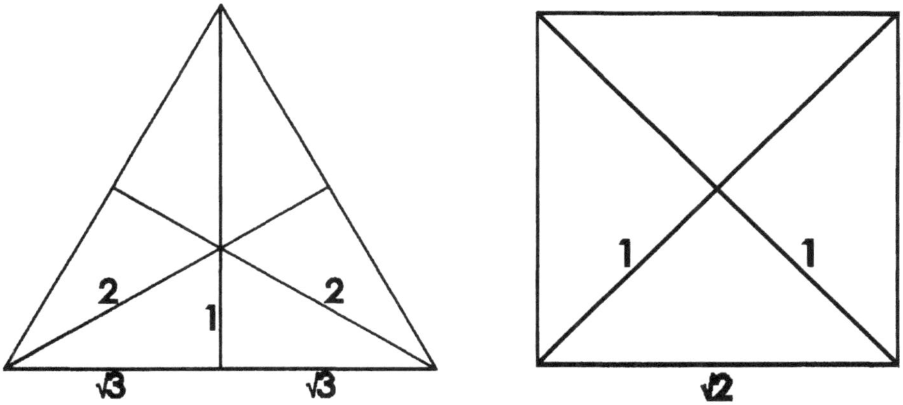

Fig. 51. Complex triangles and squares. In the specific arrangements discussed in the *Timaeus*, six half-equilateral stoicheic triangles form one complex triangle, and four isosceles stoicheic triangles form one square.

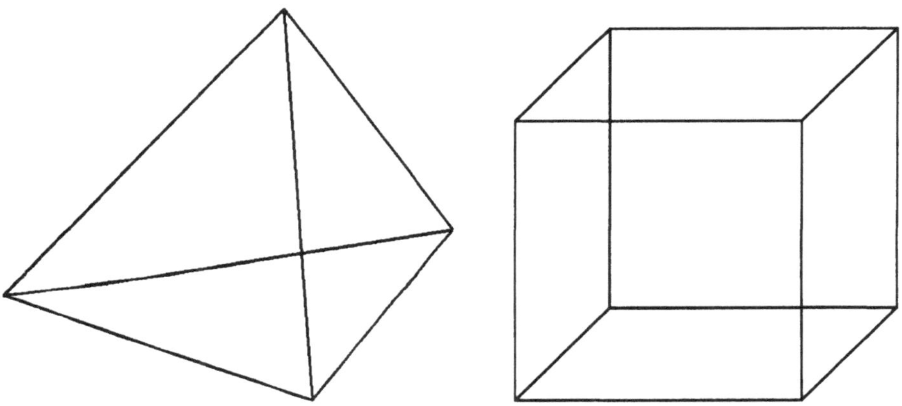

Fig. 52. Fire atom and earth atom. To form atoms, the complex triangles and the squares form into solid figures. Here we have fire, the tetrahedron and earth, a cube. Icosahedra of water an octahedra of air are also made from the complex triangles.

problematic for the theory of transmutation, both between types of one element and between elements.[44] Indeed it would be odd, given Plato's solution to Meno's paradox and his clear choice of specific shapes of the stoicheic triangles, if he supposed there to be many sizes of stoicheics when he has a simple means at his disposal of generating different varieties of the four elements. So the stoicheic triangles can form up in different ways to create different sizes of complex triangles and squares and so different sizes of atom, corresponding to the different classes of each element, and we have the further factor that there can be different mixtures of elements and grades of each element in order to explain the diversity of the phenomena that we encounter (see Figs 53 and 54).

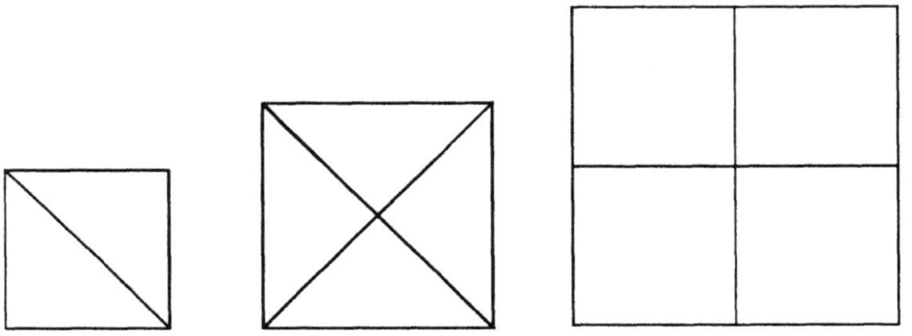

Fig. 53. 'Isotopes' – more complex planes, squares.

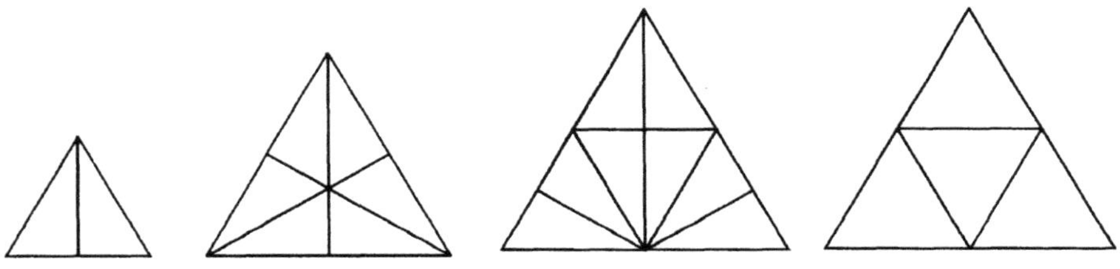

Fig. 54. 'Isotopes' – more complex planes, triangles.

The stoicheic triangles then undergo no change of character, no change in relation to themselves and no s-change, or to put it another way, incorporating the single size thesis for stoicheics, all half-equilateral triangles are congruent and remain so, and all half-square triangles are congruent and remain so.

V. The evidence of Aristotle

For the theory of change of *Timaeus* 53c ff. to work, no change in the stoicheic triangles is required and none is mentioned. Change is analysed solely in terms of the composition of stoicheic into complex triangles and these into atoms and aggregates of atoms, and the converse decomposition. At 59c, *Timaeus* is quite confident that all phenomena can be fully explained in this manner. It is perhaps important that while we are given an account of the relative abilities of the atoms to cut up what they encounter (cf. *Timaeus* 56b, 56d), we are not given such an account of the stoicheic triangles. *Timaeus* 54c ff. is important in this context. We are told that while air, fire and water may transmute into each other, they may not transmute into earth, and vice versa, as the first three are constructed from half-equilateral triangles while earth is made from the half-square.[45] If the stoicheics come apart in some way, further transmutation would be possible. While the *Timaeus* specifies how atoms and complex triangles decompose, it does not do so for the stoicheic triangles. We might take that as an indication that they are not supposed to decompose into further physical elements.

If it is the case that the stoicheic triangles of the *Timaeus* are stable, then there are some interesting consequences. The theory of human ageing tells us that the bonds which hold the constituents of our bodies together have the power to last only for a certain time, and that as these begin to fail, so we slip into decrepitude.[46] The sort of flux to which the atoms and complex triangles are subject is not then the radical flux which would have them possessing a quality at time t and not possessing it at t + 1, but a rather gentler variant. This is important because it allows the physical world a large measure of stability, how much depending on how strong the 'power to last' is. In relation to this, if we examine what the *Timaeus* has to say on how human beings and the heavens are produced, we find something very interesting indeed. *Timaeus* 42d ff. tells us that the task appointed to the demigods is to fashion the bodies of human beings. At *Timaeus* 42e we are then told:

> Taking the immortal element of mortal creatures, imitating their own fashioner, and borrowing from the cosmos portions of fire, earth, water and air on the condition that this loan could be repaid, they cemented together what they had taken, though not with the indissoluble bonds (*alutois* ... *desmois*) with which they themselves were held together, but with well constructed bonds invisible because of their smallness, they fused them into one mass, creating out of all these one body for each, thus clothing the revolutions of the immortal soul in bodies subject to egress and ingress.

So while the constituents of our bodies are put together with dissoluble bonds, those of the demigods have indissoluble bonds. Our bodies will be subject to a gentle flux, while those of the demigods will be entirely exempt from flux in terms of qualitative change. The embodiment of these demigods though is in the heavenly bodies. Earlier in his discourse, at 38e *Timaeus* told us of the generation of the planets, and stated:

> When each of the bodies required to help in the production of time had been brought into its proper motion, as generated living creatures having their bodies bound with living bonds and having learnt their duties, they moved according to the motion of the different.

Now one might attempt to draw a distinction here between the planets as living creatures and the fixed stars as demigods, but in fact the two terms are used interchangeably of each. *Timaeus* 40b talks of the fixed stars as living creatures while 40cd talks of the planets as demigods. The stability of the heavens is guaranteed by the qualitative stability of the stoicheic triangles allied with the 'indissoluble bonds' with which they are bound together. Through the mechanism of the variable durability of the bonds between the stoicheic triangles and between more complex entities,[47] the *Timaeus* theory can then account for any degree of qualitative flux at the perceptual level, up to and including the total stability of the heavens.

Aristotle's comments in *De Caelo* and *De Generatione et Corruptione* are interesting in relation to the nature of the stoicheic triangles. At *De Caelo* 306a9, Aristotle gives us an important general principle:

> Sensible things require sensible principles, eternal things eternal, and perishable things perishable, and generally something of the same kind underlies.

That of course is quite different from the atomism of Leucippus and Democritus, and, as the context makes clear, quite different in Aristotle's view from Plato as well. That geometrical atomism is Aristotle's target here at 306a ff. cannot be doubted. No one else held a theory of analysis into surfaces (306a1-4), that some but not all the elements could transmute into each other (306a4-6), and that earth was the odd man out (306a18-21). Those who 'resolve into surfaces', according to Aristotle have incorrect first principles (306a8-9).[48] Presumably those are that sensible things do not have sensible *archai* (though this is not stressed) and that perishable things have stable *archai*. These people stand on the truth of their *archai* and will not admit that their principles should be criticised in the light of their consequences.

Having discussed geometrical atomism in the latter part of *De Caelo* III.7, Aristotle opens III.8 by saying that in general it is unsound to try to give shapes to the simple bodies (*ta hapla somata*, 306b3 ff.). That is a very interesting phrase which would suggest that 'those who resolve into surfaces' do posit specifically shaped simple bodies. Aristotle goes on to raise the 'packing' problem, that such shapes could not form a plenum as, of the solids, only the square and the pyramid can do this and more shapes than these are required.

De Generatione et Corruptione is even more emphatic on these matters. At 315b25 ff. Aristotle poses what he feels is the key question, do things undergo change and growth (*alloioutai kai auxantai*) because of indivisible magnitudes (*megethos adiaireton*)? Those advocates of indivisible magnitudes are Leucippus and Democritus who suppose them to be bodies, and the *Timaeus* which supposes them to be surfaces. 316a2-4 is also important here, as Aristotle denies that putting surfaces together can account for any change of quality. Clearly he does not believe that the Platonic surfaces undergo any sort of *alloiosis*. *De Generatione et Corruptione* 325b25 ff., as we saw in Chapter 1 states that Leucippus holds that the indivisibles (*ta adiaireta*) are solids and are unlimited in shape, and Plato holds that they are surfaces and definite in number. The evidence of Aristotle then is very much in favour of the idea that the stoicheic triangles do not undergo any sort of *alloiosis*.

VI. The ageing process

One objection to the idea that the stoicheic triangles undergo *phora* but not *alloiosis* might be based on the theory of ageing of *Timaeus* 81 ff. According to this, ageing and natural death are to be explained by the gradual deterioration and consequent loss of function of the particles which constitute the body. One interpretation is that it is change in (and indeed the loss of) the intrinsic order of the stoicheic triangles that is at the root of this sort of deterioration. There is an alternative interpretation, though, which is at least as plausible. This is that while the bonds between stoicheic triangles which form complexes may deteriorate, as may those between the complexes which form atoms, as may the complexes themselves, there is no deterioration in the intrinsic order of the stoicheic triangles. At 81b5-9 the *Timaeus* begins its discussion of ageing by saying:

> When the organisation (*sustasis*) of the whole animal is new, the triangles (*trigona*) which constitute the atoms being 'fresh off the stocks' (*ek druochon*), they have strong joins between one another (*pros allela*).

It is important that the bonds that the *Timaeus* describes here are *pros allela*. These must be bonds which bind either stoicheic or complex triangles to each other, rather then any bonds internal to the stoicheic triangles. That the stoicheic triangles undergo change in their bonding relations does not entail that they undergo any change of intrinsic order. At *Timaeus* 81c6-e1 we are subsequently told:

> When the sides of the triangles loosen (*he hriza ton trigonon chala*) due to having contested many battles in a great amount of time ... finally the bonds (*desmoi*) fitting the triangles of the marrow together no longer hold out against toil, but separate.

At this point death occurs. If these bonds are *pros allela*, then the ageing process can take place without any qualitative change of the stoicheic triangles. *he hriza ton trigonon chala* is undoubtedly a rather strange phrase. I would agree with Cornford that this 'curious metaphor' must contrast with *sunkleisin* at Timaeus 81b7 and that this strongly favours Taylor's view that *hriza* means side, in the sense of the side of a complex which will form one face of an atom.[49] I would agree with Taylor against Martin and Archer-Hind that *chala* means loosens (the sides) rather than blunts (the angles).[50] Again, there is no need to suppose any change in the intrinsic order of the stoicheic triangles.

Returning to 81b5-9, one might take the view that as the triangles are *ek druochen* 'fresh off the stocks' they have been newly created.[51] However, if we build a wooden boat the wood does not undergo any change qua wood. It is merely reorganised from loose timber to a boat.[52] The wood may then be 'fresh off the stocks' without having undergone any qualitative change in itself, and so too perhaps with animals and stoicheic triangles. The verb that Plato uses in this context may support this approach. This is *sunistemi*, which has more of a sense of ordering that which is already existent than of creation de novo.[53] It is notable that this verb is used in the earlier discussion of atomism for the formation of atoms and complex triangles out of complex and stoicheic triangles respectively,[54] where no change in the stoicheic triangles is required or mentioned. So too at *Timaeus* 57c9, where as we saw there was the question of whether there were different sizes of stoicheics, the most reasonable interpretation would have *sunistemi* referring to how stoicheics are arranged in relation to each other, rather than their internal arrangement.

Even if we were to take *ek druochon* as signifying some qualitative change, it is not clear that it is stoicheic triangles that do the changing. If the referent of *trigona* at 81b6 is complex triangles, these can pass into and out of existence without prejudice to the stability of the stoicheics.[55] The correct referent of *trigona* is a problem which runs through the entire discussion of ageing. Certainly there are some uses here that can only be references to complex triangles, e.g. the *Timaeus* talks of 'the purest type of triangles, smoothest and oiliest' (82d). As there are only two types of stoicheics, and they could not be described as smoother or oilier than one another, this must be a reference to complex triangles.[56] So too at 81d, where we are told that the process of nutrition involves the triangles of the body cutting and assimilating incoming triangles. Conversely, when we age and the bonds between our triangles become weak, the incoming triangles cut up our own.[57]

Now it may be that the reference of *trigona* slips between stoicheic and complex triangles. However, in the earlier discussion of how the complex triangles are formed from the stoicheics at *Timaeus* 53c ff., Plato always has *Timaeus* carefully mark the distinction between the two, opening each discussion of the stoicheics with an explicit reference to their nature.[58] In this context *Timaeus* 57d and 61a are important, as in both of these passages there is a distinction made between the stoicheic and the complex triangles, the stoicheics being referred to in each case as *to stoicheion* triangles and the complex triangles simply as *trigona*. As some of the references of *trigona* in the discussion of ageing are clearly to complex triangles, and there is at no point any distinction drawn between these references and any other, one might conclude that all the references of *trigona* in the discussion of ageing are to complex triangles. We now have two tools which work in harmony for analysing statements such as *Timaeus* 89c:

> For plainly in the beginning the triangles (*trigona*) of each animal are organised (*sunistatai*) with the power to last only up to a certain time.

In all likelihood *sunistemi* refers to the organisation of existent particles, and *trigona* to complex triangles. So it is complexes of stoicheics which undergo intrinsic change, which is consonant with the idea that it is the *pros allela* bonds that are subject to temporal limitations. That the triangles can 'last up to a certain time' (*Timaeus* 89d) might be a temporal limitation, or more likely in view of *Timaeus* 81b ff., a 'wear and tear' limitation. If we take the reference to be to complex triangles, neither require change in the stoicheic triangles. On this interpretation, the 'biochemical' processes of ageing and nutrition are merely specialised cases of the more general 'chemistry' that is discussed at *Timaeus* 53c ff. They require no more than the division and reconstitution of atoms and complex triangles. In this respect one might note the similarity of the battling imagery between *Timaeus* 53c ff. and 81b ff., as well as the similar uses of *sunistemi*. Compare *Timaeus* 56e where we are told:

> Whenever a small amount of fire is enclosed in a large amount of air, water or earth, moving within their movements, battling and breaking up upon being defeated, two units of fire are assembled to make one form of air.[59]

If we take the two images of battling to refer to the same sort of process, then it is the atoms being broken into their faces that constitutes old age, rather than any deterioration in the stoicheic triangles. If the theories of ageing and nutrition require something beyond the general theory of chemistry, one would expect this to be marked and discussed. There is no mention of change in the stoicheic triangles and good reason in the ban on the transmutation of the scalene and isosceles triangles to believe that no such change occurs.

VII. The gold example

Against this general line of thought, one might argue that the gold example explicitly denies any sort of permanence to shapes, and to triangles in particular. There are several strategies which one might employ against this view, though. The first is to argue, as I have above, that geometrical atomism introduces a quite different theory of the nature of the world and how we might solve the problem of referring to changing phenomena to that of *Timaeus* 48e-53b. It may well be the case that everything is transient except the receptacle in 48e-53b. That seems to run into difficulties though, and perhaps the determinate and stable shapes of the *stoicheia* are introduced specifically to circumvent such problems. Ultimately these are different and incompatible theories which the readers may have to choose between, if they wish to have a coherent theory, and I shall say some more on the nature of that choice and other possible similar choices in the *Timaeus* in Chapter 9.

If one is seriously worried about what the gold example says about shapes, then it is possible to abandon the idea of stable shapes while still holding there to be some stable stoicheia, these being higher archai, perhaps lines, perhaps points. I would rather emphasise the conditional nature of the gold example though. So perhaps the key clause is at *Timaeus* 50b5-6:

> If (*ei gar*) someone were to mould all the shapes out of gold and without stopping remodel each of these into the rest ...

The conditional here may indicate that this is a thought experiment rather than a direct statement of cosmology. If someone were to do this remodelling, then we would appear to have something akin to a Heraclitean flux, where the only permanence would be found in the gold itself, and this is supposed to explicate the flux of the elements and their relation to the receptacle, and so the need for a reform of language. With that I have no quarrel, though one might wonder though what would be the case if this hyperactive character was not perpetually engaged in this remodelling. One reason for introducing this in conditional manner may be akin to the various descriptions of chaos and of the material nature of the receptacle we are given. Plato may wish to make some general comments on how we can talk about the world if we suppose there to be a receptacle, whatever degree of flux we suppose the world to be in, up to and including total flux. That does not commit Plato to the view that everything in the world is in such a radical flux, though if some things are we can still refer to them in a certain manner. Even with geometrical atomism it is quite possible that some things (including the shapes in the gold example, but not all things) may be in a rapid flux, but we are still able to refer to them though in a slightly different manner to that suggested in *Timaeus* 48e-53b.

The gold example presents us with a hypothetical manic remodeller of gold. We are not told within 48e-53b what natural processes this remodeller is an analogue for. Does he find a correlate in geometrical atomism? With the four-element theory, it is clear that there is a remodelling effect even if the cause of this is not specified. With geometrical atomism, though, we can perhaps be rather more specific. First, there is the effect which the atoms of each element have on each other, either as a general function of their shape (fire cuts more because of its size and sharp angles, 61d ff.), or the effect a much greater amount of an element has when it surrounds a smaller amount of another element (it assimilates it to its own kind, 57a-c). If we ask why these changes should come about, *Timaeus* has two answers. At 57c the receptacle is still in motion and still separating out the elements by a like-to-like principle, so there is motion of the elements and hence contact and change. At 58a ff. we find that the revolution of the same has a compressing effect on the elements, such that atoms are pushed together and so again we have change. All this will produce a remodelling of the elements, but all this takes place within the standard theory of physical change, and there is no need here to suppose that anything remodels the stoicheic triangles.

Let me emphasise that I do not wish to deny that Plato has a theory of flux in the *Timaeus*. The world we see is in a constant flux with earth, water, air and fire all appearing to change into each other. The question is how radical that flux is at root. If earth, water, air and fire are the roots then that flux is indeed a radical one. Zeyl makes the interesting comment:

> All commentators agree that Plato intends to show (a) that the constituents of the physical world ('phenomena') are caught up in constant change (as is forcefully argued in the case of the 'elements' at 49b7-c7) and (b) that this fact necessitates a reform in the use of certain locutions as referring expressions.[60]

One might get the impression from Zeyl's remarks that for Plato the phenomena are all of the physical world, and that the elements of 49b ff. are the basic constituents of the phenomena. This may be so for the four-element theory of 49b ff., but is simply not the case for 53c ff. where the elements are decomposed to atoms, planes and stoicheic triangles, all of which are far below our perceptual thresholds, such that there is more to the physical world than the phenomena.[61] Surely one of the great glories of geometrical atomism is that it decouples the nature of what we see from the nature of the micro-world. Plato exploits this fact to great effect in his explanations of phenomena in the latter part of the *Timaeus*, and not least of course in the very essence of geometrical atomism itself, that the micro-world is populated by these geometrical entities. The most basic of these may well be qualitatively stable. That what persists through change should be different from that which we see changing is beyond doubt, but that does not imply what persists must be indeterminate, merely that it should be different, as geometrical atomism allows.[62] It is possible then to take a view of geometrical atomism, flux and the receptacle such that there is enough stability in the world to underwrite the complete stability of the heavens, if the stoicheic triangles are qualitatively stable and the bodies of the gods (i.e. the heavenly bodies, mostly fire) are put together with *alutois desmois*, indissoluble bonds, as at *Timaeus* 43a2, *desmoi* being what link the stoicheic triangles into planes and the planes into solid figures in geometrical atomism.

VIII. Geometrical atomism and Socrates' dream

One objection that has been raised against the idea that the stoicheic triangles are the basic physical entities for Plato is that this would deprive him of any analysis of matter.[63] In my view this is not so. While the stoicheic triangles may be the end of an analysis by physical decomposition, which might be taken as one sense of giving an account of a physical entity, there may be other sorts of account which we can have of the stoicheic triangles. Although *Timaeus* talks of the stoicheic triangles as the *archai* of the more complex triangles, he also goes on to talk of higher *archai*. At 48c3 he speaks of 'the *arche* or *archai* of everything', and later at *Timaeus* 53d, referring to the theory of stoicheic triangles he says:

> These we hypothesise (*hupotithemetha*) as the principle (*archen*) of fire and of the other bodies ... but the principles (*archas*) of these which are higher (*anothen*) are known only to God and whoever is friendly to him.

These higher principles need not refer to some more basic physical or mathematical entities, however. It is notable that *Timaeus* refers here to higher *archai*, principles or starting points, rather than to further *stoicheia*. At the termination of explanation in terms of analysis by decomposition into physical elements, one option for Plato is to switch to a teleological analysis. These higher *archai* then may be teleological explanations of why the stoicheic triangles are as they are. One interesting aspect of the passage at *Timaeus* 53d is that the use of *hupotithemi* and *anothen* is highly reminiscent of *Phaedo* 101d, the line allegory of the *Republic*, and the defence of hypotheses by higher hypotheses. One might compare here the uses of *anothen* at *Phaedo* 101d7, *Republic* 514b2 and 518b4, and the uses of *hupotithemi* at *Phaedo* 101d ff. and *Republic* 510b ff., and the *Republic* on line and cave, especially 511a5 where the soul must *ton hupotheseon anotero ekbainen*. One need hardly stress that the *Phaedo* and the *Republic* are not looking for further physical hypotheses at this point. The final clause concerning the man friendly to God might also suggest some intention to invoke explanations in terms of the good. When we look at the sort of arguments that *Timaeus* uses to justify his choice of stoicheic triangles this is borne out. At *Timaeus* 54a ff., we are told that while the isosceles triangle has but one form, the scalene has many and we must postulate the fairest, and teleological language permeates this entire section. It is likely then that the stoicheic triangles are the most elementary constituents of the physical world of the *Timaeus*, without being the ultimate explanatory *archai*.

The geometrical atomism of the *Timaeus* is very interesting in relation to the passage of the *Theaetetus* at 201d ff. known as 'Socrates' dream'. As with Meno's paradox, the dream poses wide and deep philosophical questions, much broader than the themes for science that I can tackle here, but one issue for the study of the physical world posed by the dream is the question of letters and syllables and the accounts we can have of the simple and the complex. We can give an account of the complex in terms of its component parts, but we cannot give a similar account of the simple.

According to Socrates' dream at *Theaetetus* 201d ff., the regress is finite and terminates with basic elements, the stoicheia, which lack accounts. If they lack accounts, then they are perhaps unknowable. One problem for science that comes out of this is as follows. We can give an account of complex physical entities, such as the elements earth, water, air and fire, in terms of their component parts. Doubtless the dream has a good deal to say on the nature of that account, via the question of Hesiod's wagon at *Theaetetus* 207a, and whether a list of component parts is adequate, or an account of the interrelation of those parts, etc. There is a problem, though, with the physical stoicheia. However successful our physical analysis/account of complex physical entities, we cannot give the same sort of analysis/account of physical stoicheia.[64] As far as the physical world is concerned, it is notable that the *Timaeus* argues that there are basic elements, the stoicheic triangles, but at the termination of analysis from complex to simple these can be given teleological accounts.[65] It is interesting, and surely no coincidence, that the discussion of the *Timaeus* is presented with the same letters and syllables terminology. While letters and syllables may be inappropriate if we take the receptacle to be the key basic item, as argued in section I of this chapter that is not so if the stoicheic triangles are basic. They are the letters, the complexes of stoicheics the syllables, and the elements are neither letters nor syllables in line with *Timaeus* 48b.[66] But the *Timaeus* would want to deny that the elements of the physical world are perceptible, the contrary being claimed at *Theaetetus* 202b. It is important then that the *Timaeus* allows us some account of these stoicheia as otherwise we would have no access to them. There is not then a specific problem here with the physical world; the physical 'atoms', here taking 'atom' in its more literal sense as indivisible fundamental entity, can have accounts as well as names, though of course there may still be a problem with logical atomism, which would beset any claim to knowledge.[67] I do not wish to suggest that this is Plato's solution to the more general philosophical problems raised by the dream which run deeper than this, although it may point us in certain directions. What I do wish to bring out is that here we see a further epistemological reason for teleology, to add to the cosmological and nomological reasons we have seen, and of course the further epistemological reasons related to underdetermination and Meno's paradox.[68] In this sense, Plato's geometrical atomism is a long way removed from that of the Presocratics. So I would disagree with Vlastos when he compares Plato to Empedocles, Anaxagoras and the atomists, and says: 'So in offering us his own stoicheia Plato shows that he has enrolled in the same programme of physical elementarism.'[69] At one level that is so, but Plato's atomism has a far richer structure both physically and philosophically.

The *Timaeus*, then, may have a teleological account of the stoicheic triangles at the end of an account of physical entities in terms of decomposition. To look forward to the *Philebus* for a moment, another sort of account of the stoicheic triangles may be available there. In the *Philebus*, we can think of points on a continuum as the imposition of limit on the unlimited.[70] If we take two points on a continuum, we can then define a specific quantity or distance between them. If we do this in two dimensions, we can then have an account of all the plane figures. So we arrive at a point where the *Timaeus* could take over with an account of how the physical world is constituted from two specific types of plane figures, the half-equilateral and the isosceles stoicheic triangles. I put this as generally as possible as there is a question as to what the triangles of the *Timaeus* are triangles of and so too a question in the *Philebus* concerning whether *peras* and *apeiron* are constitutive of things or an analysis of things. Whichever way we take these questions, the *Philebus* would be able to give a further account of the stoicheic triangles. It is worth noting that the *Philebus* accepts four-element theory, and that at *Timaeus* 53b we are told that when the demiurge orders the universe from its primordial chaos, where all is devoid of 'reason and measure' (*alogos kai ametros*) (53a9), he first 'created shapes by means of forms and numbers' (*eidesi te kai arithmois*) (53b4-5).[71]

Geometrical atomism would seem to leave us in a rather better position than the receptacle in terms of explanation. The stoicheic triangles are themselves explicable in terms of a further teleological account, in a way in which the characterless and basic receptacle was not. So too the stoicheic triangles are explanatory of the phenomena, as the *Timaeus* from 53c ff. goes on to show. They are constitutive of other things without imperilling their own stability in a way that the receptacle found difficult. We can have an account of how they fit together to make more complex entities, and how the particles of our bodies interact with other particles to create sensations.

IX. Geometrical atomism and language

One of the key consequences of there being some basic, stable entities in the *Timaeus* is that the problem of a collapse of discourse concerning the world around us is avoided. As Gill puts it, whatever turns out to be stable gives the world sufficient permanence for language to get a grip on it.[72] The receptacle appears to be fatally flawed as a candidate for this role. It cannot be both constitutive of the phenomena and unchanging as it is required to be in order to license *to toiouton* references to changing phenomena. If the stoicheic triangles are the basic, stable entities it would appear that they can be properly constitutive of changing phenomena, and this does not imperil their stability. Thus reference can be made to the changing phenomena. That will be so for anything else which is qualitatively stable, which might well include the heavenly bodies. So as *Sophist* 262e demands, our sentences will have subjects, and I will say something more about false judgement in the *Timaeus* in Chapter 9.[73]

This may underestimate the grip that language can get on the world of the *Timaeus*, though. In several late works (*Theaetetus* 201d ff., *Politicus* 277d ff., *Sophist* 253a ff., *Philebus* 18b ff.) Plato employs the letters and syllables analogy to illustrate some points about the nature of language, and arguably in some or all of these cases about the nature of the world.[74] Plato is often interested in which letters do or do not combine to make syllables, and is concerned with the bonds (*desmoi*) between letters.[75]

Consider then what sort of problem geometrical atomism is introduced to solve, and how geometrical atomism is expressed. The original problem is a puzzle about the correct use of language and the structure of the world, and it is introduced using the letters and syllables analogy. Geometrical atomism is expressed in terms of *stoicheia* – we are given an account of which stoicheic triangles will combine with others, which will not, and how this combining occurs – and the bonds between the triangles are referred to as *desmoi*.[76] Indeed the *Timaeus* throughout seems very interested in the question of mathematical bonds, the bonds which hold men and gods together, the bonds which hold soul and body together and the bonds which hold the physical stoicheia together.[77]

The receptacle of 48e-53b might be thought to provide the ontological counterpart to a subject-predicate analysis of the world. Qualities appear in the receptacle and the proper description of these is that x is f-like, where x is the receptacle, f-like is the quality and F the form that this part of the receptacle is participating in. The receptacle always appears in the subject position and has no qualities in itself.[78] It halts the regress of Socrates' dream because it is ultimately basic ontologically, and is a pure subject linguistically (everything is predicated to the receptacle, it is predicated of nothing). The cost though is high, as we have seen. The receptacle is inexplicable (we can say nothing positive about the receptacle) and non-explanatory (its relation to what is in it is highly problematic).

With geometrical atomism though we have a different model of the world and perhaps a different conception of the way that language relates to the world.[79] What is not often recognised is how strong a reductive account of the world the *Timaeus* gives us. *Timaeus* 61c-68d gives an analysis of each of our senses in terms of geometrical atomism. Thus for instance white is the effect of the larger fire particles which dilate the visual stream (67e), and hot is due to the sharpness and rapid motion of the fire particles (61e). This then is no longer a simple subject-predicate arrangement as it might be if we were taking the receptacle to be the sole basic entity (i.e. (this part of) the receptacle is white, is hot). Hot and white are now related to the motion and arrangement of certain types of atom, which in turn can be given analyses in terms of (the motion and arrangement of) complexes of stoicheics and ultimately stoicheic triangles themselves.

This brings us back to letters and syllables again. Instead of an isomorphism between a subject/predicate analysis of language and a substance/attribute ontology, we might look to an isomorphism between the way that the basic components of language can be arranged and structured into something meaningful and the basic components of the world can be arranged and structured into something beautiful and good. Can language pick out the basic components and relations of the world? That would seem eminently possible. In the same way that we might say that 'Theaetetus flies', and be wrong or right about that, we might say that '(this atom of) fire cuts (this earth atom)' or 'this stoicheic triangle forms a bond (with that one)'.

It is interesting to note that relative to the discussions of language in the *Theaetetus* and *Sophist* which are keen to stress that some words, letters and syllables do, or do not, fit together properly, so the *Timaeus* is keen to stress that some of the components of the world and some complexes of those components do, or do not, fit together properly. The *Philebus* is interested in the relationship between language and music at 17a ff. In both what is important in producing a meaningful result is how the components are linked and arranged together. An interesting link between the *Timaeus* and this passage is that *Philebus* 17e warns us, after the example of music but in a quite general manner, that an indeterminate plurality leaves you with an indeterminate account, while *Timaeus* 55c makes the same sort of remarks about the number of worlds and the number of elements.[80]

The links to other late works are of course rather speculative. However, if we are to link the various discussions of letters and syllables together, and that in itself seems a reasonable thing to do, then I see no reason why the *Timaeus* should not be included in that project, especially as the results are so interesting. Even if we stay within the *Timaeus*, geometrical atomism, if we take the stoicheic triangles to be the basic entities, can be seen to provide a different conception of the world and its relation to language to the receptacle passages, if we take the receptacle to be the basic entity. As indicated, this avoids the receptacle paradox and the problem of whether the receptacle is either explicable or explanatory, and language appears to be given a reasonable grip on the world. If we take this view, then geometrical atomism either replaces radically modifies the receptacle passages of 48e-53b. The further implications and advantages of that move, and whether we can justify it as a reading of this section of the *Timaeus*, will be the subject of the next chapter. If the *Timaeus* postulates the stoicheic triangles as stable, basic entities though, this has considerable ramifications for the nature of flux envisaged (in relation to both the phenomenal and the micro worlds), and for nature of the relationship between language and the world.

8

Geometrical Atomism – Matter and Space

In the previous chapter, we examined the relationship between geometrical atomism, flux and language. I argued that the receptacle, in an ontology of forms, receptacle and phenomena, is incapable of providing a licence for *to toiouton* references to the phenomena. If the receptacle is in some way constitutive of the phenomena, its own unchanging nature is compromised and it cannot license such references, nor can it license such references if it is not constitutive of the phenomena. Nor can theories which postulate visible and changing ultimate constituents of the universe succeed here. Plato's solution appears to be geometrical atomism, a radically different theory which postulates stable micro-entities in a new ontology of forms, receptacle, micro-entities and phenomena. As these micro-entities can combine and then dissociate to create the phenomena without imperilling their own stability, the problem of *to toiouton* references to changing phenomena seems to be solved.

Geometrical atomism, however, does not dispose of the receptacle, and so we still have the problem of the nature of the receptacle and how it relates to what is in it. This chapter then will deal with the relation of geometrical atomism, matter and space. As the stoicheic triangles do not undergo any change, they do not need something more basic, such as the receptacle, to constitute them. So perhaps the best conception of the receptacle for geometrical atomism is a theory of the space in which the entities of geometrical atomism move, combine and dissociate, rather than some theory of a material substrate. However, the evidence for considering the receptacle to be in some way material is considerable. That should be no surprise, if at some stage we are being invited to consider the receptacle as in some way constitutive of the phenomena. In this chapter, though, I argue that we are presented with a different conception of the receptacle after the introduction of geometrical atomism. Before the introduction of geometrical atomism we are given descriptions which would suggest either that the receptacle is a material substrate, or that it has both material and spatial features. After the introduction of geometrical atomism, we find spatial descriptions only, and no material analogues. Aristotle's evidence is interesting in regard to this, as he appears to have a change in attitude between the *Physics* and *De Caelo* and *De Generatione et Corruptione*.

A second theme of this chapter will be the importance accorded to geometrical atomism in the *Timaeus*, something that is often under-appreciated. The *Timaeus* from 53c is given over to the nature of geometrical atomism and a geometrical atomist analysis of the phenomena. We are also given a detailed analysis of qualities and perception entirely in terms of the geometrical atomist constitution of our bodies and the external world. That seems to be of considerable significance, especially against any theory that qualities occur directly in the receptacle. There are some serious difficulties with the qualities idea anyway relating to the localisation of qualities, the coherence of entities and the nature of causation. Perception also raises the question of the relation of the *Timaeus* and the *Theaetetus* again. I argue that there are considerable differences between these works on this matter, and that the position offered by the *Theaetetus* is unlikely to be Plato's own.

I. The receptacle and metaphors

There is then a difficulty concerning whether the receptacle is supposed to be space, or matter. Certainly at first glance, it would seem that we are not given an entirely consistent account, as there appears to be a mixture of strongly material and strongly spatial descriptions of the receptacle.[1] The evidence for the receptacle as matter is that in the gold analogy the receptacle is referred to as that 'out of which' things are formed (specifically *ek krusos*, 'out of gold', 50a6). So too it is referred to as a moulding-stuff or wax (*ekmageion*, 50c2) or soft material (*malakon*, 50e8), and as mother (*meter*, 50d2, 51a4) and nurse (*tithene*, 49a6, 52d6).[2] It is like an odourless base for perfumes (50e). The receptacle is also that of which part is ignified and liquefied (*to pepuromenon meros*, 51b4, cf. 52d5) to produce phenomenal fire and water, and so would seem to be a material constituent.[3] This would at least seem to invite us to consider the receptacle to be material, especially since, as emphasised in the previous chapter, the receptacle appears to need to be constitutive of phenomena in some manner in order to license *to toiouton* references to changing phenomena. A further question here is whether this collection of material metaphors can be made to yield a consistent account of a material receptacle.

On the other hand, the evidence for the receptacle as space is that the receptacle is specifically or implicitly referred to as space (*chora*, 52b4, 52d3, 53a6, cf. 58a7) and as a seat or place (*hedra*, 52b1, 53a2) and as that 'in which' things occur (*en hoi*, 49e7, 50d1, *en hautoi* 50e5, *en tini topoi* 52a6, 52b4).[4] It is also referred to as a winnowing-basket (*plokanon*, 52e6, cf. 57c. 88de), the motion of which sorts the particles it contains.[5] This is a particularly strong spatial metaphor, as it would suggest that the particles are independent of but contained by the winnowing-basket.

So perhaps Plato treats the receptacle as both space and matter? I would agree with Algra that it is unwarranted to suppose, without some investigation and argument, that Plato held essentialist assumptions about space and matter (i.e. the essential nature of space precludes it being matter and/or the essential nature of matter precludes it being space). Especially so, one might feel, as Descartes held that there could be no body without extension and no extension without body, and so provides a good example of a non-essentialist.[6] Here then is one strategy for attempting to give Plato a consistent account of the receptacle.

As we shall see, a good many of the metaphors Plato uses are open to a wide variety of interpretation. The list of material and spatial metaphors I gave above should be regarded as a preliminary one based on first impressions of the metaphors, as some metaphors can be taken to be either material or spatial or material and spatial. We might also be able to argue that different metaphors are meant to bring out different aspects of the receptacle and so are not inimicable to each other.

Having said this, I do not believe that treating the receptacle as matter is consistent with some of the spatial metaphors which are employed. Here I am not overly concerned about whether it is possible to reconcile the 'in which' (*en hoi*) and the 'out of which' (*ex hou*) expressions.[7] We might have a theory where the receptacle is matter and is space and these expressions answer different questions, say where is it and what is it. I am rather more concerned that the theory of geometrical atomism requires there to be empty space and that Plato seems quite aware of this requirement. The winnowing-basket metaphor, which is critical to the functioning of the cosmos, is not only one that would seem to separate matter and space but on a straightforward interpretation would require there to be empty space between the particles so that whatever is winnowed will not pack to fit the space in the basket fully. I will adopt a different strategy on the nature of the receptacle, which is that Plato presents us with a material or a material/space theory of the receptacle prior to the introduction of geometrical atomism, and a space theory after its introduction.

As commented in the previous chapter, it is of the utmost importance to be able to relate 48e-53b to geometrical atomism.[8] That is so on the general interpretational principle that Plato wrote his works as wholes were we need to be able to see the relation of each part to the whole. So too there appear to be several cues in 48e-53b which prompt such a link. There are comments about shapes, shaping and triangles, there is the promise to explain the receptacle/contents more fully (50c6) and there is the link between 48e-53b at 53b7-c3 which states that now we will look at the disposition and origin of the elements.

II. The *Timaeus* and empty space

If Plato recognises the existence of empty space, this will be important evidence on the relation of space and matter and the nature of the receptacle. On this matter I am largely in agreement with O'Brien.[9] At *Timaeus* 58a ff., Plato appears to say both that there is no void, as the revolution of the all compresses the elements, and 'allows no void space to be left' (*kenen choran oudemian eai leipesthai*), and then that there are voids (*kenonta*) and emptinesses (*diakena*) between the atoms of the elements. This is far from being the only instance where we have such a discrepancy. One can also look to *Timaeus* 59a1-2, 60c1-2, 79b1, 79b3, 80c3 for a denial of a void, and *Timaeus* 60e5, 61a1, 61b1 and 61b4 for the existence of gaps between atoms. As O'Brien points out, the context here is important.[10] At *Timaeus* 58a4 ff., Plato is seeking to explain why the elements do not separate out completely, but are continually in motion and are changing into one another. There are two options for giving Plato a consistent account. With Cherniss, one can deny the existence of gaps between the atoms.[11] As O'Brien points out, though, this is mathematically false, even if one assumes many sizes of atoms, and Plato seems quite sure there are such gaps,[12] and that breaking up particles will result in empty spaces. *Timaeus* 58b is most easily read as recognition of a packing problem as smaller atoms force their way into gaps left by larger ones. It is interesting that at *Timaeus* 81b6 we are told:

> The blood particles, being chopped up within us and being surrounded by the living creature as by a heaven, are compelled to imitate the motion of the universe. So when each of these divided entities moves towards its kin, it fills again the emptied space (*kenothen*).

I find it hard to believe that Plato was not aware that his theory of atoms and how they move and change implied that there must be gaps between atoms and spaces when they change. Certainly Aristotle is aware that of this sort of problem, and he asserts:

> It is entirely irrational to attempt to assign shapes to the simple bodies, primarily because when placed together they will not fill the whole. It is supposed that there are three plane figures which can fill space, the triangle, the square and the hexagon, and among solids only two, the pyramid and the cube. Of necessity they need more than these as they consider there to be more elements.[13]

Aristotle is of course aware of Plato's denial of a void, e.g. at *De Generatione et Corruptione* 325b30 ff., though it should be noted that this is in the context of a comparison with Leucippus.

There are two points I would add to O'Brien's critique of Cherniss, both based on the fact that this is supposed to be a dynamic theory attempting to account for motion and change. First, it is one thing to provide an arrangement whereby the atoms can be made to tessellate in three dimensions. It is another to be able to account for the motion of the atoms in that way, as the arrangement may well be one that locks rather than allows motion.[14] Secondly, when atoms decompose they do so from three-dimensional entities into two-dimensional entities, and however brief the time taken to recompose, during that time there must be gaps. One might add that even if the first arrangement of atoms tessellates in three dimensions, there is no guarantee that after a transmutation the new arrangement of atoms will continue to tessellate.

The alternative approach is to question Plato's denial of a void. At 58a7 we ought to take *kenen choran oudemian eai leipesthai* to mean that the revolution of the whole tends to leave no empty space rather than precludes the existence of empty space.[15] Why is this so? Because the revolution of the whole has a compressive effect (*sphingei*, 58a7) and so brings the atoms together as far as possible, so there are no large-scale voids and motion but there may be gaps between atoms. I agree with O'Brien against Taylor and Cornford, who argue that as simple circular motion does not produce such an effect, Plato cannot have countenanced it, that Plato supposes just such an effect.[16] This is the natural reading of *sphingei*, 'bind tight, press together' at 58a7. Why should he suppose that to be so? We might look to Empedocles Fr. 38.4 where there is a possibly similar usage of *sphingon*.[17] The ancients though were aware of the action of a whirlpool in moving heavy bodies to its centre. At *De Caelo* 295a ff. Aristotle states that those who argue that the earth is in the centre of the cosmos because of the action of a vortex around it argue from observations of whirlpools.[18] Indeed, gravitational theories based on this idea were common in the seventeenth century, notably that of Descartes. Plato, on this basis, might well believe that circular motion in certain circumstances does indeed create some form of compressive effect. However Plato envisages the compressive effect working, the conclusion must be that while there are no large-scale voids for Plato, there are gaps between atoms and when they decompose, and so for Plato there is empty space. That there is empty space for Plato goes very much against the view of the receptacle as matter, or space and matter.

III. Material metaphors

As Plato appears to have empty space as a consequence of geometrical atomism, it would seem that the receptacle can be only space. That though would seem to run up against a good deal of the evidence cited in section one which would seem to indicate that the receptacle is material. Let me suggest two strategies for allowing Plato a consistent account. First, one might try to argue that the receptacle is at no stage intended to be material. As noted earlier, there is a good deal of flexibility in how we might interpret many of the material-looking descriptions, so perhaps it is possible to interpret all of these in spatial rather than material terms. Secondly, one might argue that just as geometrical atomism provides a quite different solution to the problem of how we can licence references to changing phenomena, so too we are given a different conception of the receptacle with geometrical atomism. So perhaps we are given material or spatial/material descriptions of the receptacle up to the introduction of geometrical atomism, but only spatial metaphors after its introduction.

Let us begin by examining the metaphors which give rise to the idea of the receptacle as matter, and see if they are susceptible to some other interpretation. The receptacle is described as the 'nurse of all becoming'. Certainly one could take this as a material metaphor, taking the nurse to be a wet nurse giving material sustenance to what comes to be in the receptacle. There is though nothing which compels us to take this metaphor in such a manner. Space as much as matter might be described as a nurse (everything has to be somewhere, 52b1), and the occurrence of *tithene* at 52d5 is sandwiched between and linked to the occurrence of *chora* at 52d2 and the description of the receptacle as a *plokanon* at 52e6, the strongest of the spatial metaphors. Another possibility is that the receptacle is a nurse in the sense of rocking its charges. *Tithene* at 52d5 leads up to the description of the receptacle shaking the things that are in it, and 88de has the *trophon kai tithene tou pantos*, the 'rearer and nurse of the universe' always moving and agitating the cosmos. This will help keep the cosmos in good order, just as proper exercise will help to keep the human body in good health. We are by no means compelled to take the description of the receptacle as a nurse to mean that it is in some way material.

That the receptacle is described as a mother (*meter*), and at 50d there is also a father and some offspring looks more of a difficulty. However, this metaphor is introduced by *Timaeus* stating that we must conceive of three things, that which becomes, that in which it becomes (*en hoi*, 50d1) and the model in whose likeness that which becomes is born. So the receptacle as mother is supposed to explicate the *en hoi* relation.[19] Cornford argues that a common view of conception in the ancient world was that the male supplied the seed and the female supplied the place.[20] Such a view is mentioned by Aristotle, who debates the issue.[21] Algra argues that Cornford's references to other views do not decide the issue.[22] What must go a long way to settling the issue though is the *Timaeus*' account of conception. If we look to *Timaeus* 91a ff., we find that the male indeed supplies the seed (like the fruit from a tree, *apo dendron karpon*, 91d) while the woman supplies the place in which it grows (it is sown as in a tilled field, *arouran*, 91d2), but does not supply any matter. The *meter* metaphor on this interpretation then looks more spatial than material. It may also be the case that neither the nurse or mother descriptions are meant to imply that the receptacle is spatial or material, but are merely intended to bring out the fact that the phenomena are to be thought of as some form of union between forms and the receptacle.

If we take *ekmageion* (50c2) to be a moulding-stuff, out of which we make things then this looks very much a material metaphor. So too one might draw parallels with the *malakos*, the soft material of 50e8, though if we do draw this parallel it is important to note that the strongly spatial *en hautoi* at 50e5 precedes malakos, as does *en hoi* at 50d6. *Ekmageion* is not necessarily a moulding stuff out of which we model things but may also be wax in which one marks a seal. This is certainly the sense of *ekmageion* in the wax block model of the mind in the Theaetetus.[23] The verbs here are important as well. The *ekmageion* is moved and shaped, and in the next sentence the verb *tupo* is used.[24] Now *tupo* can have the sense of to stamp or impress as well as to model. *Tupo* is again used in the context of the soft material (50d4, 50d6), but so is *apomatto* at 50e8, which has more of a sense of stamp or impress. So we might take this as a wax in which we stamp or impress figures, when the metaphor again looks rather more spatial. One might again argue that the point of these metaphors is neither spatial nor material, but merely to emphasise the characterlessness of the receptacle.

The gold example is seen by many to be the strongest material metaphor, as here we have shapes being moulded and re-moulded out of gold (*ek chrusou*, 50a6). Mohr has argued that it is the 'exigency of the metaphor'[25] that produces the impression of a material substrate, but I would tend to agree with Algra that we can hardly take the gold analogy any other way.[26] One strategy we might employ is to argue that some metaphors appear to be material until the introduction of geometrical atomism when a non-material explanation becomes available. Consider the references to the receptacle being ignified and liquefied (*Timaeus* 51b4, cf. 52d5). We are told that 'that part which is made fiery each time appears as fire' (51b4-5). As Algra comments, it is hard to make sense of the idea that phenomenal fire is an ignified part of the receptacle unless the receptacle is some form of material constituent.[27] Indeed, it is hard to see how one might ignify or liquefy space, and so this looks strongly material. With the introduction of geometrical atomism, though, one might argue that what this means is that fire atoms have either moved into or been formed in this area of space.

However, with the gold example there is no question of waiting for geometrical atomism for a full explanation. Here we are told that on account of this analogy we can have the safe answer of gold and the other answer of *to toiouton*. This can only be the case if the gold is constitutive of the phenomena in some manner, and so we are led to some form of material conception of the receptacle. Earlier I argued that it is important to recognise that the gold example is prefaced by a conditional (*ei gar*, 50a5), and is a thought experiment rather than a direct cosmological statement. Thus the radical flux supposed here need not be supposed elsewhere. But the gold example is supposed to explain something, and that surely is why we can refer to changing things as *to toiouton*, and the answer to that clearly enough is if there is an unchanging material substrate, then perhaps we can.

So too the idea of the receptacle as being like an odourless base for perfumes seems to invite a material interpretation. Whether that is the same relation of receptacle to form and phenomena as the gold example could be debated at length. Here one might argue though, as with the wax and soft material metaphors that the point here is neither spatial nor material, but to emphasises the characterlessness of the receptacle as a suitable receiver.

It is hard to avoid impression reading *Timaeus* 48e-53b that we are at least being invited to think about the possibility of the receptacle as a material entity. If there is a transition to spatial metaphors, is it possible that is relatively gentle? It is noticeable that material metaphors predominate at the beginning of 48e-53b, and that there is a transition to the predominance of spatial metaphors near the end. We might try then to reformulate the earlier material metaphors in acceptable spatial terms. I would argue that the choice is rather more stark. I do not think we can successfully reformulate all the material metaphors. Rather, 48e-53b presents us with a tripartite ontology of receptacle, forms and phenomena, with a theory of the relation between flux and language and an appropriate conception of a material/spatial receptacle. 53c ff. presents us with a new ontology with the addition of the entities of geometrical atomism, a new theory of the relation of flux and language and, as I argue in the next section, an appropriate conception of the receptacle as space.

IV. Spatial metaphors

As a second strategy, we might accept that there are metaphors which indicate that the receptacle is material, but offer the following argument. If we conduct the discussion of the constituents of the world solely at the perceptual level, then there are no physical *stoicheia*, as earth, water air and fire appear to change into each other. If so, then in order to make any proper references to the phenomena, there must be a relationship between the receptacle and the elements which supports such references, such that the receptacle must be in some way constitutive of the elements, or in other words is matter, or is both matter and space. So we get a mixture of metaphors between *Timaeus* 48e and 53c.

Now consider the situation when geometrical atomism is introduced at *Timaeus* 53c ff. We now have physical *stoicheia* which are non-perceptible, the stoicheic triangles. It is quite possible that it is their stability which underwrites proper references to the visible and changing elements, and the stoicheic triangles are certainly constitutive of the elements. After the introduction of geometrical atomism then, there is no need for the receptacle to be constitutive of anything. It can be just space. The way in which geometrical atomism is set up, with two-dimensional entities forming up to create three-dimensional entities, and three-dimensional entities which cannot be fitted together without gaps between, seems to demand such a view. It is also interesting to note that after the introduction of geometrical atomism, we find plenty of space-type descriptions, but no matter-type descriptions at all. We can find a continuation of the *plokanon* metaphor from 52e at 57c, with the only mention of the receptacle after the introduction of geometrical atomism at 57c7 (*tes dechomenes*) and 88de where it is nurse and rocks the universe. As noted above, the winnowing-basket is perhaps the strongest of the spatial metaphors as the particles in it are independent of it and it is their container, and looks particularly apposite for geometrical atomism. So too we can find mention of *chora* in many places after *Timaeus* 53c,[28] and indeed *hedra*.[29] We can also find mention of *kenos* at 58b3 asserting gaps, at 59a1-2, 60c1-2, 79b1, 79b3, 80c3 denying a void, and *diakenos* at 58b5, 60e5, 61a1, 61b1 and 61b4 asserting gaps. *Meter* is not used again although *tithene* is once at 88d6, where as we have seen, the nurse agitates and keeps the universe in motion for the sake of its order. There is no need to take 'nurse' as at any time indicating that the receptacle is material. *Ekmageion* is used once at 72c for the relation of spleen and liver, *malakos* is used several times either as 'soft' in contrast to hard,[30] or in relation to objects or as describing soft parts of the human body.[31]

It is also possible to combine these two strategies. Perhaps this interpretation of 53c ff. does not have to struggle against a multitude of material metaphors in 48e-53c, if we take mothering as spatial due to the *Timaeus*' account of conception, nursing as rocking, and *ekmageion/malakos* to be wax in which we stamp impressions, or to be really about characterlessness rather than matter or space. In which case one would be left with the gold example, and ignification and liquefaction. If so, that might well focus our attentions on what I have argued are the key issue here, which is whether the receptacle can be unchanging itself and yet be constitutive of changing things.[32]

As emphasised earlier, it is of critical importance to be aware of the differences between *Timaeus* 48e-53b and 53c ff. The first theory operates only at the level of perception for physical entities, while he second postulates micro-structure. The second theory postulates more entities than is permitted by the tripartite ontology of the first. The second theory corrects the first on transmutation. It is quite possible then that the second theory involves a reformulation of the relation between matter and space.

To say that this is the position of the *Timaeus*, or indeed Plato's position, would perhaps be going too far, however, as this reformulation is not made explicit, even if some of the other differences between 48e-53b and 53c ff. are evident and marked. But Plato may wish his readers to be active here and consider their own attitudes to the theories presented, and the possible problems and solutions.

It may be that the reader is given (at least) three theories here, elements on their own, elements and receptacle as matter/space, and geometrical atomism. I say at least three, as it may be the case that *Timaeus* 49d6 is genuinely ambiguous between whether we call fire this or the suchlike, or we call this or the suchlike fire, and asks the reader to make the choice of how to interpret it. Clearly the elements on their own is an inadequate theory, as it does not license references to the phenomena. What of the receptacle as matter and space? *Timaeus* 50bc, as we have seen, seems flatly contradictory. The receptacle does not depart from its own nature nor does it take on any shape, yet in the next sentence it is changed and shaped by the things which enter it, and this is 'hard to explain and marvellous' (*dusphraston kai thaumaston*, 50c6), and there is a promise to follow this up later. Yet the second attempt at describing the receptacle ends with the conclusion that the receptacle is invisible and shapeless (*anoraton kai amorphon*, 51a8) and its relation with the intelligible is 'most difficult and hard to grasp' (*aporotata* ... *kai dusalotaton*, 51b1). *Timaeus* 48e-53b then leaves us in a radical *aporia* (impasse) concerning the receptacle. The receptacle also appears to be inexplicable and non-explanatory.[33] Perhaps then we are meant to at least question whether the receptacle cannot do everything that is asked of it, and if this is an adequate account of the relation between matter, space and the phenomena. *Timaeus* 53c ff. then gives us the materials for a reformulation.

If we are being generous to Plato, he may well have seen some of the problems inherent in an undifferentiated theory of matter where matter and space are identified, and may also have seen something along the lines of Berkeley's criticisms of Locke on matter, that if what is under discussion has no characteristics it is otiose, and if it has characteristics it is not matter. That geometrical atomism can provide coherent answers to the major problems with the receptacle as matter theory seems more than coincidental.

V. The receptacle and qualities

There is a very important but often underappreciated fact about the *Timaeus*. This is that 53c through to 89d is given over to an analysis of how geometrical atomism can account for all the phenomena which we perceive. It is not about how qualities occur in the receptacle, but about how qualities might be thought of in terms of a geometrical atomist analysis of our bodies and of external bodies. Given what I have argued about the relationship of the theories of *Timaeus* 48e-53b to geometrical atomism, that would seem to be highly significant. This is especially so as there are certain problems with the view that qualities occur directly in the receptacle that geometrical atomism might be thought to solve.

There is a problem, which I shall call the localisation problem, if qualities occur directly in the receptacle. The receptacle, it is emphasised is *amorphos* and although this is not explicitly stated, given its characterlessness we might reasonably expect it to be partless (*ameros*) as well. Forms are not spatial entities, so they too are *amorphos* and we would reasonably expect them to be *ameros* as well. So when the receptacle partakes of the intelligible why is it that only a part of the receptacle is affected rather than the whole receptacle? What is it that spatially limits and localises qualities which occur in the receptacle? The only other resource here is the demiurge, but that looks highly problematic. It is not in the least bit clear how the demiurge might achieve localisation unless we go to *Timaeus* 53b4-5 where the demiurge begins to generate shapes by means of forms and numbers. That though would seem to point us directly at a geometrical atomism account of localisation. As the demiurge retires after the ordering of the cosmos, in order for localisation of phenomena to continue what he does at this stage must be permanent. Again, that looks problematic unless we go to *Timaeus* 53b.

On a geometrical atomism analysis of the basis of qualities, which appears to be what we are given, the localisation of phenomena is straightforward. The demiurge creates spatially limited entities, the stoicheic triangles, which move around in space and combine with each other to form more complex entities such as the four elements. So this area can be described as fiery because in this area the stoicheic triangles have formed up into tetrahedra, and the interaction of these tetrahedra with our bodies gives rise to the qualities which we associate with fire.

One might question the relation of the stoicheic triangles to the receptacle here. If one were desperate to hang on to the receptacle as matter theory, then one might postulate the stoicheic triangles as triangles in the receptacle whose matter is provided by the receptacle. As they move, their shape slips across this matter. Thus the stoicheic triangles would be merely configurations of the receptacle. This would not solve the original problem of the receptacle (if the receptacle constitutes the stoicheic triangles and so the phenomena, it changes as the stoicheics move/the phenomena change, if it does not constitute them, it cannot license stable references), and there is a further serious objection. There is a problem in differentiating between 'proper' and 'phantom' shapes on the configurations view. Here, in two dimensions, we have some regular shapes which have not tessellated. On the geometrical entities and space view ('proper' shapes shaded) this is unproblematic. We can easily differentiate between the entities and the empty space, as in Fig. 55 (left). So we have two types of regular entities and four regular but small spaces and an irregular central space. On the right though, with the shading removed, where we have a configured receptacle, what then makes the regular shapes any more (or less) configurations than the small regular shapes or the central irregular shape?

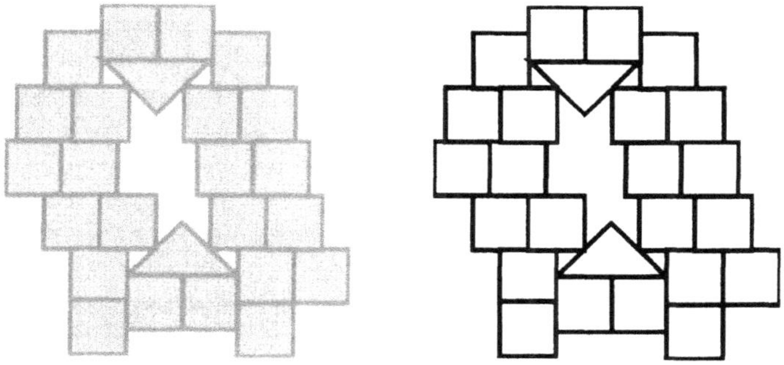

Fig. 55. A selection of two-dimensional shapes in the receptacle, on the entities and space view, and on the configurations view.

There is no mention of any irregular shapes in the receptacle once the demiurge has done his work. Such shapes would have to be as causally efficacious in their own way as the 'proper' shapes yet we are given no account of them whatsoever.

A second problem along these lines is as follows. In Fig. 56 on the left, working on the geometrical entities and space model, we have, unproblematically, eight squares and some space. On the right though, using the configurations view, suddenly we have 9 squares, as the central square is no more or less a square than any of the others. The idea that the entities of geometrical atomism are no more than configurations of the receptacle is untenable.

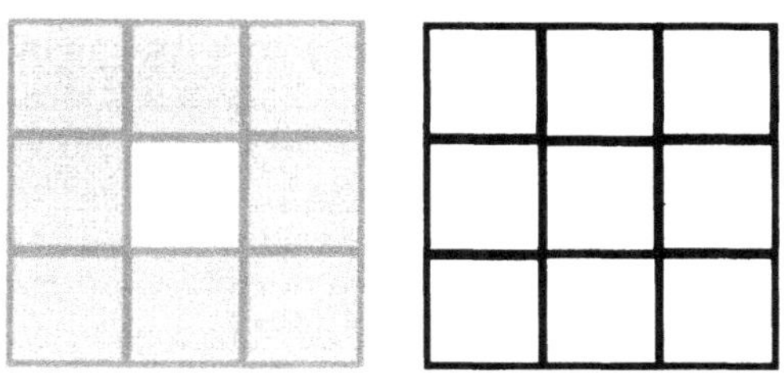

Fig. 56. A counting problem on the entities and space view and the configurations view.

There is also a problem with the coherence of entities and their movement in the receptacle, if we are thinking only in terms of qualities. If an entity is only a collection of qualities, one might be worried as to why this collection of qualities stays grouped together as the entity moves. Again, the demiurge is not going to be an adequate answer here, for while he may set up the order of the cosmos, he does not maintain it. Another resource available is the like-to-like principle, perhaps in alliance with the compressive principle generated by the motion of the same. As these principles are expressed quite generally, one might seek to apply them to qualities, though one must note that this is not how these principles are applied in the *Timaeus* and it does look odd to have such a principle operate with qualities. If the like-to-like principle does work operate with qualities, one might worry why one gets complex entities with many qualities rather than simple patches of like qualities. One might also worry about how a tiger keeps its stripes, not only when it is in motion and we are concerned as to why all these qualities move together, but also when it is stationary. Surely the black of the black stripes and the gold of the gold stripes will be affected by the like-to-like principle and the tiger will dissociate into patches of like qualities, if there is nothing to prevent this happening? At the level of qualities, there does not appear to be anything which will prevent such a dissociation.

On a geometrical atomist account, the coherence and movement of bodies is not a problem. The stoicheic triangles have coherence and permanence which is not affected by their motion. The complex entities which they form have varied permanence depending on the strength of the bonds, as we saw in the last chapter.

One might also consider the question of the nature of causal interactions if we have only qualities in the receptacle. Perhaps one could put together a theory of causation with qualities, though one would have to be careful that this was not merely a regularity theory and that it would satisfy the demand of the *Republic*'s cave allegory that there should be something more than the phenomena such that we can escape the cave. Again, though, this is not the way the account of the *Timaeus* goes. We do not have qualities interacting with qualities, but causation appears to take place between the various entities of geometrical atomism which give rise to qualities.

Certain interpretations of the receptacle run into trouble with causation, and the problems seem most severe for Cornford's images in a mirror view of the receptacle. Take two billiard balls on a table and a mirror image of this. I roll one ball towards the other one, and it bumps into and moves it. The billiard balls have a causal relation with one another, but do the images in the mirror? If the images are images of forms, surely we don't want to say that forms are causally related to one another? On a geometrical atomist account, causal interaction is not a problem. The entities of geometrical atomism can quite happily interact with one another (come into contact, decompose each other), and provide a basis for changing qualities.

A good deal of the motivation for the view that the receptacle contains qualities, I suggest, comes from an attempt to align the *Timaeus* with the first part of the *Theaetetus*.[34] The discussion of whether knowledge is perception generates the Theaetetus/Protagoras/Heraclitus position of a radical flux of qualities. Whether that is Plato's own position is a matter of debate.[35] The *Timaeus*, as we have seen, differs from the *Theaetetus* in permitting us to refer to changing entities as *to toiouton*. We might with Owen consider that to be a 'lame plea' and so date the *Timaeus* earlier, or investigate why the *Timaeus* might allow such references. The least deviation from the *Theaetetus* is the idea of the receptacle containing qualities, but on many grounds that appears highly problematic. The introduction of geometrical atomism looks much more promising but then the *Timaeus* looks much less like the *Theaetetus*. There is more to come, as there are radical and irreconcilable differences between the theories of perception advanced in the *Timaeus* and *Theaetetus*.

VI. Perception in the *Timaeus* and *Theaetetus*

There are then problems with the idea that the receptacle contains qualities, and indeed this is not the account of perception that we are given at *Timaeus* 61c ff. Perception in the *Timaeus* is generated from the interaction of geometrical atomist entities with our bodies (which themselves are constituted from geometrical atomist entities). This account is developed in some detail for each of the senses. Plato though also presents us with a theory of perception in the *Theaetetus*. This raises two questions. Is the *Theaetetus* view of perception the same as the *Timaeus* account? Is the *Theaetetus* view of perception Plato's view? My view is that the account given in the *Theaetetus* is not Plato's view, and that the account given in the *Timaeus* is radically different. Readers familiar with Burnyeat's work on the *Theaetetus* will recognise that I am sympathetic to his 'B' reading of the first part of the *Theaetetus* in what follows (Plato does not accept the Theaetetus/Protagoras/Heraclitus account of the world). I shall say more about the relation of readings of the *Theaetetus* to how we might approach interpreting the *Timaeus* in the next chapter. There are of course alternative approaches. Taylor believes there are significant differences between the *Theaetetus* and *Timaeus* but believes the theory to be Timaeus' and not Plato's.[36] Cornford believes the theory of perception in the *Theaetetus* to be Plato's own and not to have significant differences with that of the *Timaeus*.[37] McDowell argues that while there are differences between the theories, these are not insuperable.[38]

What then are the differences on perception between the *Theaetetus* and the *Timaeus*? To begin with, I agree with Fine on some points of difference between the two theories.[39] In the *Theaetetus*, there are no perceiver-independent properties.[40] An object is white if and only if it is perceived to be white. In the *Timaeus* though, the shapes and sizes of the entities postulated by geometrical atomism are not perceiver-dependent.[41] Indeed, the very postulation of such entities would seem to run counter to the *Theaetetus*. If we put this in terms of later terminology, the *Timaeus* postulates primary qualities where the *Theaetetus* does not.

What of what might be called secondary qualities, such as colours and tastes? Here it would seem that there is some measure of agreement between the *Timaeus* and the *Theaetetus*, at least on the idea that perception is due to an interaction.[42] If Plato identifies colours as 'flames streaming off bodies having particles so proportioned to the visual ray as to yield perception',[43] then one might argue that objects can (and do) do this without being perceived, that colours are thus perceiver-independent, and so conclude with Fine that:

> The *Timaeus* allows, what the *Theaetetus* disallows, that something is white when it is not being perceived at all.[44]

However, this is not all that Plato has to say on the matter, and we are subsequently told that particles of the same size as the visual ray are transparent, while those that are larger dilate it and those that are smaller contract it. Then at *Timaeus* 67e4-6 we are told:

> Thus we should assign the name 'white' to that which dilates the visual ray, and 'black' to the opposite.[45]

This to some extent makes colours perceiver-relative again.[46] Objects are not white in themselves, but emit fire particles of greater size than those of the human visual ray, and because of their effect on that ray we perceive white. One could easily imagine a creature whose visual ray consisted of particles either slightly larger or slightly smaller than those in the visual ray of humans. There would then be some objects, emitting particles close in size to those of these visual rays, which it would see as white and we would see as black, or vice versa.

This is not to say that objects do not have perceiver-independent properties. An object may well have the property of 'emitting fire particles which have the ability to dilate the human visual stream' or more simply 'emitting fire particles of a specific size', whether observed by humans or not. Similarly hotness may only occur when a sentient being comes into contact with fire atoms, but fire atoms have sharp angles and move swiftly whether they are in contact with humans or not, as the *Timaeus* has them interacting with other atoms unperceived by humans.[47] So too objects will be sorted by the shaking of the receptacle and the compressive effect of the same by a like-to-like principle whether they are perceived to be by humans or not. Even if we distinguish between weight as a perceptual quality generated by an interaction between ourselves and objects and weight as an effect of physical processes, clearly objects (such as atoms) will have weight in the second sense independent of perception.[48]

There is though a more fundamental difference between the *Theaetetus* and *Timaeus* on perception. *Theaetetus* 152c lays down two conditions for knowledge. Knowledge must be unerring, and must be of what is, such that perception, if it is to be knowledge, must both be unerring and of what is.[49] One way of summarising the argument of the first part of the *Theaetetus* is as follows.[50] In order for perception to be unerring, and for the 'man is the measure' doctrine of Protagoras,[51] perceptions need to be relativised to persons and instances. A suitable ontology here would be one of Heraclitean flux, where everything is changing in every respect all the time.[52] However, there is a problem with such a flux, as we saw in the previous chapter. The *Theaetetus* is adamant that we cannot refer to entities in total flux, and indeed at 157a ff. insists that we cannot use the verb 'to be', or any other name that makes things stand still of these changing entities.[53] So if perception is unerring then it is not of what is, and so perception cannot be knowledge.

The *Timaeus* might well agree on the prerequisites for knowledge, but disagrees on why perception is not knowledge, and indeed reverses the approach of the *Theaetetus*. The *Timaeus* is quite clear that human perception (in contrast to that of the world soul) is fallible.[54] When the revolutions which constitute our minds are disturbed, then we are prone to error.[55] Perception in the *Timaeus*, though, is of what is, at least to a greater extent than is allowed in the *Theaetetus*. Unless we conclude that both the receptacle and geometrical atomism fail in providing a licence for *to toiouton* references to changing phenomena, then the *Timaeus* allows references to changing phenomena which the *Theaetetus* proscribes. It is interesting to note that perception is not infallible in other works. Certainly the *Republic* does not treat it so at 602d with the example of things which appear bent when placed in water, nor indeed does the *Sophist* in relation to mirror images at 266c ff., nor the *Philebus* in relation to seeing at a distance at 38c.

The *Theaetetus* theory of perception then is unlikely to be Plato's own. It seems more likely that it is a consequence of trying to make sense of the idea that knowledge is perception along with the prerequisite that knowledge must be infallible. The further consequence of that is that perception cannot be of what is and so the attempt to define knowledge as perception fails. The theory of perception in the first part of the *Theaetetus* is radically different to that of the *Timaeus*.

VII. The evidence of Aristotle

The theory of the entities of geometrical atomism moving around in space seems to me to be an excellent one for someone in fourth-century Greece to hold. It has the advantage of attributing to Plato a coherent theory of the relation of flux, language, matter and space, and also suggests that Plato sees some of the problems with the receptacle as matter theory. Here I will examine the evidence of Aristotle, which is commonly thought to interpret the receptacle as a material substrate. The evidence of Aristotle, however, is not consistent, and may undergo a significant change between the *Physics* and *De Caelo* and *De Generatione et Corruptione*. Certainly he appears to give a classic statement of the receptacle as both matter and space in *Physics* IV.2, where he says:

> Plato identifies matter (*hulen*) and space (*choran*) in the *Timaeus*. For that which participates (*metaleptikon*) and space are one and the same. His account of that which participates differs in the unwritten opinions, but he is consistent in identifying place (*topon*) and space.[56]

While this looks straightforward, one might raise two doubts here. Plato does not use the word *hule* at all in the discussion of the *Timaeus*, so this is an Aristotelian interpretation, subject to the usual worries about such interpretations, rather than a direct statement. The structure of this passage is also interesting. One might read it that while Plato is consistent between the *Timaeus* and the unwritten opinions on the question of space place he is not on the question of matter, place and space. Aristotle goes on to make the key criticism against an identification of matter and place, that while place can be separated from an object its matter cannot be.[57]

De Caelo II.7 presents a rather different picture. Aristotle is of course well aware of the gold example (see the passage from *De Generatione et Corruptione* below). He discusses whether the transition from one element to another might be effected by separation, and having dismissed this possibility, suggests two other theories:

> Change of shape, just as a sphere or a cube might be generated out of the same piece of wax (*ek tou autou kerou*), or the analysis into surfaces, as some have said.[58]

It is clear from subsequent discussion of triangles that he associates the second theory, and not the first, with Plato. That is very interesting because the first theory here looks very much like the gold example (note the *ek tou* expression) and the subsequent wax and soft material analogies. Nor is there any question here of identifying matter and space. Especially so, on might say, as Aristotle recognises that the transition between elements will leave some individual triangles 'suspended'. Slightly later, Aristotle does say that:

> It is necessary for that which underlies (*to hupokeimenon*) to be devoid of forms and shapes, so that it will be more capable of modification, just like the all receiving (*pandeches*), as is written in the *Timaeus*.[59]

That, though, is in the context of putting forward his own theory of the *hupokeimenon*, and that Plato describes the receptacle as amorphos is uncontentious. More important is what use Plato may put the receptacle to. *De Generatione et Corruptione* is interesting on these matters. Aristotle tells us:

> What is written in the *Timaeus* is not well defined. It is not at all clear if the allreceiving (*pandeches*) is separate from the stoicheia (*chorizetai ton stoicheion*) and he makes no use of this, after saying it is a substratum

> (*hupokeimenon*) prior to the so-called elements, just as gold things are made out of gold.[60]

That is significant on several counts. Aristotle now seems to have serious doubts as to whether Plato equates matter and space if the elements might be separate from the receptacle. Aristotle has also picked up the point that indeed Plato does not make any further use of the receptacle where one might expect him to.[61] Aristotle goes on to say:

> He produces an analysis of the elements which are solid as far as surfaces, though it is not possible for the nurse, the primary matter, to be the surfaces.[62]

So Aristotle recognises that there is a problem with the relation between the receptacle considered to be material and the stoicheic triangles, and we might note in this context the evidence we saw in the previous chapter on Aristotle's view that the stoicheic triangles are basic, primary and undergo no *alloiosis*.[63]

Aristotle's evidence then, from *De Caelo* and *De Generatione et Corruptione*, at least recognises that:

a. There is a problem with the relation of the receptacle and the stoicheic triangles.
b. Plato's primary analysis of the elements is into planes and not into matter like the gold example.
c. Plato describes but does not employ a material receptacle.

One can see how the view Aristotle puts forward in the *Physics* might have arisen, especially given some of the similarities between the receptacle of *Timaeus* 48e-53b and the analysis of *Metaphysics* V.8 and VII.3, and the *Physics* view seems to be a reasonable criticism of the receptacle as matter/space theory. Identifying matter, space and place, or even just matter and space, is likely to lead to difficulties. That assumption on its own might not be so problematic, but it seems to be coupled in the receptacle as matter/space theory with the idea of a single, undifferentiated material substrate dealing with change. It is the combination of the assumptions that leads to the problem that the receptacle cannot be constitutive of phenomena without imperilling its own status as an unchanging entity. Geometrical atomism of is very nature breaks with these assumptions, something Aristotle seems to come to realise.

VIII. Plato, Democritus and Descartes

I might be accused of producing a 'Democritean' account of the receptacle and its contents, something, which Cornford was vigorously opposed to and, is held by some to have produced a definitive refutation.[64] What follows is an analysis of Cornford's arguments in *Plato's Cosmology* pp. 198-210, highlighting where my views differ from the Democritean account which Cornford attacks, and where they are similar, rebutting his objections. On some points I shall be fairly brief as we have already dealt with the some of the relevant issues.

Contra Cornford (*PC* 199-200), as we have seen, shape is not a mistranslation of *morphe* at *Timaeus* 50c1. In any case I am indifferent to what is in the receptacle at this stage of the *Timaeus*. I do not read Democritus' account into Plato on the basis of the *plokanon*, the winnowing basket (*PC* 200-3). I agree with Cornford that there are considerable differences between a *plokanon* which is shaken and other instruments which are given a rotatory motion and so might be though to be like a Democritean vortex. As argued in Chapter 1, this may be significant as a move questioning the atomists right to assume a specific sort of motion. Certainly I do not think that Plato accepts the Democritean account of all shapes and sizes of atoms (*PC* 200).

Cornford puts forward his favourite analogy of the receptacle as a mirror (*PC* 200 ff.).[65] As I have argued, there is no evidence for the receptacle as a mirror, an analogy which Plato does not use for the receptacle despite using many descriptions of the receptacle and despite discussing mirrors and mirror images elsewhere in the *Timaeus*.[66] We have also seen some of the problems Plato would have been saddled with had he thought the receptacle to be like a mirror. Kung has argued that the receptacle is not a mirror on three grounds:[67]

1. For Plato, a mirror is causally acted upon by the objects it images, while forms do not act and the receptacle is not acted upon.

2. A mirror inherently distorts an image, making left appear right, while the receptacle ought not to.

3. For Plato, a mirror neither contains nor constitutes its images.

Algra quite rightly argues that if we consider the receptacle to be like a mirror, we are not necessarily not committed to it being like a mirror (either as we understand mirrors, or as Plato did) in all respects.[68] This to some extent disarms Strange's comment that it is odd that Plato should describe the relation between receptacle and forms as *dusphrastos* and *thaumastos* when he has already explained quite succinctly how mirrors work.[69] He might wish to deny that the receptacle is like a mirror in this respect, but then one must ask in what respect is the receptacle like a mirror? The other major concern has to be that if the receptacle does not constitute what is imaged in it, as with a mirror, then the receptacle will not provide a basis for licensing *to toiouton* references to changing phenomena, while if the receptacle does constitute the phenomena in some manner then it would seem to be most unlike a mirror.

According to Cornford (*PC* 204ff.), the *Theaetetus* gives us Plato's own theory of sensible qualities and physical bodies, and we can use such theories to help us interpret the *Timaeus*. This actually goes to the heart of one of the key issues in how we should look to interpret the *Timaeus* and has some wide reaching implications, as I shall discuss in the next chapter. We have seen, however, that there are considerable differences in the treatment of perception between the two works, and it is by no means clear that the theories to which Cornford refers (*Theaetetus* 155d ff.) are Plato's own.

Cornford has some very odd views on motion within the receptacle (*PC* 205 ff.). If I understand his view, it is that if Plato postulated particles like those of Democritus, they would be inanimate and so would not be capable of motion, 'since no bodily change can occur without the self-motion of the soul' (*PC* 205). This seems straightforwardly false. It may be true that soul is required to originate motion, and perhaps is required to maintain teleologically ordered motion (i.e. the heavens) but it is not required to maintain motion itself nor is it required for change to occur from such motion. There is, on a literal reading of the cosmogony, a problem about the origin of motion prior to the world soul but not about the maintenance of motion. In relation to this, it is interesting to note that Cornford seeks to deny two things. As we have seen, he denies that the revolution of the whole creates a compressive effect which forces the particles together (*PC* 243-6). He also wishes to deny that the receptacle shakes or is shaken by its contents. He says that 'Plato cannot mean that Space really shakes or is shaken by the qualities' (*PC* 209),[70] and so (wrongly) cuts off the two critical sources of motion for inanimate particles.

Cornford also attempts to draw on the *Politicus* to support his view (*PC* 206 ff.). I think I have said enough in Chapter 4 concerning the fundamental cosmological differences between the *Politicus* and the *Timaeus*. One simply cannot use the *Politicus* to help interpret the *Timaeus* without recognition of these critical differences.

Having dissociated myself from Cornford's Democritean position, I do of course recognise several similarities between Plato and the atomists, notably on atomism, the void and perception.[71] The critical differences between Plato and the atomists which I would like to highlight, however, are that where the atomists postulate unlimited shapes and sizes of ultimate particles, Plato postulates a small number restricted by teleological principles, postulates structured particles, and sees the need for the teleological ordering of both the extrinsic and intrinsic organisation of the ultimate particles.

My account of Plato has some similarities, but also some important differences with the views of Descartes on the nature of the physical world. Both Plato and Descartes treat matter as shapes with specifiable geometrical properties and nothing else. Descartes, however, identifies matter and extension, such that empty space is inconceivable.[72] Plato's view is rather of extended entities moving in space allowing for some small-scale empty space, as we have seen. For Descartes the ultimate particles of matter are three-dimensional, come in an indefinite number of shapes and sizes and are indefinitely divisible, all contrary to Plato's account.[73] While some hold that Descartes geometrised space,[74] we cannot say the same for Plato. The receptacle is specifically shapeless (*amorphon*, 51b7).

Plato and Descartes appear to be in strong agreement on the inherent properties of matter. Matter has geometrical properties and motion, and no other properties. So for Descartes interaction between particles is by contact action only, and this serves to explain all attractive and repulsive effects that might otherwise be thought to operate at a distance, including the hard cases of gravity, electricity and magnetism.[75] Plato is keen to stress this as well, at *Timaeus* 80b8-c4:

> And moreover the flowing of waters, the fall of thunderbolts and the wonderful attraction effects of electricity[76] and the magnet,[77] all these are not due to any power of attraction, but to the fact that there is no void and that the particles push into each other.

He has *Timaeus* go on to say that as particles disintegrate or combine they will move to their own regions, and by means of these complex processes what appeared marvellous 'will become clear to someone investigating in the proper manner', a sentiment with which Descartes would wholeheartedly agree.

So both Descartes and Plato have a strongly reductive account of physical phenomena. The only question here is whether Plato considers the like-to-like principle to be an inherent property of bodies (as with Newtonian gravitation operating at a distance), or an effect of certain types of motion on bodies (as with Cartesian gravitation which is an effect on bodies generated by vortices). The latter seems more likely as we do not hear of this principle in the absence of motion, and electrical and magnetic effects are explained along the same of lines. Plato's views have some interesting similarities with those of Leucippus and Democritus, and indeed with those of Descartes, but they have important differences as well.

The comparison with Descartes is worthwhile, not only to bring out the strongly reductive character of geometrical atomism but also to illustrate the viability of a strongly geometrical conception of the ultimate constituents of the world. We might ask whether the stoicheic triangles are planes of matter or are more like coathangers with material boundaries enclosing space. Coathangers would seem unlikely due to the possibility of stoicheia or atoms passing through one another. Cornford objects to planes on the grounds that atoms would then be hollow (PC 205), but as long as the structure is rigid I do not see a problem here. An alternative is that these are simply geometrical entities. One might agree with Crombie that:

> If we are tempted to ask: 'Triangles of what?' we must resist the question. They are simply triangles.[78]

Crombie though has the triangles as triangles of space. So too Descartes' particles are particles of extension/matter, there being no empty space.[79] A more radical view would have the triangles as geometrical entities, which are extended in space but not constituted of space. Thus they might move in space without changing space. One would then have to think about the analogues Plato uses for the interactions between atoms in a more sophisticated and mathematical manner (fire particles cutting because they have sharp edges looks distinctly material), but that is possible. The simplest view, though, is that the triangles are matter in space.

Here, as elsewhere, Plato's proto-realism would seem to be an important driving force in postulating real but unobservable micro-entities which have specific inherent geometrical properties. That in turn forms the basis for a realist account of the laws by which they interact. Such laws are not empirical generalisations but are based on the inherent geometrical properties of, say, fire atoms, which of necessity have certain effects on other atoms when they meet them in certain ways. So too Plato seems to postulate a relation between the entities of geometrical atomism and language such that we can get a good grip on these entities. He also seems driven to postulate ultimate constituents of the natural world which are both explicable and explanatory.

IX. Teleology and bonding

Some further comments on the nature of geometrical atomism. I see no reason to suppose there to be irregular figures, at the level of atoms, complexes or the stoicheic triangles in the cosmos. There is an issue relating to the analysis of the sense of smell at *Timaeus* 66d ff. If the difficulty is that the olfactory veins are too small to admit individual atoms of earth and water and too large to sense individual atoms of fire and air, and none of the elements can be smelled, only the intermediate states where water is turning into air or air into water, then perhaps there are intermediate atoms of some sort.[80] However, Vlastos argues convincingly that the olfactory veins cannot be so small, and that the problem is to do with the viscosity of the smoke or mist inhaled and not the sizes of individual particles.[81] As we can account for smell in this manner, there is no need for irregular particles which would not fit in with the scheme of beauty devised by the demiurge.

One question that is frequently asked of any account of the receptacle and its contents is what is supposed to be in the pre-cosmic chaos? Are there irregular figures here, especially as at *Timaeus* 53a the elements only possess traces of their proper nature and at 69a nothing partakes of proportion in the pre-cosmos except by chance? Here I return to some points made in Chapter 1. Are we to take the ordering of the pre-cosmic chaos into a cosmos literally or metaphorically? There I argued that if we take this literally, the point of the assorted passages on the pre-cosmos is not to describe a specific state of chaos but rather to make the more general point that whatever less ordered state preceded the currently ordered state, the transition between the two is not something that could have come about by necessity and chance alone. If we take these passages as metaphorical, or as I put it, a piece of counter-factual analysis, then the emphasis is more epistemological on what we need to assume in terms of the intrinsic and extrinsic order of the elements such that we can have a reasonable account of the current cosmos. If Plato did believe in a complete chaos prior to the work of the demiurge, then the stoicheic triangles might lack intrinsic order, as indeed would everything else, though it would be hard to see why the elements would have traces of order. Even if this were so, I do not see why this would prejudice the stability of the *stoicheia* in the cosmos, although one would be able to draw a distinction between the receptacle which seems to exist in the same manner in the pre-cosmos and the *stoicheia* which do not.

Why does Plato select the half-square and the half-equilateral as his fairest triangles? Taylor cites the ratios of the angles, 1:1:2 (45:45:90) and 1:2:3 (30:60:90),[82] while Cornford suggests that such triangles allow for a sequence of 'isotopes' which are relatively closely spaced in terms of triangles used,[83] though as Artman and Schafer argue it is not clear why that should be an important criterion.[84] Correctly, in my view, they relate this matter to that of the fairest of bonds discussed at *Timaeus* 31c-32a, and the question of doubling a square and tripling a triangle.[85] The process of doubling a square is something which is well known from the *Meno*.[86] From *Timaeus* 31c ff. the best bonds exhibit the ratio a:b = b:c. For the square complex, where a is an arm and b the hypotenuse of a half-square stoicheic, and c the diagonal of the square, a:b = b:c, as $1:\sqrt{2} = 2:\sqrt{2}$, and c = 2a (Fig. 57). For the complex triangle, where a is the short arm and b the long of a half-equilateral stoicheic, again a:b = b:c as $1:\sqrt{3} = \sqrt{3}:3$, and in this case c = 3a (Fig. 58). As Artman and Schafer comment,[88] we thus get the fairest of bonds between 1, 2 and 3, and we also get 1, 2 and 3 in the ratios

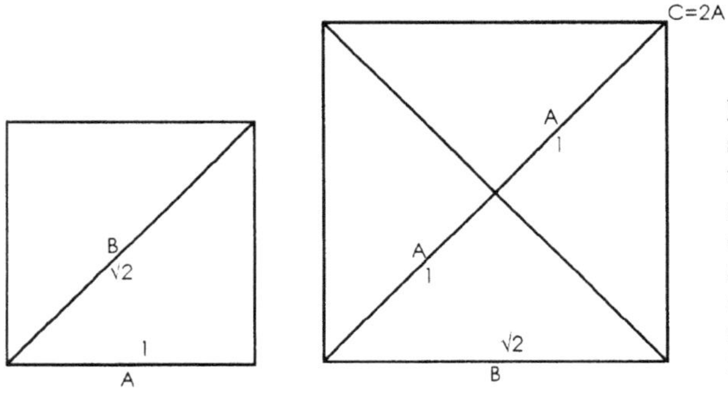

Fig. 57. Doubling a square.[87] A:B = B:C, where A is the side of the small square, B the diagonal of the small square and C the diagonal of the large square, or 1:2 = 2:2, and C = 2A.

Fig. 58. Tripling a triangle. A:B = B:C, where A is half the base of the small triangle, B is the height of the small triangles, and C is the height of the large triangle, or 1:3 = 3:3, and C = 3A.

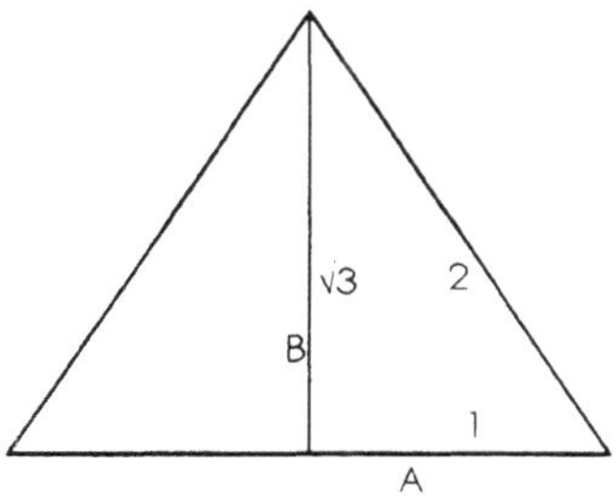

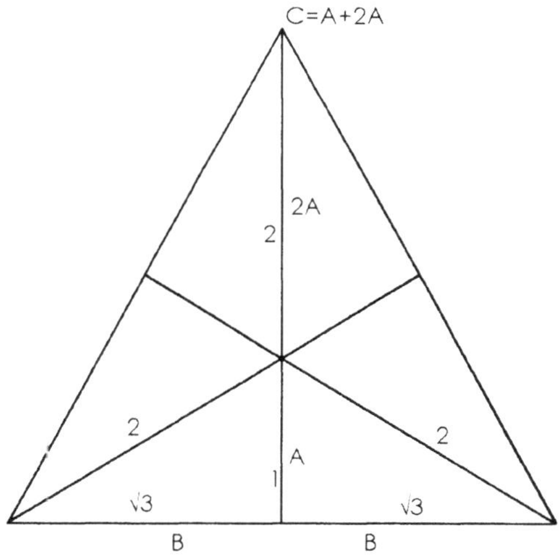

of the angles as Taylor observes. This works very well for the constitution of the standard atomic faces, though it is less clear why on these grounds the faces should join to make the perfect solids. If the like-to-like principle of the *Timaeus* is quite general rather than merely for types of atoms, and *Timaeus* 81a2-3 would suggest it is, then similar stoicheic triangles can be brought together by this principle.

Aristotle does point out a major weakness in geometrical atomism, though he perhaps does not make the most of it. At *De Caelo* 299b25 ff. he argues that as lines can be placed next to each other, end to end, or on top of one another, it is too much to ask that the surfaces only contact each other side by side. Aristotle could considerably strengthen his case by pointing out that there are indefinitely many ways in which surfaces might be put together and we are being asked to believe this always happens in one way without any reason, and that there are two layers of this, the stoicheics joining to make complex triangles and the complexes joining to make atoms.

One answer to this might be that both do so in order to instantiate the best. Stoicheic triangles join up in the specified manner to produce the best bonds, complexes join up to produce the best three-dimensional figures. One might accept that in the initial ordering of the cosmos the demiurge arranges everything just so. The problem though is the theory of change. Atoms decompose, but how do they reorganise into other atoms? In Chapter 1 I suggested that we might consider each atom to be a mini-cosmos. I would stand by that, but there is a significant difference with the cosmos in that the cosmos is alive and shows the ongoing action of intelligence, as do the heavenly bodies, and thus the heavens retain their order. There is no suggestion that stoicheics, complexes or atoms are intelligent entities. A further difficulty here is that of isotopes. If there are different complexes which might be formed, and different ways in which complexes might join together to create different sizes of atomic faces, then how the stoicheics and the complexes join up seems underdetermined.

X. The significance of geometrical atomism

I shall finish this chapter with some more general comments on the historical significance of Plato's geometrical atomism. I would agree with Vlastos that one of the finest legacies left by the Greek atomist tradition (and I include Plato and some of the physiologoi in this) is the development of two-level theories.[89] The idea that there is a micro-world beyond our perception, which while it underpins our sense perceptions can be radically different in its nature from the macro-world, is an enormously significant step forward in the history of science. Although Vlastos puts up some conjectural equations for physical processes involving analysis into component parts, it would be misleading to suggest that Plato ever made use of the possibilities for mathematisation opened up by geometrical atomism.[90] While Plato does make use of such possibilities in astronomy (or at least urges others to!), there is no evidence of any similar programme for physics or chemistry. Vlastos makes the important point that while in astronomy important 'hard facts' were being gathered to which theories would have to adhere, no similar process was happening for terrestrial disciplines, which had to rely on commonplace observations.[91] Thus it was hard for the Greeks to choose between Plato and the atomists, and indeed other matter theories. Indeed some issues here appear radically underdetermined, which is why we see Plato employing his teleological solution to Meno's paradox and so hypothesising some ideal basic entities, which in some ways places him closer to modern atomism (all particles as perfect spheres, atoms for the earlier atomists, sub-atomic particles for the later ones).

It is worth considering some of the ways in which Plato's geometrical atomism is rather more like modern atomism than that of Leucippus and Democritus. Plato is strong on the idea that there are a small number of types of ultimate particles, which are mathematically well defined, against the indefinite number of shapes and sizes of the atomists. Plato's discussion of geometrical atomism stresses that matter has deep structure in the sense of ultimate particles forming structures which in turn themselves form further structures, and so on. That is an idea which is absent from Leucippus and Democritus. Although Plato is not specific about how bonding between particles occurs, and indeed this is the major theoretical flaw in the scheme, he is right that this is not an accidental matter based on mechanical interaction but happens in a specific and well defined manner.[92] His theory of 'isotopes' is also interesting, not as a precursor to any modern ideas about isotopes but rather as another emphasis on the intricacies of structure at the sub-atomic level. This of course is a somewhat Whiggish approach to assessing Plato's geometrical atomism, but given the praise that the ancient atomists have received precisely along those lines it is worth pointing out what we might be able to say about Plato too. If we are going to give the ancient atomists credit for in this manner, then we must also give credit to Plato. Perhaps more importantly, Plato's atomism, like his cosmology, is perfectly viable in the context of fourth-century Greece. It is aimed at solving certain problems in the accounts of the atomists and physiologoi, and is no less plausible as an account of the physical world.

As I argued in Chapter 1, it would be wrong to see the seventeenth century as solely reviving the atomism of Leucippus and Democritus to the exclusion of Plato's geometrical atomism. That the seventeenth century felt the need to cure atomism of atheism was not merely a religious predilection. The philosophical problems with Presocratic atomism, of why atoms should have certain shapes and combine in certain ways, which Plato addressed with teleology and the demiourgos, were now addresses with a Christian deity. So we find Boyle saying that:

> The provident demiourgos wisely suited the fabric of the parts to the uses that were to be made of them.[93]

Similarly, we find Newton saying that:

> It seems probable to me that God in the beginning formed matter into solid, massy, hard, impenetrable movable particles, of such sizes and figures and with such other properties and in such proportion in space, as most conduced to the end for which he formed them.[94]

So too there was a considerable debate between those who advocated a plenum of particles and those who favoured atoms and a void. It is notable that those who favoured the plenum and rejected the idea of action at a distance, such as Descartes, adopted a similar solution to Plato for the awkward cases of gravity and magnetism. *Timaeus* 80c argues that the 'attractive' powers of electricity and magnetism are not due to any action at a distance, but can be explained by the fact that there is no void and the atoms jostle each other and move to their own region. Descartes, while he used vortices to explain gravitational effects did so without a void, and used screw-shaped particles moving among smaller particles to explain magnetism.[95]

A further important aspect of Plato's thought related to geometrical atomism is the question of irrational numbers and measurement. The Pythagoreans treated geometry arithmetically, by attempting to treat geometrical problems as part of the theory of natural numbers, that is as numbers composed of indivisible monads.[96] Thus every geometrical length ought to be expressible as the ratio of two natural numbers. If these numbers represent a length, then if we ask how long something is, rather than measure the distance we count the number of monadic lengths involved. So too, according to Aristotle the Pythagoreans treated physical entities as in some way constituted out of number.[97] The great problem for these projects comes with the discovery of the irrationality of the square root of two, for here we have a number/length that cannot be expressed as a ratio of two natural numbers, or as a multiple of a monadic length.[98] That Plato was aware of this for not only lengths but for areas and volumes too is made clear by the Athenian's explicit comments at *Laws* 819d ff.[99]

It has been argued, most notably by Popper,[100] that in response to the difficulties of the Pythagorean programme Plato advocated geometrical rather than arithmetical means for the description and explanation of the world. The part of the *Timaeus* in this was of proposing a specifically geometrical atomism. As if to emphasise the overcoming of the difficulties dogging the Pythagoreans, the stoicheic triangles have sides of root two and root three. It may also be the case that Euclid's Elements were more than an exercise in pure geometry, and were intended as an Organon of the Platonic cosmos.[101] Proclus for instance comments:

> Some have thought that the subject matter of the various books [of Euclid] pertains to the cosmos, and that they are intended to help us in our contemplation of, and theorizing about, the universe.[102]

Popper comments that one of Plato's main contributions is that

> Ever since, but not before, geometry (rather than arithmetic) appears as the fundamental instrument of all physical explanations and descriptions, in the theory of matter as well as cosmology.[103]

Plato's geometrical atomism then is of some importance. It seeks to rectify specific philosophical problems in the ideas of the atomists and the physiologoi. It provides in the stoicheic triangles a stable basis both for the world and for our references to the world. It allows for diverse phenomena, ranging from phenomenal flux of the elements to the total qualitative stability of the heavens, and so underpins the regularity and stability of the motions of the heavens. In the latter part of the *Timaeus* it is employed to provide outline explanations for all the major phenomena, including, significantly, an account of how all our senses work and how phenomena can be analysed in terms of the shape and motion of the atoms.

A final point on the seriousness of the *Timaeus*. On my view, the *Timaeus* highlights problems with atomism, with the receptacle and with flux theories. It advocates (or wishes its readers to consider seriously) the merits of a teleological geometrical atomism. Clearly Plato objects to the philosophy of the atomists and physiologoi. One point of the *Timaeus* may be to show that their positions on the physical world are inadequate too, while views (newly) developed out of Plato's philosophy are adequate. I see no reason on this view why Plato should not be considered a serious cosmologist or the *Timaeus* be considered a serious work of cosmology.

9

Epistemology in the *Timaeus* and *Philebus*

It would seem then that we can produce a coherent and viable set of interpretations of important passages in the *Timaeus* based on the ideas that the motions of the heavens are regular and stable and that the basic entities of geometrical atomism, the stoicheic triangles, do not undergo any *alloiosis*. This might be thought to create some tension with *Timaeus* 27c-29d. We might accept that tension, either as unrecognised by Plato, or as inherent in any attempt to give a full account of the world of becoming, which according to Timaeus we cannot give a fully consistent account of. What follows is a rather more radical way of attempting to deal with this tension.

The *Timaeus* develops an important analogy between the world soul and the human soul. This is done in detail, and is critical to the conception of humans and their place in the cosmos in the *Timaeus*. A model of mind is developed, as is an account of the mechanics of false judgement. Yet the epistemology that is inherent in this view does not sit well with 27c-29d, and seems considerably more optimistic. So too there are other intimations scattered through the *Timaeus* that we can do rather better than an initial reading of 27c-29d might lead us to believe, especially in relation to astronomy. The point of this is to show that there are tensions in the *Timaeus* irrespective of the view we take on the heavens and the stoicheia. So too there are some distinct oddities about the views we are given at 27c-29d, both inherent to those views and in relation to views given in other works. Should we then believe the views given at *Timaeus* 27c-29d to be Plato's own without further consideration?

I argue that there are good grounds for believing that the *Timaeus* is not a repository of Platonic doctrine. We have already seen that the astronomy of the *Timaeus* is best viewed as a prototype rather than as a completed and definitive view. So perhaps the *Timaeus* (at least in part) sets out to pose its readers some problems. One of them may be this. What do we hold to be fundamental when we investigate nature? Some version of a two worlds (TW) theory? The stability of the heavens or the stoicheia? Can these be held together without reformulation? If we take the stability of the heavens and the stoicheia as fundamental, then we need to rethink

TW somewhat, and there seem to be grounds for much greater epistemological optimism toward the investigation of nature. That is highly significant for the sort of results Plato might expect from such an investigation. One problem here of course is that if Plato is challenging his readers to examine their own views, it becomes a tricky matter to discern Plato's own views. I think there are some indications in the *Timaeus* on this matter though, and certainly in other later works, notably the *Philebus*, *Laws* and *Epinomis*, there appears to be greater optimism toward the idea that the investigation of nature and the *technai* will actually generate some knowledge. Clearly this approach to the *Timaeus* is a conjectural one, but it does provide an interesting alternative to the view that Plato believes we simply cannot produce a consistent account of the physical world and so should not be overly concerned about tensions and inconsistencies in the investigation of nature.

I. World soul and human souls

Let us begin with the question of what the world soul can know and what it can form opinions about, and how it acquires these states. Having described the construction of the world soul, at 37a *Timaeus* goes on to say:

> Therefore, as soul was blended out of its three constituent parts, sameness, difference and *ousias* ... whenever she has contact with something that has either dispersed or undivided *ousian*, she is set in motion throughout the whole of herself and declares what this thing is the same as and what it is different from, and how it affects and is affected by each thing, in which manner and how and when, both in relation to the things which come to be and those which always are. When this gives rise to discourse which is equally true whether concerning that which is different or that which is the same, it is carried within the self moved without speech or sound.[1] Whenever this occurs concerning the sensible (*to aistheton*) the circle of the different running correctly announces this to the whole of the soul and opinions and beliefs which are secure and true occur (*doxai kai pisteis gignontai bebaioi kai aletheis*). On the other hand, where the rational (*to logistikon*)[2] is concerned and the smoothly running circle of the same declares this, intelligence and knowledge of necessity result.

The epistemological implications of this passage have been underestimated. If the world soul can have opinions concerning the sensible which are both secure and true, then there cannot be anything intractable in the nature of the sensible, or the nature of opinion, which prevents the formation of such opinions. There may be difficulties for we mere mortals, but that is a different matter, and one that locates the problems with us rather than with the sensible world or with opinion. Timaeus goes on to give us an account of the formation of the human soul and its capacities, and a description of how it is that human beings are able to form false judgements.

At 41c ff. having created a race of demigods to be the souls of the heavenly bodies, the demiurge then gave them the job of helping in the creation of mortals. He once more blends together sameness, difference and *ousia*, though this time not as uniformly as before,[3] and divides the resulting mixture into the appropriate number of souls. Though not explicitly stated, it would seem that he also performs the same acts of dividing into strips and binding into circles that he does with the world soul, such that our souls consist of the two revolutions. He then turns these over to the demigods, who proceed to fashion a body for each. Having been implanted in physical bodies subject to influx and efflux, mortal souls are then subject to sensations, to desire mingled with pleasure and pain, to fear, anger and other emotions.[4] The fundamental moral injunction is then to master rather than be mastered by our emotions. The crucial difference between our souls and that of the world soul can be put like this. The revolutions of the same and the different that make up the world soul are regular, stable and imperturbable.[5] This makes the world soul perfectly rational, and all of its judgements of sameness and difference are correct. However, when our souls are bound into our bodies, the revolutions are disrupted, they partake in all six motions,[6] and the influx of sensations further disorders their motions, such that initially the two revolutions barely hold together, as at *Timaeus* 43d:

> They produced all manner of twisted circlings, and all manner of fractures and corruptions of the circles in as many places as was possible, with the result that they barely held together, moving irrationally (*alogos*), being at one time reversed, at another sideways, and at some times upside down.

The world soul immediately begins intelligent life, as its revolutions are initially perfect, unlike those of newly born humans.[7] At *Timaeus* 43e we are told that this is because of the imperfection of our mental orbits:

> This is like when someone sets themselves upside down, with their head against the ground, holding their feet aloft by pushing them against something. In this case right and left both of the sufferer and of any onlookers will appear to be reversed to the other party. Whenever they encounter some external object, whether it be of the class of the same or the different, then they label it the same as or different from something contrary to the truth, showing themselves to be incorrect and foolish.

On this basis, we make some incorrect attributions of sameness and difference. There are two important parallels with other late works here. At *Sophist* 263d we are told:

> When statements about you state as different what is the same and what is not as what is, it is very likely that the occurrence of such a combination of verbs and nouns really and truly constitutes false statement.

It is interesting that the *Theaetetus* alludes to the same sort of example as the images in its discussion of false judgement, though without drawing the same conclusion. In the discussion of the wax block in the *Theaetetus*, at 193c Socrates tells us that:

> The possibility of false judgement remains in the following manner ... it may be that I am in error suffering as sight is affected by a mirror, left changing into right. In such a case other judging and false judging happen.[8]

At *Timaeus* 46a ff., we are given an analysis in terms of fire particles for why mirrors produce this reversal of left and right. What is remarkable about this model of mind and false judgement is how optimistic *Timaeus* is about the possibility of human beings bettering themselves. At 44a ff. *Timaeus* tells us that at first we are hardly judge anything correctly, our mental revolutions having been disrupted at birth, though with time and effort as the orbits settle:

> Those revolutions which are assailed by these external sensations are drawn along by them together with whole vessel of the soul, and at such times they appear to have the mastery but are in fact mastered. On account of these sufferings, now as in the beginning the soul is at first without intelligence, as when it was bound into a mortal body. But when the stream of growth and nutrients flows less strongly, then the revolutions become calmer, pursuing their own path and becoming more stable as time progresses, and eventually, as each of the circles comes to travel its own pattern, the revolutions correct themselves, and announce both the same and the different correctly, so rendering their possessor intelligent. If he now takes part in some correct nurture of education, he becomes perfect and entirely sound in mind, having escaped the greatest of maladies.[9]

The *Timaeus* suggests both that there is a natural tendency for the orbits to settle down (44b) and that it is our job to help them do so as far as is possible (44c), by mastering rather than being mastered by our perceptions. Note how this relates to the moral injunction of *Timaeus* 42b. If we can master our emotions we can become just, if we can master our perceptions, we can become rational.

So too *Timaeus* 90a-d is important here. In order to become like god,[10] we need to rectify the revolutions in our own heads by studying those of the universe. This all looks very promising. If we can train ourselves properly, then it would seem that we too can have some opinions concerning the sensible which will be like those of the world soul. All the difficulties here for humans are tractable ones, and given what the world soul is capable of, there are no intractable problems with sensible entities or the nature of opinion. We shall come back to the important role astronomy has to play in this in a little while. For the moment though I will merely note that this all looks distinctly more optimistic than *Timaeus* 27c-29d.

II. Souls and common concepts

One might gather from the account of the construction of the world soul, in the way that the demiurge mixes together portions of sameness, difference and being (*ousia*) that each of these are being treated as complete predicates. That is, it would appear that it is legitimate to say of something that 'it is same' perhaps in virtue of the sameness it possesses, rather than saying 'it is the same as x'. However, other passages seem to treat sameness and difference as incomplete predicates. At *Timaeus* 37a9, when the world soul interacts with anything, it announces internally 'what it is the same as and different from, in what respect'. So too at *Timaeus* 44a2 where humans judge things to be *tauton toi kai thateron tou*, 'the same as and different from'. Why should there be this apparent disparity? It might be that Plato has not yet properly distinguished properties and relations, and only apparently treats sameness and difference as relations.[11] Another suggestion is this. As all of the complete uses are in the passages concerning the construction of the world soul and human souls,[12] perhaps this is a metaphorical means of attributing something to those souls.[13] As *Timaeus* goes on to tell us that whenever they encounter anything, both the world soul and humans make judgements of what that thing is the same as and what it is different from, perhaps what is being attributed to the soul is an innate ability to make judgements of sameness and difference.[14] What though of the use of *einai* here? If its treatment is similar to that of sameness and difference, then perhaps Plato reduces what he considers to be a two-place operator (the 'is' of identity and predication) to a one-place operator (which then looks like but is not the 'is' of existence) in order to attribute another innate ability, perhaps that of forming statements of identity and predication. This is even more fundamental, for in order to be able to form judgements of relations of sameness and difference we require the ability to form such propositions.[15] If this is so, it would explain why, though souls are constituted from sameness difference and *ousia* that *Timaeus* only speaks of making judgements of sameness and difference, as in doing so we are also employing the ability attributed by *einai*.

Plato may have had several reasons for introducing such innate abilities. If, by this time he has rejected *anamnesis* with its recollected knowledge, he may feel the need to postulate some fundamental innate abilities in order to be able to explain learning ab initio. The *Timaeus* model avoids some of the notorious difficulties concerning what the newborn child can know or perceive according to the account of recollection at *Phaedo* 73b-75c. According to the *Timaeus*, we are able to make judgements about our sense perceptions straight away, although these will be rather poor due to the confused initial state of our mental revolutions. One advantage of the *Timaeus* account is that it allows for intellectual development, in the sense of an increase in our cognitive powers associated with the steadying of our mental revolutions as well as the mere gain of knowledge allowed by recollection. Plato is perhaps not giving us an exhaustive account of functioning of the mind, especially as in later works he allows that we may think pictorially.[16] Rather, he is allowing us to think propositionally about anything that is brought before the mind. One might also note in relation to my discussion of the *deuteros plous* in Chapter 3 that Plato develops a very active and independent account of the mind. Perhaps the clearest statement of this is at *Timaeus* 77bc, where we find that the third kind of soul seated in the midriff, in distinction to the revolutions in our heads:

> Inasmuch as it continues to be entirely passive, and does not revolve of itself or repel external motion it has no capacity for reasoning about its own experiences.

As we have seen, the two key injunctions in the *Timaeus* are that our minds should have mastery over our perceptions so that we can become wise, and over our emotions so that we can become good. We might also consider this theory of innate abilities in relation to the 'common ideas' of the *Theaetetus*. At 185c Socrates and Theaetetus have the following interchange:[17]

> *Socrates*: Through what does the ability function which reveals to you what is common (*koinon*) to all things, that which you name is or is not, and others which we were asking about just now? What organs will you assign for all of these, through what does the perceptive part of us perceive each of these?
>
> *Theaetetus*: You mean being and not being, likeness and unlikeness, sameness and difference, and one and the others numbers as applied to them ... I don't know what to say, Socrates, except that it seems that there isn't a special organ for these as there is for others, but the soul is its own instrument examining what is common in all things.

The theory of recollection would undoubtedly have a tale to tell here about the acquisition of these common ideas. One of the interesting aspects of the *Timaeus* is that Plato presents us with a fairly parsimonious theory of the innate abilities of the mind. The *Theaetetus* mentions several common ideas, such as unity and plurality,[18] likeness and unlikeness,[19] beautiful and ugly, good and bad,[20] as well as sameness and difference. Now, a more generous innatist might make the ability to apply all of these innate capabilities. The *Timaeus* though grants only the last two, which in conjunction with the ability to think propositionally, might be thought to be foundational. The *Timaeus* suggests that we form some of our abilities to make judgements from abstraction from what is given in perceptual experience. For instance, at *Timaeus* 47a we are told that:

> Sight of day and night, of months and the cycle of the years has procured for us number and the concept of time, and has also led us to seek the nature of

the universe. From these we have derived philosophy, the greatest gift of the Gods to mortals.[21]

It is presumably by applying our judgements of sameness and difference to our observations of the heavens that we acquire the concepts of number and time that we can subsequently apply to other perceptions. Similarly perhaps with our knowledge of sameness and difference. It is presumably by applying our innate abilities to form judgements of sameness and difference that we come to some conscious knowledge of sameness and difference.

A further important feature of the account of mind in the *Timaeus* is the way that the structure and abilities of the mind relate to the structure of the world. The world is made up of things that are the same as and different from other things. The world of the *Timaeus* is also structured in time and space, and is organised by the demiurge according to number (*kat' arithmos*). Our minds have a fallible ability to judge sameness and difference, and are able to acquire the concepts of time, space and number. So too the structure of the world can be though of in terms of letters and syllables as the arrangement of the stoicheic triangles giving an interesting isomorphism between language and the world. All this might indicate some greater optimism to the cognition of the world than is suggested at 27c-29d. It would seem that Aristotle took the main thrust of Plato's account of the soul to be along these lines too. In his critique of Plato's view of the soul in *De Anima*, he states that:

In the same manner Plato in the *Timaeus* constructs the soul out of the elements (*ek ton stoicheion*). For he believes that like is known by like, and that things are formed from these elements (*ek ton archon*).[22]

So too Aristotle at *De Anima* 406b26:

In the same way the *Timaeus* gives a physical account of how the soul moves the body. The self-motion of the soul is responsible for the motion of the body on account of its intimate relationship with it. For it had been created out of the elements, and was divided according to harmonious ratios, in order that it might have some innate (*sumphuton*) perception of this harmony and the universe might move with harmonious motions, and the straight was bent into a circle, and having divided one into two, meeting at two points, he further divided one of these into seven circles, thus making the motions of the heavens identical with the motions of the soul.

III. Souls and epistemology

Let us now compare *Timaeus* 37a ff. with 29b ff. At 29b6 ff. the implication was of a *sungenes* matching of secure (*bebaios*, 29b6), accounts with secure entities, and of 'likely' accounts with likenesses. At 37a ff. though, there exist beliefs and opinions (*doxai kai pisteis*) which are both secure and true (*bebaios kai alethes*) in contradiction of both the letter and spirit of 27c-29d.[23] At 37a ff. there is no *sungenes* stricture on explanation. Instead of the idea that any account of a likeness can be only a likely account (29c1), we have an analysis of the psychological mechanics of error in terms of the revolutions of the human mind, and of false judgement in terms of sameness and difference. As the world soul can have secure and true opinions of the sensible (*to aistheton*) there cannot be anything intractable with the nature of the sensible which prevents the formation of such opinions. So there is a contrast of intractable problems with the world of becoming to tractable problems with human beings concerning the sensible between 27c-29d and 37a ff. Given the analogy developed between the world soul and human souls, and the analysis of error in terms of sameness and difference, it is possible for us, if we train ourselves properly and have mastery over our perceptions, to have this sort of opinion concerning the sensible. That too is contrary to the spirit of the closing part of *Timaeus*' opening speech which tells us that we who are human in nature should be ready to accept a likely account and seek no further. How much tension there is between 27c-29d and 37a ff. of course depends on how strongly we interpret these passages, but there is no doubt that such tension exists. It is important to stress here that the passage at *Timaeus* 37a ff. is not an isolated example which we can explain away as a slip by the author. The human psychology developed in analogy with the world soul is carefully worked out and developed over several pages and is one of the central themes of the *Timaeus*. It is important to note that *Timaeus* makes this a significant motif when he is concluding his speech at 89d ff. At 90d we should supply what is fitting for the divine part of us, and what is *sungenes* with that are the revolutions of the heavens. Then at *Timaeus* 90d we are told that we ought to be:

> Correcting those revolutions in us that were disturbed at birth, by learning thoroughly the revolutions and harmonies of the universe, by making our intelligent part like to that intelligence according to its original nature.

Timaeus 37a ff. is also interesting in relation to TW. *Timaeus* 28a contrasts the intelligible with sensible, but according to *Timaeus* 37bc, the domains of opinion and knowledge for the world soul are *to aistheton*, 'the sensible' and *to logistikon*, 'the accountable/calculable' respectively.[24] While one might try arguing that the domains of the sensible and the intelligible are quite separate, there is no prima facie reason why some things which are sensible should not be amenable to calculations. In order to be amenable to calculations, these sensible entities might have to be stable, or their motions might need to be describable in a manner which is in itself stable. If there were such things that would go against the idea that the world can be simply divided into sensible things which are always changing and intelligible things which are never changing. Prime candidates here of course are the heavenly bodies, which if their fundamental particles are qualitatively stable and are bound with indissoluble bonds, and their motions are entirely regular, are a set of stable and mathematically precise entities. Timaeus has some very interesting things to say about how we might use our observations here. *Timaeus* 47bc tells us:

> God devised and gave to us vision in order that we might observe the rational revolutions of the heavens and use them against the revolutions of thought that are in us, which are like them, though those are clear and ours confused, and by learning thoroughly and partaking in calculations correct according to nature (*logismon kata phusin orthotetos*), by imitation of the entirely unwandering revolutions of God we might stabilise the wandering revolutions in ourselves (*tas en hemin peplanememas*).[25]

The use of the plural throughout this passage strongly suggests that both types of our mental revolutions (same and different) are under discussion here. If so, and we can make calculation correct according to nature then the implication would seem to be that we can have some knowledge of the heavenly bodies. That would again be contrary to the views put forward at 27c-29d. The importance laid on astronomy here as the subject which will bring our minds closer to the ideal of the world soul may go a long way to explaining the central and otherwise somewhat disproportionate role that astronomy is given in education in the later works. One might compare this passage with *Timaeus* 90cd, *Laws* 966e ff. and *Epinomis* 978b ff.

We need to exercise caution with all of this material. The language of 37a ff. is highly figurative, and the analogy between the world soul and the human soul, while highly interesting, to some degree remains at the level of metaphor and analogy. A great number of questions remain concerning just what the world soul can know and what sort of stability its opinions might have. Plato does not spell out the full consequences of this analogy for epistemology, nor does he make any direct comparison with *Timaeus* 27c-29d. We must be careful then not to press this material too hard.

What we can say, even if we interpret 37a ff. in a fairly minimal manner, is that there are significant tensions between the epistemology suggested by the analogy of the world soul and the human soul and that which Timaeus outlines at 27c-29d, both in detail and in general tenor, that of 37a ff. seeming to be considerably more optimistic. *Timaeus* 37a ff. also provides a supplementary argument for the regularity and stability of the heavens. If the world soul is entirely rational, then one would expect its motions to be entirely regular.[26] As its motions are those of the heavens, the heavens must be regular and stable.

IV. Timaeus 27c-29d and two worlds (TW) theory

There are certain oddities about Timaeus' speech at 27c-29d, and there are difficulties in making the speech sit well with several other passages and comments in the *Timaeus*. Having invoked the deities, at 27d Timaeus begins by saying:

> It is possible, in my opinion, firstly to make the following distinction. What is it that always is and never becomes, and what is it that always becomes and never is? The one is apprehended by the intellect with reasoning, and is always the same, the other in turn comes to be believed by opinion with unreasoning sensation, always coming to be and passing away, but never really existing.

This can be construed as a very strong version of TW. As Frede has argued, coming to be here cannot mean coming into being if what becomes is always becoming and never is. Nor is it easy to take 'always' as meaning 'always changing in some respect, but perhaps not all' as that would mean that what becomes in some respect could also be said to be in some respect, and that appears contrary to the sharp bifurcation of what always is and what never is.[27] This passage has an interesting similarity to *Theaetetus* 152de, where Socrates says:

> It becomes all the things which we, incorrectly, say it to be. For it never is anything, but always becomes.[28]

The *Theaetetus* passage occurs in the description of a radical Heraclitean flux. One issue we will have to address soon is what Plato's attitude to such a flux is in the *Theaetetus* and how 27c-29d relates to geometrical atomism and the stability of the stoicheic triangles. Here I merely want to note that *Timaeus* 27d is (or at least appears at first sight to be) an inordinately strong statement of TW.

Clearly, if we take *Timaeus* 27d at face value there are problems in arguing either that the heavens are stable or that the stoicheic triangles are qualitatively stable. Let us leave that on one side for a moment, and look at whether this passage sits well with other aspects of the *Timaeus*. First, there is going to be a problem with status of the entities supposed by geometrical atomism. Whether the stoicheic triangles are stable or not, complex triangles and atoms undergo generation and destruction. None of these are sensible as all are well below human perceptual thresholds. Yet the category of intelligible and undergoing change is not catered for at 27d. So how can there be any account of these at all? These problems are exacerbated by what Timaeus says at 56b, that at least part of what he has to say about geometrical atomism is *kata ton orthon logon kai kata ton eikota*, in accordance with the correct account as well as the likely. That, although little attention is paid to it in the literature, is a quite remarkable statement. If we believe 27c-29d a likely account is the best we can hope for, but here is Timaeus in possession of the correct account, and saying that in this case the likely account is in agreement with it. One then wants to know what this correct account is, how Timaeus has access to it and what the implications are for the epistemology which Timaeus espouses at 27c-29d. Even if we take the subjunctive *esto* which introduces this phrase to mean something like 'let this be in accordance with ...', such that Timaeus is less certain that this is the correct account, still he asserts that there is a correct account to be had. I would reiterate that in relation to the heavens (whether they are stable or not) at *Timaeus* 47c we can make *logismon kata phusin orthotetos*, calculations correct according to nature, in order to help stabilise our mental revolutions, and at 37b when the world soul encounters anything sensible *doxai kai pisteis gignontai bebaioi kai alethes*, beliefs and opinions which are stable and true occur. There are then oddities both in the nature of *Timaeus* 27d and in its relation to other parts of the *Timaeus*. At 29b Timaeus states:

> In every matter it is of great importance to start at the natural beginning. Thus concerning both the likeness and the paradigm, one must make the following distinction, that accounts of them will be of the same family (*sungeneis*) as the things they expound. Therefore accounts of that which is steadfast, firm, and is clearly apprehended with the mind are steadfast and unchanging, and so far as it is possible for them to be irrefutable and undefeatable as befits such accounts, they must not fall short of this. On the other hand, that which has been copied from the model, being a likeness, will have accounts that possess likelihood, standing in an analogous relation to the previous accounts. Whatever is related as becoming is to being, is also related as opinion is to truth.

Here the *sungenes* ('of the same kind or family'), stricture seems somewhat odd. One of the main points of the theory of forms, one might think, is to postulate some entities which are not *sungenes* with what they are supposed to help explain. This is especially so if one of the major motivations behind the theory of forms is to postulate cognitively stable entities in order to resolve the compresence of opposites. There is a sense here in which the intelligible must be related to the physical, as the explanans must fit the explanandum, but the forms could hardly be thought to be *sungenes* with what they are supposed to be explaining. Indeed, some of the arguments of the first part of the *Parmenides* might be thought to play precisely on that. We also have the word play of *eikones* and *eikotas*, 'likeness' and 'likely' at 29c2. Presumably that is a consequence of the *sungenes* stricture, but it is a rather strange one. There is no suggestion in the *Phaedo* that any account of why one thing is larger than another will be only a 'likely account' because it deals with the physical world. The cases of two men and two horses are treated equally with two numbers, ten and eight.[29] While there are obvious echoes of the *Republic* here, the association of likeness and likely account is a new departure. Geometrical atomism might be thought to breach this stricture, again whether the stoicheic triangles are qualitatively stable or not. The entities postulated by geometrical atomism are the *archai* of sensible things, without being sensible themselves, and of course if the stoicheic triangles are stable the *archai* of changing things are stable. One of Aristotle's complaints, as we saw, was precisely this, that *archai* should be *homogenes* with what they are *archai* of, but in geometrical atomism they are not.[30]

One might also ask who, prior to Plato, believed that explanans and explananda are *sungenes*, or more accurately, who Plato believed to hold such views. Two prime candidates here are Parmenides and Heraclitus. There are indeed certain affinities between Parmenides' poem and Timaeus' speech. Both separate the objects of reason and sensation, reckoning these to be co-ordinate with what is knowable and what is opinable, and both require explanations to be *sungenes* with what they explain.[31] We might also compare Timaeus' repeated use of *eikos* to describe the status of any account of the physical with Parmenides' similar usage at Fr. 8/60.[32] Both Parmenides and Timaeus draw a distinction between the veracity of what they have just said and the nature of the account they are about to give,[33] and both aim to provide the best account of the physical world that it is possible for mortals.[34] A second figure who Plato believes thinks that accounts are *sungenes* with their explananda is Heraclitus.[35] Whether this is an accurate picture of the historical Heraclitus is another matter, but the *Theaetetus* certainly takes this line about Heraclitean flux.[36] So perhaps Timaeus' speech at 27c-29e sets up a Parmenidean account of reason and the intelligible realm and a Heraclitean account of sensation and the physical realm.

That ought to make us pause though. If the *Timaeus* comes after the *Parmenides* (and in particular after the final argument of the first part of the *Parmenides*, 133b-134e), what is a Parmenidean account of the intelligible doing here, especially with the *sungenes* stricture in play? With Heracliteanism, again there is a question of the relation to the *Theaetetus*. Is a radical flux Plato's position there, and is it Plato's position in the *Timaeus*? If so, how does it relate to geometrical atomism and the passages cited above? So once more, *Timaeus* 29b is somewhat strange in itself and in its relation to the rest of the *Timaeus*. At the close of Timaeus' introductory speech at 29c, he says:

> If then, Socrates, on a great many matters concerning the Gods and the generation of the universe we should not be able to give what is due (*apodounai*) in terms of accounts which are in every way entirely in agreement with themselves and perfectly exact, do not be surprised. But if our rendering is inferior to none in likelihood, this must please, remembering that both I who speak and you the judges are human, such that concerning these things we should gladly accept the likely story, as befits this, and not seek any further.

Socrates immediately acquiesces to *Timaeus*. Resonances with Plato's earlier works should also make us highly suspicious here. Should we 'gladly accept a likely account', or would Plato's Socrates of *Meno* 81d and 86bc castigate this as a 'lazy' approach, likely to stop men searching? Would the Socrates of *Gorgias* 482b so glibly accept any account that does not agree with itself? Wouldn't the Socrates of *Phaedo* 90b ff. reject Timaeus as an *antilogikos*, one who does not put sufficient trust in the power of argument? What of the Socrates who has emphasised consistency so strongly in the *Theaetetus*?[37] Are we meant to acquiesce so tamely with Timaeus on these matters?

V. Two worlds and stability

So far in this chapter I hope to have shown two things. First, the *Timaeus* has an account of epistemology based on the analogy of the world soul and the human soul which is worked out in some detail, with an account of human intellectual development and of the mechanics of false judgement. This epistemology not only seems rather more optimistic than that of 27c-29d, in places it also appears blatantly to contradict it. Secondly, whether or not one considers the heavens and the *stoicheia* to be stable in the *Timaeus*, there are certain oddities internal to 27c-29d, and in the way it relates to some other passages in the *Timaeus*. There is I believe an important conclusion for the interpretation of the *Timaeus* which comes out of this. There are several key passages in the *Timaeus* which, if taken in isolation, are indeterminate between different readings. A classic case of this is how we construe *Timaeus* 49d ff., and which of fire, this and suchlike are subjects and which are predicates. Naturally we then look beyond the immediate context to see which interpretation will harmonise best with the rest of what the *Timaeus* has to say, and perhaps with what other works have to say too. On the grounds given above, though, we cannot use the strong version of TW given at 27c-29d as an infallible guide to the interpretation of such passages. Put more colloquially, we cannot use it as a stick to beat the rest of the *Timaeus* into submission. Of course this cuts both ways. Nor can we use the stability of the heavens and the *stoicheia* as infallible guides. Throughout I have chosen the options which I believe harmonise best with the stability of the heavens and stoicheia. That has been worth doing to show the viability of such an approach in dealing with a wide range of passages, and to show how the interpretations this strategy generates can be linked together into a coherent picture. This remains a strategy for dealing with indeterminate passages though, and does not demonstrate how we must interpret any of them, and of course ultimately this reading comes up against the apparent barrier of *Timaeus* 27c-29d.

What then are we to do with the *Timaeus*? Perhaps it is incompetently written and Plato was unaware of the tensions in the account he has

Timaeus give. Perhaps he is aware of at least some of the tensions, but accepts that we cannot have an entirely consistent account of these matters. That though would mean that he accepts 27c-29d, at least to the extent of believing we cannot have a fully consistent account of the world of becoming, in which case one might expect several of the more indeterminate passages to be more clearly delineated in their alignment with 27c-29d, and there to be less evident contrasts with the account of the heavens and the epistemology of 37a ff. A more radical suggestion is that the choices being offered for the interpretation of the indeterminate passages are systematic, and that we are presented with the following dilemma:

1. If we accept the strong version of TW given at 27c-29d, and we apply the principles involved there to other issues, then there are problems with postulating stable heavens and *stoicheia*, and ultimately with referring to changing phenomena.

2. If the heavens and *stoicheia* are stable, and we apply the principles involved here to other issues, then there are problems with sticking to a strong version of TW.

One of the difficulties with (1), as I have sought to show in the previous two chapters, is that it might well leave us without the resources to deal with the problems thrown up by the receptacle. Perhaps Plato wishes to elicit some sort of a response from his readers. The idea of the tension outlined above may be to provoke the readers into working their way through the indeterminate passages and coming to their own conclusion on the matter.[38] They must decide for themselves what is fundamental in an account of the world about us and which consequences they can accept. What I shall do in the next few sections is suggest some ways in which we might seek to undermine, at least to some extent, the idea that the *Timaeus* simply gives us Platonic doctrine. Here I shall attempt to counter a common view, succinctly expressed by Sayre, that:

> The major dialogues of the later period (with the exception perhaps of the *Timaeus* and the *Laws*) were constructed to serve as dialectical instruments rather than as repositories of doctrine.[39]

This approach has certain advantages. If we can read the *Timaeus* in this manner, it narrows the gap in style between the *Timaeus* and other post-*Republic* and *Phaedrus* works. In particular it throws up some interesting contrasts and parallels with the *Theaetetus*, particularly on the question of whether Plato is committed to the idea of some form of Heraclitean flux of the physical world. One difficulty with this approach is that it then becomes a rather slippery matter to determine Plato's own position. Perhaps that is so anyway once we take the matter of Platonic anonymity into account. In some ways that is a difficulty for some of what I have argued in this book, as I have sought to give a developmental account of Plato's views on science. One can argue that even if Plato wishes to provoke his readers to think for themselves, he still has some ideas of his own, whether positive or negative, to get across. The account of falsehood in the *Sophist* and some refutations of positions, particularly of Protagoras in the *Theaetetus* would be the prime candidates here. The problem though is in generating any hard and fast rules for distinguishing between a position given to a character for some further purpose, and Plato's own views.

I stand by my developmental account of Plato's astronomy and cosmology, though if we are to take this approach to Plato it has to be put in somewhat more conjectural terms than if we were dealing with someone who simply wrote doctrine. Certainly I would argue that the view I have pursued here is quite a charitable one of Plato. It sees him build on the work of the Presocratics and make some important advances in both astronomy and cosmology. It is important that a great number of those views are repeated in both the *Laws* and *Epinomis*, in both cases in ways in which would seem to indicate that they are Plato's views at this stage.

VI. Writing and *paidia*

I take it that Plato's usual aim in writing is not so much to impart knowledge, but rather to attempt to generate understanding, and so in the main his purpose is dialectical rather than dogmatic and that he attempts to create some sort of interaction between the reader and the text.[40] Merely reading an explanation will not produce understanding in the same way that reading a fact will produce knowledge.[41] So at *Phaedrus* 275d Socrates tells us:

> Writing has this strange property, in which it is very like painting. For the products of painting stand before us as if alive, yet if you ask them something, they are altogether solemnly silent. And so it is with written words. You might believe them to speak with some understanding, but if you should question them wishing to learn their account, they always point out one and the same thing.[42]

Is writing then a singularly unhelpful medium for Plato? The fact that he wrote so much, and with so much care, suggests not.[43] Perhaps then there is a via media for Plato between treatise writing and oral dialectic, a style that attempts to create a form of dialectic between reader and text, making writing a worthwhile philosophical activity. Plato employs various literary/philosophic tools to achieve this objective, such as aporia, often presenting extreme and unacceptable alternatives as an apparently exhaustive dilemma,[44] paradox,[45] and complex irony.[46] A common factor is their ability to act as a challenge or a philosophical irritant to the readers, forcing them to do some thinking for themselves, and to return to read the text at a different level of sophistication.

We also need to consider the contrast which Plato draws between *spoude* (earnestness, serious) and *paidia* (child's play, sport, pastime). The key material here begins at *Phaedrus* 276b ff. with a contrast between what a good gardener would do when he is serious and what he would do in a more playful mood. Then at *Phaedrus* 276d ff. Plato has Socrates say:

> Fittingly, it is for the sake of a pastime (*paidias*) that that he will sow seeds and write in the garden of letters, when he writes, this being laid up as a store of reminders, both against the onset of forgetful old age, and as a track for all those who follow him, and he will be pleased to watch the growth of these tender things ... but in my opinion, much better is the serious treatment (*spoude*) of these matters, that is whenever someone employs the method of dialectic, and taking a suitable soul, plants and sows words of wisdom, which are able to defend themselves and their planter, and are not barren but are pregnant with seed, whence other truths are brought forth in other minds, rendering these seeds immortal, and making their possessor happy, as far as is possible for humans.

One might begin to worry here how seriously we are to take Plato's writing itself. This is exacerbated at *Phaedrus* 277e-278a, where we are told:

> The man who supposes that there is much that is playful (*paidian*) in the written word concerning any subject, and that nothing (*oudena*) ever yet written, whether in metre or in prose, is worthy of very serious (*megales spoudes*) consideration ... Only (*monois*) in oral teaching for the sake of learning are really written in the soul things which are distinct, complete and worthy of great trouble concerning justice, beauty and goodness.

So while Plato takes great care in composing his works he states, in writing, that nothing written is worthy of great pains, and is but a *paidia*. Can this paradox be resolved? Consider the way in which Plato's theory of writing philosophy ought to be presented in writing. Its principles should apply reflexively, and in line with the dilemma strategy, Plato overdraws the contrast between oral dialectic and writing to the point where paradox occurs, leaving the reader in aporia, and that this contrast is thoroughly steeped in the *paidia* and *spoude* terminology. The resulting book paradox is self-referential and heavily ironic. In order to resolve it we need to shift levels, and this should provoke us into considering the nature of Plato's writing. The challenge to the reader is to sort out the relationship between the written and the spoken word such that the written word can have some merit in the right circumstances, and so avoid paradox, while not being the equal of oral dialectic.

So the theory which says not to trust the written word but to examine the thought that lies behind it is itself not to be trusted as a written theory but only as the thought behind the resolution of the written paradox that expresses it.[47] We must also consider the *paidia* description and the *paidia* and *spoude* contrast. Once writing escapes the book paradox, it is no longer a worthless amusement. Certainly a work should be written playfully in order to provoke us to investigate the thought that lies behind it, and writing is still inferior to oral dialectic, but it has real value and serious intent. Thus the *paidia* description is one that carries a complex irony, guaranteeing both the playfulness and the seriousness of a piece of writing, highlighting both its merits and its drawbacks as writing.

This is of some importance, for while Plato does not have a fixed philosophical terminology, some terms build up resonances, and the contrast between *spoude* and *paidia* and the *paidia* description are two of these. Important examples of the *paidia* and *spoude* contrast are at *Republic* 545e, 602b, 608a, *Phaedrus* 276b, 276e, 277e, 278a, *Sophist* 237b, *Politicus* 288c, *Philebus* 30e, and *Laws* 688b.[48] Important examples of the *paidia* description in relation to the status of a piece of writing are at *Republic* 536c, *Parmenides* 137b, *Theaetetus* 161e, *Politicus* 268d, *Timaeus* 59cd and *Philebus* 28c. This widespread and consistent usage of *paidia* and *spoude* would suggest a coherent strategy on Plato's part of reminding his readers of his views on writing.

One perennial worry about Plato's philosophy of science is whether he believed that the investigation of nature could yield any worthwhile results. On some accounts, the *Timaeus* states that such work is a mere childish amusement, that here we can only have a likely account, and can aim no higher than to have unstable opinions concerning the physical, sensible and changeable things that make up the world about us. This is sometimes argued on the grounds that at 59d *Timaeus* tells us that providing accounts of the phenomena will be pleasant, and that anyone doing so 'will be creating in his life a pastime (*paidian*) both suitable and prudent'. However the use of *paidia* and its cognates has certain resonances in later Plato, and if this is so here, then this is a very positive thing for Plato to say. Rather than relegating the *Timaeus* to a mere harmless diversion, the *paidia* description elevates it to a serious work where we perhaps ought to be looking beyond the playful facade to the arguments and debates running beneath the surface. So perhaps we should not take the *Timaeus* as a direct statement of cosmology, but as an interesting and playful series of hypotheses that will lead us to the serious thought that lies behind them. An important comparison here for the *Timaeus* is with the *Philebus*. One would hardly wish to say that the *Philebus* is not a philosophically serious work on the grounds that *Philebus* 28c and 30e7 use *paidia*, in both cases Socrates commenting on the nature of the discussion. I see no reason, at the outset, to suppose that the *Timaeus* is not a philosophically serious work or that Plato's general theory of writing does not operate in it. If the *Timaeus* is a joke, or some other form of amusement, it is no more and no less so than other late period works.[49]

VII. Plato and myth

Let me try to undermine the idea that we should take what *Timaeus* has to say at face value in another way. The account of the *Timaeus* is presented as a myth. How should we deal with that? For a good deal of Plato's later cosmology, this is not a major concern as much of the critical material is presented in non-mythical form in other works (the *Philebus*, and especially the *Laws* and *Epinomis*). However, the attitude we take to Plato's use of myth will affect the status of he account of the *Timaeus*. There is a problem with Plato's use of myth in the *Phaedrus*, in that he is apparently critical of it but still employs it. Socrates is not interested in discussing the myth that Boreas abducted Orethuia, because this will not help him with the Delphic injunction to know himself.[50] Nor is he interested in the tales of monsters that are part of this myth. At *Phaedrus* 230a he says:

> I investigate myself rather than these things, to see if I am a more complex and fearsome beast than Typhon, or a gentler and more single-minded creature, who has by nature a less puffed up fate appointed by God.[51]

Given that Plato overdraws the contrast here to worry his readers about the status of myth, how might some myths be philosophically interesting and able to help with the Delphic injunction, and others not? One might draw a distinction between those myths which are in a sense purely historical, where the only judgements we might make are whether to accept the general form and particular details of that myth,[52] or how to rationalise that myth,[53] and those that call upon the reader to make some philosophical judgements, perhaps even offering various criteria on which they might be judged. The problem with the myth of Boreas and Orethuia is that it falls in the first category here. This is in sharp contrast to the myth that Socrates offers at 246a ff., which deals with the human soul and commentators have found to be a mine of information, and will help the readers with the Delphic injunction, as in coming to a general understanding of the human soul they will know themselves better.[54]

With that in mind, consider the treatment of the myth of Solon. Socrates wishes to hear an account of the state of the *Republic* in action.[55] At 19b ff. Socrates compares his description of the ideal state with a painting depicting animals. In both cases, he says one wants to see the subject in action. This seems to be an allusion to *Phaedrus* 275d and the discussion of the nature of writing. The function of this, I suggest, is to throw doubt on both myth and writing as vehicles for philosophy.[56] *Timaeus* 19b ff. suggests that some sorts of myth are useful but others are not. The problem here is exacerbated by the fact that at the beginning and the end of it, Solon's myth is described as wholly true.[57] What though is the nature of the warrant for its truth? Several criteria are suggested, all of which are external to the readers, all calling upon them to accept the myth on someone or something else's authority. All the criteria alluded to here are ones that Plato has given us grounds to be suspicious of in earlier works. At *Timaeus* 20d, we are asked to accept the myth because Solon is wise and old. Yet surely we have seen enough of the fate of the 'experts' in the early dialogues not to put our trust in the testimony of an eminent man, whatever his credentials. In the *Phaedrus*, Socrates upbraids *Phaedrus* for being interested in the pedigree of a myth rather than its content.[58] At 21d Solon's sources are questioned, and at 23a we find that the source is the ancient writings of the Egyptians. The *Phaedrus*, however, has warned us not simply to accept the written word, whatever its provenance, and the reference here to Egypt may be a further allusion to the *Phaedrus* in addition to *Timaeus* 19b.[59] At 26a ff. a different tack is taken. Perhaps the story is trustworthy because it is well remembered. However, at 19a Socrates finishes his synopsis of the *Republic*, on which he should be the expert, having discoursed the previous day, by saying that this completes the contents, and asks if Timaeus agrees that this is a complete list of the contents, which he does. This synopsis though is far from complete. Memory, then, is not to be trusted. The overall effect of *Timaeus* 19b-27a is then to make us suspicious of all external authorities in judging a myth and to make us rely on our own internal criteria. In contrast to the myth of Solon, which deals with historical matters, Timaeus' main account not only deals with the human soul and its place in the universe, but also discusses the criteria by which we should be judging the contents of the myth, one that invites philosophical rather than mere historical judgement. That in turn may affect debates about whether we should take the tale of creation literally, or more allegorically, setting up a framework of postulates about the order of the universe within which to discuss cosmogony and cosmology. Our suspicions ought also to be aroused by the fact that Timaeus is described as an expert in astronomy,[60] remembering the fate of other experts in earlier works.

VIII. Interpreting the *Timaeus*

Two further thoughts on undermining the view that the *Timaeus* presents us with dogma. In Chapter 5 we saw that the best way to view the model of the heavens that Plato has Timaeus put forward is not as a literal piece of cosmology, but rather as something which embodies important cosmological principles while having flaws in its astronomy. In short, it is a prototype of this sort of cosmology and astronomy for others to develop. I have also argued that if we take Plato be presenting us with a straightforward cosmology and cosmogony, his views on chaos, flux and the material nature of the receptacle appear confused if not contradictory. A more generous way of viewing these issues is that Plato presents us with a range of options and a general way of dealing with them. So whatever chaos is supposed a teleological ordering is required to produce the current cosmos, and no version of a material receptacle can do what is being asked of it. Thus Plato's treatment of these issues is rather more flexible and more powerful, and one might note that although he does not give a full, direct cosmology there are principles which he wishes to get across.

I will now put forward two suggestions for how we might go about reading the *Timaeus*. Both are tentative, but both lead to some interesting perspectives on what the *Timaeus* may be trying to do. The *Timaeus*, if one reads it carefully, can be a very difficult work to pin down. Sometimes one might feel that reading the *Timaeus* is rather (frustratingly!) like looking at a kaleidoscope. Each time you look at it anew, taking a fresh starting point, a different pattern emerges. Unlike a kaleidoscope, though, certain patterns keep emerging. Depending on what you hold to be fundamental in the *Timaeus*, and how one interprets certain key passages in order to accord with those fundamentals, different, non-compatible but recurring patterns emerge. That is somewhat disorienting, and might cause one to wonder what precisely is going on in the *Timaeus*. Perhaps, though, that is a deliberate effect that Plato wishes the *Timaeus* to have on its readers. Why might that be so? In some works, the nature of the discourse appears to be related to the nature of the subject matter. Thus one might argue that in the *Philebus* the key ideas of limit and unlimited and their relation to philosophy in part shape the nature of the discourse between Socrates and his interlocutors. What I want to suggest here is that the problems the reader is presented with in dealing with the account that *Timaeus* gives are like those that Plato believes are critical in beginning to give a proper account of the physical world. Quite simply, what do we take as fundamental? If we make certain assumptions about the physical world, then how well do they accord with our other beliefs and how coherent and philosophically satisfactory is the broader picture account that we can generate? Perhaps this is also why the *Timaeus* is in the main a monologue. We cannot have dialectic and elenchus directly with the physical world, so we need to make assumptions and see how these pan out, to see whether these account for what we see and whether they are philosophically acceptable. So too with the monologue that we are given by *Timaeus*. If I am right about Plato's strategy here, then he is asking his readers to do quite a lot in terms of thinking for themselves and interacting with the various problems that the *Timaeus* might set.[61] That though would make the *Timaeus* stylistically rather closer to other later period works (and here I mean philosophical and literary style rather than stylometry). This brings us back to the dilemma posed between TW and the stability of the heavens and stoicheia which I suggested earlier. The choice between the two is critical to the understanding of the world about us, and critical to the account which *Timaeus* gives.

What relation does the *Timaeus* hold to other late works? Here I shall introduce my second suggestion for reading the *Timaeus*. Burnyeat, in his book on the *Theaetetus*, proposes two readings, A and B, for the first section on the question of whether knowledge is perception.[62] On reading A, Plato himself accepts the views of Protagoras and Heraclitus provided that they are restricted to perception and the sensible world. Sensible things are then in a 'perpetual flux of becoming',[63] and each person is the judge of what sensible qualities they perceive in this flux. So according to Burnyeat:

> Thus the argument, taken as a whole, supports the view set forth at length in earlier dialogues like the Phaedo and the Republic, the celebrated Platonic doctrine that true reality is a nonsensible realm of changeless being, the Forms, and it is in these alone that knowledge can find its objects.[64]

On reading B, however, Plato does not accept the views of Protagoras and Heraclitus. Rather, if one wants to hold that knowledge is perception (Theaetetus' thesis) then one must also hold a Protagorean epistemology and a Heraclitean account of the world. If these are necessary and sufficient conditions for Theaetetus' view, then all three stand, and ultimately fall, together.[65]

That one can generate something akin to the position of the *Theaetetus* reading A from the *Timaeus* is well known. We simply accept the statement of TW at *Timaeus* 27c-29e as Plato's own view, the discussion of the elements and the receptacle gives us a flux, which meshes with the partial imposition of reason on necessity where necessity can be taken as chaos. So Plato holds a Heraclitean view of one realm and a Parmenidean view of the other. One might take the *Theaetetus* and *Timaeus* to be in rough agreement here, if one goes for the A reading of the *Theaetetus*, or argue that the *Timaeus* is early (should be taken with the *Phaedo* and *Republic*, and prior to the *Parmenides* and *Theaetetus*) if one goes for the B reading of the *Theaetetus*. Some commentators, notably Cornford, use what is effectively an A reading of the *Theaetetus* as a reference point in attempting to establish an A reading of the *Timaeus*.[66]

What I want to suggest is that, as in the *Theaetetus*, we have an alternative to the A reading in the *Timaeus*.[67] This is based around the regularity and stability of the heavens, the stable stoicheia and the alternative epistemology of the analogy with the world soul. We need not take it that Plato is committed to a Heraclitean account of the physical world in either work. The *Theaetetus* is aporetic, and consciously so. While there is no successful definition of knowledge, *Theaetetus* is better for realising he does not know what he though he knew.[68] One task of the *Timaeus* may be to offer us the TW/Heraclitean account again, this time talking in greater detail about the nature of the physical world, but offering something more constructive as an alternative account of that world.[69] There are some interesting parallels between the *Theaetetus* and the *Timaeus*. The *Timaeus* gives us a model of the mind, tied to a discussion of the psychology of false judgement (using similar examples to the *Theaetetus*), a discussion of flux, of perception, of the relation of language and the world and perhaps most importantly, a discussion of letters and syllables.[70] If I am right that at least for the physical world, at the end of analysis by decomposition we can switch to a teleological account of the physical stoicheia, then the *Timaeus* provides at least a partial answer to one of the key aporia of the *Theaetetus*. The implications of that need investigation on a wide stage, and that, arguably, is what the *Timaeus* provides.

IX. The *Philebus* and divine cosmology

The *Timaeus*, then, presents us with some alternatives, with the onus placed on us to make a decision about our own beliefs. If we wish to hang on to TW, then we need to adopt the commensurate views on the more indeterminate passages. If we believe that the stability of the heavens and stoicheia is fundamental, then we may need to reconsider how strong a version of TW to accept, if any. The latter may give us grounds for greater epistemological optimism. The world soul can have opinions which are true and stable, and so can we, if we work hard enough, especially at astronomy. Is this greater epistemological optimism toward the investigation of nature reflected in other late works? I believe this to be so for the *Philebus*, *Laws* and *Epinomis*. In the *Philebus*, the discussion of knowledge, *techne* and measurement from 55d on is highly optimistic and would seem to allow there to be different types of knowledge, arranged in a hierarchy from the most accurate and abstract down to those which rely very much on empirical knowledge without a great deal of measurement.

At *Philebus* 55c, Socrates considers whether some types of *cheirotechnikais*, manual crafts, are closer to knowledge than others.[71] He goes on to say that if we abstracted arithmetic, measuring and weighing from all branches of the *technai*, all we would be left with would be guesswork learned by *empeiria kai tini tribei*, an empirical knack and grinding away.[72] *Philebus* 56a ff. praises those *technai* which employ number and measurement against those which employ guesswork, and considers them to be closer to the truth. Earlier in the *Gorgias* Plato had reckoned that it was the possession of a 'rational account' that distinguished a *techne* from mere *empeiria*.[73] Here we see something which is typical of Plato's later development, which is a greater mathematical orientation in explanation. This is not to suggest that the earlier account excluded mathematics or that the later excludes more discursive accounts, the subject matter helping to determine what is appropriate, but rather there is a significant change of emphasis.

According to *Philebus* 56a, music, at least in terms of flute playing and lyre playing, uses very little of arithmetic, measuring and weighing anyway, and the actual playing of an instrument requires a good deal of practice in finding the right notes. This is also the case with medicine, farming, piloting and generalship. On the other hand, there are *technai* which use instruments, such as a compass, and a rule, typically the building crafts such as ship building and house building. Having distinguished two sorts of *technai*, those using measurement being the more exact, Socrates then proceeds at 56cd to distinguish two sorts of arithmetic, that of the many and that of the philosophers. Some people reckon with unequal units, such as two armies, two bulls, or pairing the smallest and greatest of things, while others demand that all the units are equal. So one can contrast too calculation and measurement used in building and trade with philosophical geometry and diligent calculations and distinguish two sorts of arithmetic. The conclusion then at *Philebus* 57ce is that there are different types of knowledge (*episteme*) which can be arranged in order of the clarity and accuracy they possess. Throughout all this, Socrates has been quite happy to talk of knowledge in conjunction with the *technai*. There seems to be no question here that to be in possession of a *technai* is to possess knowledge of some sort.

It is also interesting to note the order of the *technai* here, which in some ways is quite surprising. The various *technai* of building are rated higher than those of medicine and generalship on this account. Perhaps though, coming after the demiourgos, the craftsman god of the *Timaeus*, that is not such a surprise, though it does run against the famous comment from Xenophon, that 'what are called the mechanical arts carry a social stigma, and are rightly dishonoured in our cities'.[74]

Following the discussion of various types of pleasures, at *Philebus* 61d Socrates returns to talking about knowledge again, and we are told:

> So knowledge (*episteme*) differs from knowledge, on the one hand concentrating on the things that undergo generation and destruction, on the other on things that do not suffer generation and destruction and are always the same and similar.

Furthermore, these lower grades of knowledge are accorded some considerable importance for practical decision making, as the following exchange between Socrates and Protarchus at *Philebus* 62a ff. demonstrates:

> *Socrates*: Let us suppose then that there is a man who understands what justice itself is, and has an account in accordance with his understanding, and moreover that he has the same sort of thoughts concerning all other things.
>
> *Protarchus*: Let that be supposed.
>
> *Socrates*: Now will this man have sufficient knowledge, having account of the divine circle and sphere, but being ignorant of the human circle and sphere, even if he employs these circles and other rules in house building?[75]
>
> *Protarchus*: We ought to call it a ridiculous state, Socrates, for someone to have divine knowledge only.

> *Socrates*: What is this you say? Are we to throw the unstable and impure skill of the false and circle and rule into our mixture?
>
> *Protarchus*: Necessarily, if we are to be able to find the road home on each occasion.
>
> *Socrates*: And music, which we just previously said was full of guesswork and imitation and was not pure?
>
> *Protarchus*: That seems necessary to me, if our life is to be any sort of life at all.[76]

The last comment is highly significant here as one of the main themes of the *Philebus* has been the nature of the life we might have in relation to the debate between intelligence, good and pleasure. There is the suggestion that if we lived a life solely of pleasure, and so entirely lacking in intelligence, our life would in fact be like that of a shellfish. In order to have a reasonable life, it seems we need the knowledge of the practical *technai* as well as that of the philosophers. What is not in doubt again here is that to be in possession of a *techne* is to be in possession of knowledge of some form, and of useful knowledge. The *Philebus* in this respect seems to have an even greater epistemological optimism to the investigation of nature than the *Timaeus*. The whole discussion of the *technai* here is notable for its lack of any strong TW assumptions. For the *Philebus* it seems a straightforward matter that we can have some kind of useful knowledge of the world about us. It is also perhaps significant that the two leading examples of *techne* in the *Philebus*' discussion of letters and syllables, and later in the discussion of the *techne* and types of knowledge, are music and grammar, precisely those examples at *Theaetetus* 207b, where the expert's discipline differs from that of the child.

10

Conclusion

At the outset I highlighted several key issues for Plato's philosophy of science. I will now look at how the arguments of this book affect those issues and how we should assess Plato's conception of nature and the investigation of nature.

I. The role of empirical work

The anti-empirical charges against Plato's conception of science are ill founded. We have looked in detail at the two passages most often cited by Plato's detractors, *Republic* VII and *Timaeus* 68b-d. In the *Republic* passage, Plato draws a distinction between teaching astronomy as part of the education of the guardians and doing astronomy as an investigation of the heavens. While we may well 'leave the things in the heavens alone' in the former, the contrast implies that we do not in the latter. This is entirely consonant with the conception of investigation advanced in the epistemological allegories. Even if he held TW at this point, there is no demarcation for investigation at the border of the two realms.

At *Timaeus* 68b-d, Plato appears to be concerned with the proportions of different sized fire particles which constitute a colour rather than with mixing pigments. Certainly he believes that this is something that cannot be found out by practical tests, but that would appear to be due to something quite specific to this problem, and may reveal a relatively sophisticated doubt about what can be found out by such tests. It is significant that Plato's consistent conception of the *technai* allows learning of a limited sort by *empeiria kai tribe*, by trial and error and grinding away, and so is consonant with his views in the *Republic* and *Timaeus* passages.

This is not to say that there are not weaknesses in relation to empirical work in Plato's conception of science. Plato is somewhat condescending about the empirical parts of the *technai*, and in general underrates both the importance and the extent of the empirical bases for most disciplines. Against that it must be recognised that he holds that there are experts in these *technai*, men who have mastered the empirical basics and have gone on to a proper sytematised knowledge, particularly in medicine, and that he is fulsome in his praise of the accuracy of the Egyptian astronomical observations. One weakness shown up by the nature of the *Timaeus* is that

Plato does not envisage a dialectic between investigators and nature. At one level, of philosophical dialectic, that is so, but modern philosophers of science would stress very strongly the interaction between theory and experiment which is so fruitful for modern science. Plato does not ignore empirical evidence, but there is no sense of the dynamic interplay with nature which allows modern science to make rapid progress.

That, however, could be said of virtually all Presocratic science, with very few exceptions, the Hippocratics perhaps being the likeliest candidates here. Plato may have weaknesses in this area but they are not so great as to pick him out as notably weak against the background of Presocratic science. A criticism that might be levelled at much of Greek science is that many of the contributors presented their work as the final word on the matter and in a finished system. It is very rare to find anyone leaving possibilities open or indicating that further research along certain lines would be fruitful. In this sense Greek science in general tended to be dogmatic. Plato, to some degree at least, can be absolved of this, especially if we consider the *Timaeus* to be more about how to do cosmology than specific cosmology itself. One can perhaps see this in the presentation of geometrical atomism at 54a, where anyone who could give an argument for better basic entities than the two types of triangle would be welcomed as a friend rather than a foe. Plato was well aware of many of the deficiencies of the very simple astronomical model he puts forward in the *Timaeus*, and so could not have meant this model to be taken literally or dogmatically.

Observation and experiment are key issues not only because they are important for science, but also because negative assessments of Plato's views have been used in a number of scenarios to attack his conception of science and to downplay its significance in the ancient world.

II. Plato as a scientific (proto-)realist

Plato can best be seen as someone who is struggling with the foundations of scientific realism. If we adopt this strategy for looking at Plato's conception of science, it allows us to explain why he takes certain stances that might otherwise be thought to be rather odd. Regularity is something which needs to be explained for Plato, in particular the regularity of the heavens. Lacking conceptions of gravity and force, Plato makes the cosmos and all the heavenly bodies animate, intelligent deities which obey good laws because they themselves are good. In short, lacking the idea of physical law Plato uses an analogue from civil law. Plato is not alone in using biological and political metaphors to explain the integrity and behaviour of the cosmos. Plato does not revert here to supernatural or animistic explanations. He strips these living beings of all but the necessary attributes of self-motion and intelligence. These are not capricious entities, but being gods will always choose the good thus guaranteeing and explaining the regularity of the heavens. Plato makes use of the highly limited tools he has available to make sense of the regularity of the heavens. An important part of Plato's reaction to the atomists is that their multiple cosmoi will not support anything more than a *de facto* regularity theory.

We also need to understand Plato's reaction to Meno's paradox/underdetermination in relation to his proto-realism. In order to be a realist, one must be able to justify a choice from an indefinite field of possibilities where others deny the validity of any further criteria. This can be seen most clearly in choice of shapes for the ultimate particles, but the effects run deep throughout Plato's approach to problems in science and the choice of regular circular motion for the heavens is also important here.

Plato is clearly not an instrumentalist. There is no motivation towards instrumentalism in his views, as he has a means of dealing with underdetermination and is unlikely to be worried by other anti-realist arguments. Hypotheses which save the phenomena have a rather different meaning for Plato, especially as he asks for regular, circular and ordered motion. Someone who believes we can escape from the cave, and has a solution to Meno's paradox has no need to be an instrumentalist. However, neither can we cannot assimilate Plato to modern scientific realism. He is struggling with the foundations of this approach in the philosophy of science, he lacks important scientific and philosophical tools, and the ancient debates differ significantly from the modern ones. What we can do though is appreciate what Plato was trying to do, and narrow the gap with modern philosophy of science. While there may appear to be a yawning chasm between Plato and modern positivist and instrumentalist accounts of science, not to mention relativist and sociological analyses, there is a considerable affinity with modern scientific realism. That Plato was a proto-realist is also of considerable importance in understanding his motivation for attempting to solve some difficulties in the philosophy of science by the use of teleology.

III. Reasons for teleology

Plato has interesting epistemological, cosmological and nomological reasons for teleology, in addition to any other reasons he may have. These reasons are all related to his proto-realism. Doubtless Plato has other reasons for teleology as well, but it is important to recognise the role that teleology plays in allowing Plato to answer certain problems in a realist manner. Plato's answer to underdetermination/Meno's paradox is to give priority to hypotheses which cite the good. So too with the epistemological problems of Socrates' dream. At least for physical entities, at the end of an analysis by decomposition into component parts, we have a teleological account of the stoicheia.

Plato also has cosmological reasons for advocating teleology. A key question for cosmogony and cosmology is how the cosmos acquired its order and how it maintains it, given that cannot be explained in terms of matter and forces alone. A survey of the history of science, right down to the present day, indicates two recurring strategies for answering this question. One is to suppose that this cosmos is part of an array of many, perhaps indefinitely many differing cosmoi, so it is no great surprise that one of these has order. The other is to suppose further, typically teleological or theological principles to explain this unique cosmos, especially if one holds that multiple cosmoi are non-explanatory and ontologically profligate.

So too Plato has nomological reasons for employing teleology. To meet the demands of his proto-realism in the absence of theories of gravitation, force and physical law Plato employs a teleological account based on biological and political analogies. It is important here that because the 'souls' of the stars are good, they always act in the best possible manner and so in a perfectly regular manner.

Plato's employment of teleology is not based on a metaphysical fairy tale. It is an attempt to answer real and perennial problems in the philosophy of science in a realist manner, with extremely limited resources. Plato's answers have strong affinities with modern realist positions, the more so as his conception of the good becomes more technically and mathematically oriented in later works, and so his solution to underdetermination looks more modern. At the strategic level, the project of using further explanatory principles for a unique universe is still alive and well. It is less overtly teleological, but Plato's is the first attempt at a teleological cosmos, and we can forgive him for not being possessed of two millennia of hindsight. The question of what underpins regularities is helped by the modern conception of physical law, though it might well be argued this only tells us how entities obey laws rather than why they do so.

Teleology is an important issue as it is often thought to be the Achilles' heel of Plato's conception of science, along with observation and experiment. If we can understand why Plato felt the need for teleology in science, and why that might have been a reasonable move given the context of fourth-century Greek science and his realist goals, this will go a long way towards a more positive assessment of his views. Plato's teleology is a part of an attempt to answer some serious and interesting problems in the philosophy of science in an original and systematic manner. This teleology is all the more remarkable for the lack of significant predecessors. We might to some extent look to Anaxagoras, and to the perhaps Socratic conception of the craft analogy for knowledge leading ultimately to the demiurge, or a snippet or two from Presocratic literature. Plato though gives us the first comprehensive and considered principles of a teleological cosmogony, cosmology and science.

IV. The aims of the investigation of nature

Plato believes the investigation of nature will produce worthwhile results. This needs to be discussed in two parts. First, what sort of knowledge will be generated by the investigation of nature? The TW view of the middle period might give us cause for concern here, though as I have argued, investigation in the *Republic* is not without empirical content nor does it recognise a demarcation between the two realms. By the time of the *Timaeus* and the *Philebus* however, a more optimistic attitude appears to be taking over. In the *Timaeus*, *Laws* and *Epinomis* it would seem that very good results are envisaged, at least for astronomy. This might be argued to be a special case, though: the one instance where we have physical entities which are both qualitatively entirely stable and amenable to precise mathematical description. The analogy with the world soul, however, might suggest that as our false opinions are correctable, and the world soul can have true and stable opinions, there may also be more optimism towards other aspects of the investigation of nature.

The *technai* quite generally are seen as producing useful results. Plato never doubts, for instance, that there can be experts in medicine as well as the more craft oriented *technai*, and that they will be able to apply the results gained by the *technai* fruitfully, whether they are gained by trial and error or by a more sophisticated analysis of the nature of the topic. The *Philebus* gives us a very positive view of the *technai*, with a hierarchy beginning with those that rely most on *empeiria kai tribe*, to those that employ arithmetic, measuring and weighing to those which deal with theoretical matters. The *Philebus* is quite happy to see this as a hierarchy of differing types of knowledge, all of which are important if we are going to be able to live properly

The second part of the discussion of this issue is what effect the successful study of nature will have for humans. The *Timaeus* is very clear on the benefits of the study of astronomy. By studying the visible revolutions of god, i.e. the heavens, we will help to stabilise the revolutions in our own heads and so become rational and knowledgeable. That will of course help with the standard Platonic project of becoming like god. So too a standard motif in Plato is that studying theoretical sciences is immensely important for the education of the young, and quite generally as a road to the cognisance of the good.

Having said that, one great weakness of Plato's conception of science from a modern point of view is the lack of practical benefits and technology. Since that is arguably the case with virtually every conception of science until Francis Bacon at the beginning of the seventeenth century, I hardly think we can single Plato out for criticism here. It is worth noting, however, that for Plato all the *technai* will do good. That is, a proper practitioner of a *techne* will bring about order and structure in his subject matter, be he a doctor, a builder, an astronomer or a cobbler.

While Plato employs myth, this is no mere story-telling, something which is strongly criticised, but a tool with specific philosophical ends. Plato is not reverting to pre-Milesian myth telling about the nature of the cosmos, but is often challenging his readers to think critically about the mythological account. Much of the astronomy and cosmology of the *Timaeus* is anyway repeated, without mythical setting or likely account qualifiers in the *Laws* and *Epinomis*. Again, while there are weaknesses and oddities to the results Plato envisages from an investigation of nature, they do not pick him out for special criticism against the background of the Presocratics.

V. Plato's developing philosophy of science

Plato's views on the nature of science, and on astronomy and cosmology, develop and do so in interesting ways. Plato's major contributions to astronomy are the first model of the heavens with an offset ecliptic/zodiac, and the principle that the motions of all the heavenly bodies can be resolved into combinations of regular circular motions, both of which are significant advances on the middle period views. The model of the *Timaeus* is best seen as a prototype embodying the new principles rather than as a finished article. I find it highly unlikely that Plato was not aware of at least some of the severe drawbacks of his model, and utterly improbable that he hoped to deal with these with the 'contrary power', which may simply mark out that there is something odd about Mercury and Venus. Doubtless Plato's relation to Eudoxus will remain controversial, not least for lack of hard evidence, but some form of close alliance between the two seems reasonable to suppose. Certainly the standard objections to a significant relation between them, that Plato was anti-empirical, that he did not believe in regular circular motion, that there was a radical difference in philosophies of science between them fail. The Platonic challenge, if Plato did issue it, and there are no philosophical or scientific reasons why he should not have done, was of enormous significance to astronomy right down to Copernicus and Kepler.

In terms of cosmogony and cosmology, Plato gives us the first thoroughgoing teleological account. The minutiae of the development of Plato's cosmological models are also of some significance. The model of the *Republic* is supported on a pivot, requires braces, needs an external source of motion, gradually degenerates and is not alive. By the *Timaeus*, the cosmos is seen as exhaustive of all there is, requires no pivot or support, no bracing, moves itself and is an immortal entity not subject to any degeneration. Just as the Presocratics developed theories about the nature, position and stability of the earth, so Plato develops ideas about the cosmos. The move to regular circular motion, or combinations therof, and to a stable cosmos is a significant one against a background of philosophy, science and literature which assumes a degenerating cosmos. So too is the move from the Pythagorean arithmetical conception of the relation of mathematics to the world to a geometrical model.

Of great importance is the meeting of the traditions of observational astronomy and cosmological speculation that I have argued occurs in later Plato. While the model of the *Republic* is hardly any advance on that of Philolaus, that of the *Timaeus*, introducing the offset ecliptic, is the first serious attempt to match the detail of what can be observed to a model for the motions of the heavens. The cosmological sophistication is greater than the astronomical accuracy, but that is only to be expected in a prototype which indicates how cosmology and astronomy can be brought together in a fruitful manner.

It is also possible to discern changes in Plato's though at a more general methodological level as well. While there is no ban on observation in the *Republic*, there is a change of emphasis on the importance of observation by the *Timaeus*, and the *Philebus* uprates the importance of the more practical *technai*. The *Timaeus* is also more optimistic about the results we will get in astronomy, and quite possibly other disciplines as well, and the *Philebus* allows that all the *technai* can be considered to be forms of knowledge, with some being more mathematised and so more accurate. The conception of the good in the *Philebus* also seems to be more mathematically and technically oriented. Plato's thought then develops. Some of that development has considerable significance for the history of Greek astronomy, cosmology and philosophy of science. Some of the criticisms that might be aimed at a middle period work like the *Republic* are wide of the mark when it comes to the later works.

VI. Criticisms of the atomists and physiologoi

Plato has a series of important and cogent criticisms of the atomists and the other physiologoi. His position on cosmogony at one level is quite simple. If we suppose some sort of preceding chaos, we cannot account for the order of the current cosmos by necessity or chance alone, but must postulate some form of teleological ordering. That, given the paucity of mechanisms for the accretion of order available to the Presocratics, seems a perfectly reasonable position for fourth-century Greece, especially if we consider zoogony as well as cosmogony here. Plato's conception of cosmogony and chaos bears closer analysis. His predecessors presume an extrinsic disorder of the stoicheia, but an intrinsic order. Is there any argument to support that? Take away the intrinsic order and their account looks even more implausible. Some assume that a vortex, an orderly motion, will be formed. Why that motion, rather than anything else? A chaos of motion will again make such an account implausible. Meno's paradox and underdetermination loom large here. What reasons do the atomists and physiologoi have to suppose order rather than chaos?

Plato's own cosmogony can be viewed in two ways. If he did believe that the cosmos was ordered from chaos, then he wants to make a general point, that whatever the preceding chaos a teleological ordering is required, rather than commit himself to any specific picture of chaos. If we take the cosmogony as metaphorical, as a counter-factual analysis of the cosmos now, the emphasis appears more epistemological. In order to account for what we see, what sort of order do we need to suppose? What would the cosmos be like without that sort of order? However we take Plato on cosmogony (and I think we can take this in both ways), this is an important piece of thinking about the origins of the cosmos which brings to light several difficulties in the atomist account.

We have seen some of the cosmological and philosophical objections Plato has to the idea of explaining the order of our cosmos in terms of multiple entities, be they cosmoi, shapes of the stoicheia or species. Also important for Plato is an adequate explanation of the continuing order of our cosmos. That forms part of the drive towards laws which explain the continuing regularity of our cosmos rather than considering this to be a matter of luck or requiring a further multiplication of cosmoi.

Plato's own geometrical atomism, I have argued, sets out to rectify many of the problems inherent in the Presocratics. It postulates two basic shapes for the ultimate constituents of the cosmos, rather than an unlimited number. These stoicheic triangles are not subject to any qualitative changes. This is critical for a coherent theory of change. If the basic entities, such as the elements of Empedocles are subject to change, the there are serious problems with identifying references. We also need to be able to account for aspects of phenomenal flux and phenomena stability. The key to geometrical atomism is postulating stability at a level far below that of human perception, and then explaining flux and stability in terms of the bonds that are formed between the stoicheia, which can be insubstantial leading to a phenomenal flux or permanent leading to stability (the heavenly bodies) or any degree between. It is also of considerable importance that geometrical atomism provides a conception of the structure of the world on which language has a grip. Indeed, the letters and syllables analogy that geometrical atomism is expressed in would suggest an important isomorphism between language and the world, which is significant for the sort of results we might expect from the investigation of nature.

Plato's reaction to the atomists and physiologoi is not simply that of a theologian and a teleologist objecting to materialism. There are serious philosophical difficulties with the Presocratic accounts of cosmogony, cosmology and the stoicheia which Plato addresses. Certainly his account is teleological, but as we have seen that is at least in part driven by his attempt to solve some serious philosophical problems in a realist manner. The resulting accounts of these topics are at least as plausible and viable in the context of fourth-century Greece as those of the atomists and physiologoi.

VII. Plato relative to the atomists and physiologoi

Plato's views on science and atomism are undervalued relative to those of the atomists and physiologoi. Geometrical atomism is often left out of the history of science, and all to often even omitted from histories of atomism where the focus is on Leucippus, Democritus, Epicurus and Lucretius. This is a considerable injustice as geometrical atomism has a great deal to offer. It has significant criticisms of the shortcomings of its predecessors, and proposes interesting solutions to these problems. If we judge this matter on affinities to modern atomism, it is by no means clear that the atomists will have the advantage. Geometrical atomism postulates a small number of well defined ultimate particles, allows microstructure, gives prominence to mathematics and has a structured system of bonding. If we look at this in the context of fourth-century Greece, it is probably just as good at dealing with the phenomena as the atomists, and has significant philosophical advantages. It is at least as plausible and viable as the views of Leucippus and Democritus. It is also important to recognise that when atomism is revived as a significant part of the scientific revolution of the seventeenth century, this is done in a Christian context where the role of the demiurge in solving some of the problems of early atomism is taken over by a Christian god.

Preference for the views of the atomists is often based on affinities to modern science. The atomists advocate a reductionist, mechanical, atomistic, anti-teleological conception of nature. Whether their conception of nature is entirely mechanical is not entirely clear cut, as they still employ biological analogies for physical processes, and among the physiologoi Anaxagoras uses nous and Empedocles love and strife. If we look for affinities to modern science with Plato, then most strongly of all there is the use of mathematics wherever possible. What is less often recognised is that the *Timaeus* offers a significant reductionist programme, based on the geometrical properties of the atoms. While this is undoubtedly crude, it is more sophisticated than the atomists. Plato's alleged anti-empirical attitude and pessimism about the results of the investigation of nature are often cited as significant differences with modern science, but we have seen that these allegations are groundless.

There is also an important philosophy of science dimension here. Plato is going to have greater affinities with realist conceptions of modern science. We have seen that his alleged animism and supernaturalism in relation to the heavens is actually an attempt in the absence of key resources to give a realist explanation of regularities. Plato has considerable affinities, both philosophical and scientific, to a conception of science where real laws and a solution to underdetermination are important. It may ultimately be the case that there are greater affinities between atomism and modern science, but the matter is not clear cut and they are not as great as is often supposed, and a great deal hangs on the question of whether realism is the proper way to approach modern philosophy of science.

The other major question here is how plausible and viable in the context of fourth-century Greece are the views of the atomists and physiologoi relative to Plato? On a great number of issues, notably cosmogony, cosmology and zoogony, they are struggling. Like Plato, they too lack key resources that are developed later. The brute fact of the matter is though that Plato's views are at least as plausible and viable as any Presocratic conception of science in the context of fourth-century Greece. Plato's atomism, science and indeed philosophy of science stand up well in comparison to the views of the atomists and other physiologoi.

VIII. Plato's philosophy of science

There is a set of standard criticisms of Plato's views on science which I hope to have rebutted. There are weaknesses in relation to observation and experiment, but no more so than with the Presocratics. He did not substitute mathematics for physics. Certainly he employs teleology, but there are interesting epistemological, cosmological and nomological reasons for this, often related to his struggle with the foundations of scientific realism. He does believe worthwhile results will come from the investigation of nature. Plato's views also develop in interesting ways, and he makes significant contributions to astronomy, theory of matter, cosmogony and cosmology. Some of the strategies he employs in these areas, notably in the debate about universe or multiverse, are still relevant to modern debates. So too there are some interesting affinities between Plato's proto-realism and modern scientific realist positions. There is a series of cogent and important criticisms of the atomists and physiologoi. The teleological alternative he offers is designed to circumvent the serious philosophical problems he finds with his predecessors. Plato played a significant part in the development of ancient science and philosophy of science. The conception of nature and the investigation of nature which he offers was at least as viable and plausible in the context of fourth-century Greece as anything offered previously, and is one that has had considerable influence on the development of science.

Notes

Introduction

1. Though perhaps that is a personal prejudice as I do not work in this manner!
2. See Thesleff (1982) p. 203, Taran (1975).
3. See De Vogel (1955) for discussion here.
4. For the moment I avoid calling the *technai* 'sciences' here so as not to prejudge later debates, though as will become evident I believe 'science' is a perfectly reasonable translation of *techne* in some contexts. Rhetoric is also an important and often discussed *techne* for Plato.
5. Simplicius *in De Caelo* 492.3, cf. 488.23.
6. Vlastos (1975) p. xiii.
7. Vlastos (1975) p. 29.
8. Vlastos (1975) p. 65.
9. Graham (1990) p. 8.
10. Zeyl (2000) xv.
11. See e.g. Vlastos (1975) p. 63.
12. Bhaskar, in Bynum, Browne and Porter (1981) p. 362.
13. Musgrave (1991) p. 244. Musgrave also makes an important distinction between dogmatic realists who believe we can know for sure which theories are true and critical realists who deny this. It is important to recognise the distinction between critical realists and instrumentalists.
14. Leplin (2000) p. 393 comments that 'The debate between realism and instrumentalism is at an impasse'.
15. Hesse, in Edwards (1972) vol. 4 p. 404.
16. Duhem (1906a), (1908).
17. Duhem (1908) p. 28.
18. Duhem (1906a), (1908). Lloyd (1987) p. 312 describes the view that Duhem sees as typical of ancient astronomers as 'a lack of concern with the physics of the problems in favour of a preoccupation with the mathematics, the construction of models that are purely calculating devices with nothing to do with any underlying physical realities'.
19. Vlastos (1969) p. 106.
20. Knorr (1993) p. 399.
21. Hetherington (1993a) p. 85.
22. Musgrave (1991) p. 250.
23. Throughout I shall translate *aitia* as 'reason' or 'explanation' rather than 'cause'. On this see Vlastos (1969) p. 76 ff. and cf. Sedley (1998), the literature on cause/because and the relation of *aitia* to *to dia ti* in Aristotle, e.g. Hocutt (1974), Moravscik (1974), Sorabji (1980), Annas (1982a), and note that Plato too uses *aitia* and *to dia ti* interchangeably, e.g. *Phaedo* 96a 'I was very interested in the kind of wisdom known as natural science; it seemed to me to be excellent to know the

reasons (*aitias*) for each thing, why (*dia ti*) each should come to be, why (*dia ti*) it should cease to be, and why (*dia ti*) it exists', so also *Phaedo* 98c, 99a, 100c and 101b. See too Frede (1980) p. 129 ff. on the use of *aitia* generally.

1. Cosmology, cosmogony and teleology

1. See generally Rowan-Robinson (1996), Silk (1989), Barrow and Tipler (1986), Smoot (1993), Weinberg (1993), Gribbin (1986), (1993), Liddle (1999).

2. Cf. *Timaeus* 81a.

3. A *plokanon* is anything woven, a *liknon* is specifically a winnowing-basket. See here Cornford (1937) p. 201 for discussion and a (marvellous!) illustration.

4. Cf. Aristotle and the idea that the four elements would settle out into four concentric spheres if it were not for the action of the sun.

5. See Chapter 8 on whether there are small voids due to the problems of packing the geometrical atoms together and three-dimensional atoms dissociating into two-dimensional components.

6. Rowan-Robinson (1996) pp. 62 ff., Silk (1989) pp. 89 ff.

7. Barrow and Tipler (1986) pp. 367 ff., Davies (1982) pp. 60 ff., Gribbin and Rees (1990).

8. One might also point out that once we have inhomogeneities modern cosmology can give a good account of the development of the cosmos in terms of gravity. But what of the source of the homogeneities? If the answer to that is quantum mechanical, as seems most likely, and that answer is a probabilistic one, then there are many possible arrangements for a cosmos which has our values of the fundamental constants.

9. Cf. *Philebus* 64e.

10. So Cornford (1937) p. 200.

11. See Zeyl (2000) xx ff., Sorabji (1983) pp. 276 ff., Vlastos (1964) p. 401 ff., cf. McCabe (1994) p. 180.

12. Democritus Fr. 164, Sextus Empiricus *Against the Mathematicians*, VII 116-18, Barnes' translation and italics, Barnes (1987).

13. See Diogenes Laertius IX, 31.

14. Diogenes Laertius IX, 45. On the nature of necessity, 'Democritus means by it the resistance and motion and blows of matter.' Aetius I, 26, 2.

15. Hippolytus *Refutatio* I, 13, 2, cf. Aristotle *Physics* II.4, 196a24 ff.

16. Diogenes Laertius IX, 31.

17. Anaxagoras Fr. 12, Simplicius *Physics* 164, 24 and 156, 13.

18. See *Phaedo* 98b.

19. Anaxagoras Fr. 13, Simplicius *Physics* 300, 31.

20. One might compare and contrast Plato's demiurge here.

21. See here Alexander (1956) pp. xvi ff., and the Leibniz-Clarke correspondence generally.

22. This is a standard difficulty for all static cosmoi with attractive forces.

23. See astronomical appendix for an explanation of the problem here. In the eighteenth century Laplace demonstrated the solar system to be stable within certain limits, and on being asked where was the place for god in his system, is reputed to have replied 'Sir, I have no need of that hypothesis'.

24. See e.g. *Apology* 26cd, and cf. Hippolytus *Refutatio* I, 8, 3-10 and Diogenes Laertius II, 8-9.

25. Cf. the view of Kirk, Raven and Schofield (1983) p. 380.

26. Anaxagoras Fr. 4, Simplicius *Physics* 35, 3.

27. Anaxagoras Fr. 2, Simplicius *Physics* 155, 31.

28. Simplicius *Physics* 178, 25.

29. Descartes *Principia Philosophia* III.47: Cottingham, Stoothoff and Murdoch (1985) p. 257.

30. Descartes *Principia Philosophia* III.45: Cottingham, Stoothoff and Murdoch (1985) p. 256.

31. While it would be wrong to consider Newton himself to be a mechanical philosopher, in the light of his views on religion, cosmology, and in his alchemical writings the 'active' qualities of matter, later Newtonians certainly did hold a mechanical view.

32. Barrow and Tipler (1986) p. 16.

33. Barrow and Tipler (1986) p. 21.

34. See here Davies (1982) pp. 111 ff., Barrow and Tipler (1986), Leslie (1989) pp. 127 ff., Gribbin and Rees (1990).

35. Barrow and Tipler (1986) p. 11.

36. Barrow and Tipler (1986) p. 22.

37. See Davies (1992) pp. 70-1.

38. Barrow and Tipler (1986) p. 1.

39. If extra-terrestrial life is not found, it would be the first instance in a long series of historic scientific debates in which the anthropocentric hypothesis had prove even partly valid. Barrow and Tipler (1986) p. 601.

40. See here Borelli (1680) and Hankin (1985) pp. 114 ff.

41. See here Hankin (1985) pp. 115 ff., Brown (1974), Hall (1969).

42. So too in the debate concerning preformation or epigenesis – the mechanical philosophers tended to favour preformation, as then the organisation of the embryo was already given and all they had to explain was its growth. See Bodemar (1964).

43. The infamous 'Trembly's polyp'. See Hankin (1985) pp. 131 ff., Vartanian (1957).

44. Theophrastus *Hist. Plant.* 3, 1, 4.

45. Living creatures come from moisture, heat and earth – see Guthrie (1978) vol. 2 p. 315.

46. Empedocles Fr. 98, Simplicius *Physics* 32, 6, KRS translation.

47. Empedocles Fr. 98, Simplicius *Physics* 300, 21, KRS translation.

48. Aetius v, 19, 5.

49. Empedocles Fr. 57, Aristotle *De Caelo* III.2 and Simplicius *De Caelo* 587, 1.

50. Empedocles Fr. 61, Aelian *Nat. Anim.* XVI, 29.

51. Aristotle, *Physics* II.8, 198b29. There may be some Aristotelian gloss here, but the multiplication is still evident.

52. Empedocles Fr. 59, Simplicius *De Caelo* 587, 20.

53. Reading *anotato* (highest) here. Wicksteed and Cornford (1929) pp. 144-5 argue for *ano e kato* as a plausible conjecture (though Simplicius has *anotato*) to bring this passage into line with *De Generatione et Corruptione* 334a ff.

54. Aristotle *Physics* 196a ff., cf. *De Generatione et Corruptione* 334a ff.

55. One might also argue that in line with later evolutionists (cf. the nineteenth-century debate about the age of the earth), Empedocles feels the need for there to have been a great amount of time for the process of the formation of species: 'When the human race was first generated from out of the earth, the day was of the length that ten months are now, owing to the slowness of the sun's advance. As time went on, it became the length of seven months.' Aetius 5, 18, 1.

56. Simplicius *Physics* 28, 8. Cf. Aristotle *Physics* III.4 203b25, If that which lies outside is infinite, then it would seem so are bodies and cosmoi. For why should something be here rather than there in a void?

57. Dionysius ap. Eusebium *Preparatio Evangelica* XIV, 23, 3.

58. Cf. Aristotle *De Caelo* III.4, 303a3 ff.

59. Aristotle *Physics* I.4, 187a23 ff., cf. Simplicius *De Caelo* 242, 18.

60. See Aristotle *De Generatione et Corruptione* 314a8 ff., cf. *De Caelo* III.4, 302b10 ff.

61. See *Phaedo* 101d9-e1.

62. Against Leucippus and Democritus in particular, there is no evidence for very large atoms.

63. Simplicius *Physics* 28, 8, see above.

64. Cf. also *Philebus* 64e and *Theaetetus* 183b.

65. See e.g. Cornford (1937) p. 220 ff.

66. See e.g. *Timaeus* 53a7.

67. *Timaeus* 32c considers there to be only four elements, though the *Epinomis* allows aether to be the fifth element.

68. Although it is not an exactly parallel case, such explanations always remind me of an old Goon Show joke: 'Eccles – what are you doing here?' 'Everybody got to be somewhere.'

69. See Keyt (1971), and cf. Parry (1979), (1991) and Patterson (1981).

70. Aristotle at least uses taxis in relation to such theories: 'There are some who make *to automaton* the explanation of the heavens and of all the cosmoi. For from *to automaton* arose the vortex and motion which by separating and settling brought the universe into order (*taxin*)' Aristotle, *Physics* II.4 196a24 ff.

71. See Aristotle *On Democritus* ap. Simplicium *in De Caelo* 295, 11 and Simplicius *in De Caelo* 242, 21.

72. Vlastos (1969) p. 82.

73. Boyle and Newton are important thinkers here, even leaving aside Leibniz and the idea that this is the best of all possible worlds.

2. Astronomy, observation and experiment

1. See the papers by Vlastos, Mourelatos, Turnbull and Mueller, as well as Anton's own introduction in Anton (1981), Mourelatos (1981a), Bulmer-Thomas (1984) and Lloyd (1991) pp. 333-51 for the most recent contributions.

2. Most influentially Heath (1913), and Neugebauer (1957). See Lloyd (1968) p. 78 notes 1-9 for a review of the earlier literature.

3. See *Republic* 510d ff. on the line allegory. Following G. Fine (1990), pp. 85-115. There is varied opinion on what these real objects of astronomy might be; see Mourelatos (1981a) p. 63 notes 2-10 for some possibilities here.

4. Cf. G. Fine (1990) note 3 p. 3, and Stokes (1992).

5. Heath (1913) note 2 p. 135, cf. e.g. Mueller (1981) p. 104 who comments that 'Plato assumes every science to be the study of some kind of real but non-sensible objects', so too Mittlestrass (1962) p. 117, and Neugebauer (1957) p. 152; cf. Shorey (1935) vol. 2 p. 187 note c and Lloyd (1968) p. 78 notes 1-9 for further examples of this view.

6. H.P.D. Lee (1955), p. 311. This position might cite *Timaeus* 27c-29d4 in its support.

7. See Cornford (1937) p. 29, cf. De Santillana (1961) p. 195, 'As Plato insists towards the end of his life there can be no "true physics" ', and p. 197 (referring to *Republic* 529ab) 'It shows vividly how Plato's conception of science has nothing to do with anything happening in space and time.'

8. One link with the *Republic* here is *epanabasmois* at *Symposium* 211c3, matching the allusion of ascent; cf. *Republic* 511b5-6 where hypotheses are *hoion epibaseis te kai hormas*.

9. Cf. *Epistle VII* 342d.

10. When astronomy is discussed later, its methodology is compared with geometry; cf. *Republic* 511b, 529de.

11. See Vlastos (1981) pp. 1-31, p. 11 ff. here.

12. See G. Fine (1990) p. 111 ff.

13. See e.g. *Republic* 529c, 530a in relation to the heavens; the construction is the best that can be achieved. Fine (1990) p. 97 ff. argues that the form of the good is the teleological structure of the world; 'Forms are its parts and individual sensibles instantiate it'.

14. The point about the teleological structure of the heavens and its relation to explanation can be made independently of Fine's model of knowledge (or the form of the good) here.

15. Compare the *Theaetetus* on how things are brought before the mind for judgement.

16. See G. Fine (1990) p. 85, Stokes (1992) p. 132.

17. Thus we can first opine intelligibles and then know them. Cf. my horizontal/vertical and dynamic/static discussion of the line.

18. G. Fine (1990) p. 85 draws this consequence if we attribute her TW view to the *Republic*.

19. G. Fine (1990) p. 86 takes *Republic* 520c to contradict her construal of TW.

20. See especially *Republic* 510ab, where the analogue links L2 and L3. I believe there to be two issues here.

21. See *Republic* 529d8-530b4, especially *paralattein* at 530b3.

22. One might argue that for Plato there are no just acts to be known, all actions being both just and unjust, so this is the most we can expect from the guardians. See *Republic* 479a ff., and cf. G. Fine (1990) p. 87, who is worried that if the guardians only have opinions of sensibles like everyone else, then they are no more fit to rule.

23. See Vlastos (1981) p. 10 ff.

24. See Vlastos (1965b) p. 68 ff. on necessary and contingent and (1969) p. 99 ff. on a priori and a posteriori.

25. See Jordan (1984) pp. 14-16, 24-5, 70-2.

26. See Mueller (1981) p. 103, pp. 110 ff.

27. If we think back to the epistemological allegories, geometry is used as an example of how investigation in general works.

28. See also *Republic* 526de where a division is made in geometry along the intelligible/sensible distinction for the purpose of determining which part will be useful for education, suggesting that there is a sensible part of geometry, however small in relation to other disciplines. In a much later work *Philebus* 55e ff. is clear on this; when we remove the theoretical/mathematical element of any *techne* we are left with the *empeiria*, however small that may be.

29. See e.g. *Republic* 511b.

30. See *Republic* 523a ff. especially 524e, where any contradictory perception will force the mind to consider unity, plurality, individuation and counting.

31. This is possibly implied by the order of the subjects in the curriculum. Note the fuss Socrates makes to get this right at *Republic* 527d.

32. So too with mathematics at *Republic* 523a ff., cf. *Timaeus* 47ab.

33. See Crombie (1963) vol. 2 p. 191.

34. See Crombie (1963) vol. 2 p. 37.

35. Or how we might train astronomers. This is a point anticipated by Lloyd (1968) pp. 79-80, (1970) pp. 68-9 and (1991) pp. 333-4.

36. This passage affirms that a subject straddles the two worlds, as a contrast is drawn between its more and less advanced sections along this distinction, but both are parts of that discipline.

37. Cf. e.g. H.P.D. Lee (1955).

38. Both an eminent physicist and a secondary school class 'study' physics, but one researches and the other learns.

39. *Republic* 529c7-d5.

40. *Republic* 529d7-530b2. See *Republic* 510b-511c for parallels with the line and the treatment of geometry; cf. Mourelatos (1981) p. 37 ff.

41. See *Republic* 531c3 for clarification of the nature of these problems, where we ascend to consider 'which numbers are concordant, which are not and in each case why'.

42. It picks up *Republic* 521c ff. which sets the educational purpose of the whole section, as well as 526de and 529c.

43. A point originally made by Dicks (1970) p. 234 note 39; cf. Vlastos (1981) p. 3. Some other instances of *chairein ean* as meaning to renounce are at *Protagoras* 348a, *Phaedo* 63e, 101cd (see Brandwood (1990) for a full list). Even this stronger phrase is not always totally dismissive.

44. Heath (1913) p. 139, cf. Shorey (1935) p. 187, Lloyd (1968) pp. 80-1, Vlastos (1981) p. 12.

45. Cf. *Timaeus* 47a ff. on the benefits of eyesight and hearing.

46. Shorey (1935) translates 'Have a part in the true science of astronomy', and Vlastos (1981) p. 3, 'By doing real astronomy', Cavagnaro (1997) p. 357 'have a share in real astronomy'.

47. See Vlastos (1981) p. 10.

48. See Mourelatos (1981) p. 34. Mourelatos does not specifically translate *ontos astronomias metalambanontes* as 'real astronomy', but instead cites *toi onti de astronomikon* at *Republic* 530a3 and takes it that the whole of 529a-530c deals with 'real astronomy'.

49. See Mourelatos (1981a) p. 34 ff.

50. See Mourelatos (1981a) p. 34.

51. Cf. Shorey (1935) pp. 180-9, Dicks (1970) pp. 103-6.

52. See Vlastos (1981) p. 10.

53. See Vlastos (1981) p. 11.

54. See Vlastos (1981) p. 13 ff.

55. Vlastos (1981) p.16.

56. Cf. Hetherington (1996) pp. 276 ff.

57. Dicks (1970) p. 109.

58. See Bulmer-Thomas (1984) p. 108 ff.

59. See Bulmer-Thomas (1984) p. 109 ff.

60. i.e. the three fingers at *Republic* 523a ff.

61. Cf. *Timaeus* 39c ff. where this linkage is explicit.

62. Vlastos (1975) p. 50. Vlastos is referring specifically to *Timaeus* 40c4-d3.

63. Though I would agree with Lloyd (1996) that Aristotle was an 'amateur astronomer', cf. Gregory (2000).

64. See e.g. Aristotle *De Caelo* 270b13, 292a8.

65. Compare the *Epinomis* passage with *De Caelo* 292a8.

66. See *Timaeus* 22d ff. As much of the subsequent discourse concerns astronomy, this may be highly significant. This suggestion illuminates the *Timaeus* passage whether the *Epinomis* is genuine or not.

67. Shorey (1935) p. 186, cf. his (1927) pp. 172-3, and Lloyd (1968) p. 80. For Aristotle's view see *De Caelo* 270b11 ff.
68. So too the *Politicus*: see Chapter 4 for further discussion.
69. Taking a moderately realist position here.
70. See Farrington (1936) p. 127 ff.
71. See e.g. Vlastos (1981) p. 15 ff.
72. Cf. Shorey (1935) p. 184 note c and Adam (1902) on *Republic* 530b.
73. LSJ gives 'to change or alter a little, esp. for the worse ... to pass aside, turn from the path, to deviate, vary'.
74. Agreeing with Vlastos (1991) p. 22 note 17 against Cornford (1941) and Duhem (1906a) that *auton* must mean 'about them' rather than 'in them' which would require the dative *autois*, and against Shorey (1935 vol. 2) p. 186 note a that the final clause is not a 'trailing anacoluthon'.
75. Cf. Sorabji (1980) pp. 61-2.
76. Sorabji (1980) pp. 61-2.
77. That is something which I argue changes as Plato's cosmology develops – see Chapters 4 and 6.
78. Attempts to see a more complex astronomy in the actions of the daughters of necessity (specifically some account of the retrograde motions of the planets, see Knorr (1990)) will not do as they do not discriminate in their treatment of the stars, sun, moon and five planets while only the five planets show retrograde motion. I discuss this matter in greater detail in later chapters.
79. See *Philebus* 56d ff.
80. See *Philebus* 57d.
81. Sedley (1997) p. 333, cf. Dampier-Whetham (1948) p. 28, 'He roundly condemned experiment as impious or a base mechanical art', and Crombie (1963) vol. 2 pp. 228-9.
82. Lloyd (1987) p. 140 note 133, see also Cornford (1937) pp. 276-8.
83. See e.g. Cornford (1937) p. 277 note 1.
84. See *Gorgias* 485c ff. and 500e ff.
85. See *Phaedrus* 268c ff.
86. See *Philebus* 55e ff.
87. One out of many too, if we can prepare pigments from starting materials which are a mixture of colours.
88. See Aristotle *Meteorologica* 375a1 ff., and *Meteorologica* I.5 generally, see also the pseudo-Aristotelian *On Colours*, and Democritus as reported by Theophrastus *De Sensu* 73 ff.
89. See Aristotle *Meteorologica* 374a3 ff.
90. See Lloyd (1987) p. 140 note 133.
91. *Philebus* 15b ff. is a good example here.
92. Or perhaps 'self-generating' or 'self-existent' or 'spontaneous' for *autophue*.
93. Cf. the comments on the liver at *Timaeus* 71ab.
94. The one exception to this might be argued to be the Hippocratics.
95. Farrington (1936) p. 127.
96. Heath (1913) note 2 p. 135, cf. Neugebauer (1957) p. 152.

3. Meno's paradox and underdetermination

1. See e.g. the articles and commentary under 'The Duhem-Quine Thesis' in Curd and Cover (1998), pp. 255-408, and relevant articles in Newton-Smith (2000).
2. There are various positions here. For instrumentalism, see Duhem (1914), and Van Fraassen (1980). For conventionalism, see Poincaré (1952). For pragma-

tism see Quine (1953). For relativism, now generally leading to a sociology of science/epistemology/knowledge, see Hesse (1980), Bloor (1976), (1981), Collins (1981).

3. Laudan refers to this as the 'Humean Underdetermination Problem' (HUD), which he formulates as:

HUD For any finite body of evidence, there are indefinitely many mutually contrary theories, each of which logically entails the evidence.

Laudan (1998) p. 323. Cf. Newton-Smith (2000) pp. 532-6.

4. See here Harré (1972) pp. 43 ff., (1981) pp. 6 ff. and Sober (2000).

5. The scale of the axes can be chosen arbitrarily (e.g. squared, cubed, log, or something more exotic) to produce data in a rough line. There is nothing which determines this choice of function, and recalcitrant data could be brought into line by the use of ad hoc or even discontinuous functions.

6. Again one can manipulate the axes to produce such data.

7. One might also argue that the amount of error is underdetermined.

8. One might also question how the use of error bars can be justified on purely empirical criteria.

9. If our only criteria are empirical, what does it mean for an experiment not to work?

10. Millikan's oil drop experiment is a notorious example of a scientist using ad hoc criteria to exclude results he is not happy with.

11. See e.g. Curd and Cover (1998) pp. 383-5.

12. So Einstein comments: 'I do not by any means find the chief significance of the general theory of relativity in the fact that it has predicted a few minute observable facts, but rather in the simplicity of its foundation and in its logical consistency.' See Hetherington (1996) p. 273.

13. The modern underdetermination problem, and the modern problem of induction generally, are often held to be descendants of Hume's discussion of causation.

14. Duhem (1954) p. 183, cf. p. 187.

15. Quine (1969) p. 79.

16. Gillies (1998) pp. 302 ff.

17. See Duhem (1954) p. 214 ff. Duhem considered there to be a scientific 'good sense' which guided our choice of theories (but did not of course determine them), without ever saying a great deal about the nature of this 'good sense'.

18. Quine (1953) p. 42.

19. Quine (1953) p. 43. Laudan refers to this as the 'Quinean Underdetermination Problem' (QUD) which he formulates as:

QUD Any theory can be reconciled with any recalcitrant evidence by making suitable adjustments in our other assumptions about nature.

Laudan (1998) p. 328.

20. See Laudan (1998) p. 325.

21. Quine (1953) p. 44.

22. Quine (1953) p. 43.

23. Indeed he does so specifically in the context of the statement quoted above, see Quine (1951) p. 43, and (1986) pp. 618/19.

24. Quine (1953) p. 42.

25. Quine (1953) p. 45.

26. Quine (1986) pp. 618/19.

27. Quine (1953) p. 45.

28. Quine (1953) p. 46.

29. Quine (1953) p. 46.

30. Assuming that dropping the law of non-contradiction is not one of the 'drastic adjustments' to the system! If it is, anything goes.

31. I take one of the general points of the discussion of Protagoras' theory to be that one must assert something to be true in order to assert at all.

32. Quine needs either to drop these three underdetermination principles or explain why their scope is restricted (and proximity to the edge of the fabric, as Quine might put it in *Two Dogmas* would be rather too vague). Similar problems might be argued to apply to the problem about negative instances.

33. So too with negation and predication, as with the *Sophist*.

34. Often citing Quine, see e.g. Hesse (1980) p. 33.

35. Hesse (1980) p. 33.

36. i.e. as there is some matter in zero volume the density is infinite.

37. See *Meno* 82b ff., so too *Phaedo* 72e-77a, 91e-92d.

38. Quine (1960) p. 23.

39. See e.g. Barrow (1991), and cf. Hawking (1988).

40. Burnet (1911) p. 99 comments 'This is the oldest name for what we call "natural science" '; cf. Gallop (1975) p. 234 note 56 and Hackforth (1955) p. 122.

41. Three possibilities canvassed for the *protos plous* on a non-ironical interpretation have been Anaxagoras' teleology, teleology of any sort, and divinely revealed knowledge; cf. the discussions of Crombie (1963), Gallop (1975), Goodrich (1903) and (1904), Williamson (1904), Burnet (1911), Murphy (1936), Hackforth (1955), Huby (1959), Rose (1966), Sayre (1969), Shipton (1979), Bostock (1986).

42. Whether this passage has anything to do with the historical Socrates is another matter; see Hackforth (1955) p. 127 ff. for discussion.

43. Cf. G. Fine (1987) p. 88 ff., MacKenzie (1988) p. 41 ff.

44. Throughout I shall translate *nous* as 'intelligence', though it might equally well be translated as 'mind'.

45. Cf. Aristotle *Metaphysics* 984b15-17.

46. See Guthrie (1978) p. 317, cf. McCabe (1993) p. 54.

47. So Menander, fragment 241; see Hackforth (1955) p. 137.

48. Cf. MacKenzie (1986b) p. 17 ff., Vlastos (1991) ch. 1 on riddling and complex irony.

49. Reading *kardopio* (lid of a kneading trough) as against *kardopo* (kneading trough). Cf. *ad loc.* notes in Burnet (1911) and Gallop (1975). *Kardopio* is more in tune with Aristotle's report at *De Caelo* 294b15.

50. Possibly the theories of Empedocles and Anaxagoras. See Sedley (1989) p. 363.

51. Sedley (1990) p. 370. Cf. Lennox (1985) pp. 199 ff., G. Fine (1990).

52. See e.g. G. Fine (1990).

53. See Wiggins (1986) p. 3 and Vlastos (1969) p. 82 note 15.

54. Sedley (1989) p. 370.

55. G. Fine (1990) p. 386.

56. Cf. the discussions of R. Robinson (1953), Sayre (1969), Bostock (1986).

57. See Gentzler (1991) p. 266.

58. See Robinson (1950) p. 128.

59. See e.g. Robinson (1950) pp. 129-31, Bostock (1986) p. 169.

60. R. Robinson (1953) p. 127, cf. Sayre (1969) pp. 15-28.

61. Cf. *Gorgias* 479c5, 496e5, 508b3, *Parmenides* 136a8, 136b3, 142b3, *Sophist* 251e5, *Philebus* 22a7. In what follows I might be accused of assimilating the *Republic* and *Phaedo* too closely, but note that *sumphonein* and *homolegein* rather than *sumbainein* are used in the *Republic* especially in discussion of the line allegory.

62. Gentzler (1991) p. 268.

63. Cf. Gentzler (1991) p. 269 ff.

64. See Gentzler (1991) p. 270.

65. Gentzler (1991) p. 271.

66. Cf. *Philebus* 56a, *Sophist* 254a, *Phaedrus* 260e and 270b, *Gorgias* 462c ff., *Epinomis* 976b.

67. This would rule out many of the propositions that Gentzler is concerned about. Of course, there may be some accounts which are consistent, seemingly suitably teleological and have not yet been shown to stand in a suitable relation. Here Gentzler may be correct in suggesting that *sumphonein* and *diaphonein* are contraries and not contradictories, such that we only reject as false those accounts which do not suffice teleologically or do not stand in a suitable relation, reserving judgement on others.

68. Vlastos (1975) pp. 30-1. It is possible that Anaxagoras would not give an unequivocal 'No' to this.

69. Cf. *Timaeus* 55cd, *Philebus* 17e.

70. Hawking (1993) ix says: 'I have written in the belief that the universe is governed by an order that we can perceive partially now, and that we may understand fully in the not-too-distant future.'

71. See Pais (1982) p. 6. Nor, according to Einstein, does god play dice.

72. Aristoxenus, *Elements of Harmony* II, 30-1, Gaiser's translation, cf. Simplicius *Physics* p. 151, p. 453, Themistius *Oratio* 21, 245cd, Proclus, *Philebus Commentary* p. 688.

73. See Sayre (1983).

74. Gaiser (1980) p. 12.

75. See Cooper (1977) p. 155.

76. See Lennox (1985) p. 206.

77. M.R. Wright (1995) p. 47.

78. See Duhem (1969) p. 5. Several modern commentators have followed Duhem here, most recently Cushing (1998) pp. 54-5.

79. See Lloyd (1978) and Musgrave (1991).

80. Duhem (1969) pp. 5-6. Duhem (1906) p. 103 asks us to 'Pause a moment at this fundamental passage, which so clearly shows us the doctrines of Plato touching the method of astronomy.'

81. Duhem (1969) p. 7.

82. Duhem (1969) p. 28.

83. Duhem (1954) p. 299.

84. See Duhem (1954) p. 299.

85. See *Republic* 519c: 'Our job as lawgivers is to compel the best minds to attain what we have called the highest form of knowledge, and to ascend to the vision of the good as we have described' and 520e: the guardians will descend because 'We are making a just demand of just men'.

86. See *Republic* 529c ff.

87. See e.g. Duhem (1906) pp. 105-11 and cf. (1969) pp. 5-24.

88. See e.g. *Republic* 510b3, b5, c4, c6, 511a4, a6, b5, b6, c7, d2.

89. See *Republic* 511b.

90. Duhem (1906) p. 103.

91. See *Timaeus* 40a.

92. See *Laws* 897b ff.

93. Duhem (1969) p. 7. Against this he contrasts the methods of Eudoxus and Callipus, who 'Followed the method of the astronomer, as laid down by Plato'.

94. In one context explanation A is most meritworthy, in another context or from another perspective explanation B may have the greater merit, and there is no independent perspective from which to judge which is in fact the better explanation to adopt. See van Fraassen (1980).

95. See Leplin (2000) pp. 398-9.

96. Fine is influenced here by Hilbert's comments on set theory – a proof of the consistency of a set theory ought not to come from that theory itself, but be derived from other, more stringent considerations. See A. Fine (1984), Curd and Cover (1998), Newton-Smith (2000).

97. Fine's reason for doubting decision making procedures is the historical argument so Plato may well not share Fine's doubts.

98. Hilbert's criteria are often considered to be too stringent, and we might well say the same of Fine. If Fine's argument is generalised to 'we always need higher level justification' then it is regressive and may lead to the conclusion that we have no knowledge at all. Fine's ultimate position (the Natural Ontological Attitude) is arguably a realist one (see Musgrave (1989)).

99. Simplicius *in De Caelo* 488, 18 ff., cf. 492, 31 ff.

100. See Mittlestrass (1962) p. 154 note 214, von Fritz (1971) p. 179 note 375, Knorr (1990) pp. 319 ff.

101. Zhmud (1999) p. 220, cf. Vlastos (1975) Appendix L p. 111 on whether we can put so much faith in Simplicius' 'fastidiousness'.

102. At Simplicius *in De Caelo* 493, 1 it is *engkuklion kai homales kai tetagmenon*.

103. See Vlastos (1975) Appendix L pp. 110-11.

104. See Simplicius *Physics* p. 60 Diels.

105. See Zhmud (1999) p. 218 note 23.

106. Cf. Zhmud (1999) p. 218 note 23.

107. Especially so if we look to Simplicius *in De Caelo* 492, 31 ff. where a different order of presentation means that this problem does not occur.

108. Geminus, *Isagoge* I, 19-21, Heath's (1913) p. 269 translation.

109. Van der Waerden (1978).

110. See Zhmud (1999) p. 219 note 25, and Zhmud (1997).

111. Zhmud (1998) p. 211.

112. Zhmud (1999) p. 219 ff.

113. Neugebauer (1957) p. 152. Cf. Adam (1902) vol. 2 p. 125, Shorey (1935) vol. 2 p. 187 note c, Samburksy (1956) p. 45, Mourelatos (1981a) notes on pp. 63-4, and Bulmer-Thomas (1984) p. 107 for further references to assessments of Plato's astronomy.

114. I also discuss the possible relationship and direction of influence between Plato and Eudoxus in greater detail.

4. Celestial motion in the *Timaeus*

1. See most recently Cavagnaro (1997), with whom I am in broad agreement on the relation of the *Republic*, *Timaeus* and *Laws* on this matter while differing on many details.

2. See *Phaedo* 109c ff.

3. This passage supports the idea that Plato holds an imperfection view of the physical world or that he has not yet fully worked out his views on participation. It may be that the visual analogue of what a fish might see is meant to apply in intellectual terms for humans, but that too could be used to support the sort of conclusions the *Republic* draws.

4. Note the emphatic *ton nuni* instead of *ton nun* at *Politicus* 272b2.

5. *Paralattein* does occur at *Timaeus* 22d in Critias' account, which in turn is an older myth that has come down to him (*Timaeus* 21b). As Lloyd (1991) p. 342 has pointed out, there are some oddities about the nature and status of Critias' account; perhaps all we need here though is to note that Critias is entirely willing to defer to Timaeus in matters of astronomy, i.e. 'Timaeus is the best astronomer among us, and has made a special study of the nature of the universe' (*Timaeus* 27a).

6. Cf. Popper (1945) vol. 1 p. 219 note 5.

7. Lucretius 2.1172-4, M.R. Wright (1995) translation, p. 91, cf. *Politicus* myth!

8. See Brisson (1995) pp. 349-63, Rowe (1995b) pp. 186-99, and cf. Erler (1995) pp. 375-80, Ferrari (1995) pp. 389-98, especially p. 394 note 17.

9. See *Politicus* 269d5 ff., cf. McCabe (1997) pp. 100 and 103.

10. See McCabe (1997), pp. 104 ff., and see *Politicus* 272bc and 274e5 ff.

11. See Lane (1995).

12. Burkert (1972) p. 324, cf. Heath (1913) p. 171, Mittlestrass (1962) pp. 130 ff., Owen (1953) pp. 325 ff.

13. e.g. *Timaeus* 38c6, 39d1, 40b6, and 43b3. The cognate verb *planao* has associations of to mislead, or to lead into error.

14. The six motions are up, down, left, right, backwards and forwards. According to the latest evidence, the universe does not revolve. If it did, we could expect to see systematic anisotropies related to the axis of rotation in COBE maps, but we do not.

15. We are also told that sight is the greatest gift from the gods to men.

16. Aristotle comments at *Physics* 218b that 'Some have identified time with the revolution of the whole, and others with the sphere itself' and is thought to be referring to Plato and the Pythagoreans respectively (see Simplicius' references to Eudemus, Theophrastus and Alexander, at *in De Caelo* 700, 18). His general criticism of Plato at *Physics* 219b ff. is that time, while intimately related to motion, should not be identified with motion. Whether Aristotle is quite correct here is another matter. I would agree with Cherniss (1944) p. 418 and note 349, and Vlastos (1964) p. 409 ff., that the key phrase is that the stars 'mark off and preserve the numbers of time' (*Timaeus* 38c6), and so implicitly distinguish clock and time as measure and measured.

17. Adam (1902 vol. 2) pp. 297/8.

18. e.g. the uses of 'calculate' at *Timaeus* 47c2 and 40d1.

19. Cf. *Timaeus* 37d, 38a9, 38c, and note the addition of *arithmos* at 39c8 in comparison to *Republic* 530a8.

20. Cf. other uses of *planes* in the *Timaeus*, at 19e4 to describe those who 'wander' from city to city, at 43b4 to similarly describe our mental revolutions, and in some biological contexts describing the random wanderings going on in our bodies, of phlegms at 86e7, of atomic particles at 88e2, and of the womb in women at 91c4.

21. There might be a worry here about reading the views of the *Laws* back into the *Timaeus*, but the *Timaeus* advocates similar, if not quite so well expressed or developed ideas; see e.g. *Timaeus* 34a, 89a.

22. See *Timaeus* 34a.

23. Note that the motion of the whole is regular but not perfect, too; it fails to be self-generating. See *Timaeus* 34a.

24. While we would regard this as mathematically incorrect, as we would resolve the circular motion of the stars on the same into two normal linear motions, the sense is clear enough; the fixed stars have only the translational motion imparted by the same.

25. *Laws* 822a might also be taken as a clarification that the 'contrary power' has either been rejected by this stage or was never a problem for regular celestial motion.

26. See too e.g. *Laws* 897b.

27. See *Timaeus* 40a ff. and 40b ff.

28. See e.g. *Philebus* 32d.

29. At *Philebus* 43a ff. the idea that nothing is free from the processes of deterioration and restoration is raised in language that is reminiscent of the discussions of Heraclitean flux at *Theaetetus* 179-83 and *Cratylus* 439-40. Socrates avoids discussion of the truth of this theory by suggesting that there are some changes that are so small that we do not perceive them, and hence we sometimes feel that we are in a state that does not involve deterioration or restoration.

30. See e.g. *Epinomis* 982b-d, 983c, 987a, 991c.

31. See e.g. *Timaeus* 34a on motion and cf. *Politicus* 269d-270b, and *Timaeus* 32c, 38b, and 41a on indissolubility.

32. See *Politicus* 273b ff.

33. Cf. *Politicus* 270a9 and *Timaeus* 33d9, a point made by Brumbaugh (1961) p. 523, although his concern is with models rather than uniform motion. *Timaeus* 34a is a better passage for my purposes. It is worth noting that prior to the *Timaeus* Plato always has the cosmos supported; a pivot in the *Politicus*, the spindle of necessity in the myth of Er (*Republic* 616b).

34. Cf. *Timaeus* 33a9 and *Politicus* 273e4.

35. The myth is introduced to illustrate a point in the hunt for the definition of the statesman, see *Politicus* 268de.

36. See e.g. *Phaedo* 109a ff., where the 'fish's eye view' is indeed indicative of Plato's current epistemology.

37. See e.g. *Laws* 897b, 967a etc.

38. One might make the case that the *Politicus* is transitional, rather than the last use of the old cosmology. Perhaps the two cycles of the myth reflect Plato's first dissatisfaction with degeneration, and at *Politicus* 284a ff. there is the concern with mensuration that we find in both *Timaeus* and *Philebus*, and some of the Pythagorean ontological themes which with Sayre I shall argue in later chapters are typical of the *Timaeus* and *Philebus* are to be found in the *Politicus* too.

39. See McCabe (1997) p. 99.

40. Skemp (1952) p. 147 uses 'thousands of revolutions', cf. Rowe (1995b) p. 65 'many tens of thousands of revolutions'.

41. Cf. *Timaeus* 39c.

42. *Emphronos* can mean sensible both in the sense of alive and in the sense of reasonable or prudent.

43. e.g. *Timaeus* 46e.

44. See Archer-Hind (1888), A.E. Taylor (1928), Cornford (1937) and Morrow (1950) for further discussion here.

45. Cf. Long (1977).

46. There are possibly some conservative assumptions concerning art in play here; 'good' in the sense aesthetically pleasing, perhaps, or in the sense of being an accurate representation of the subject.

47. Cf. Aristotle on good and chance: 'Something good might come about by chance; but in respect of chance, and insofar as it results from chance, it is not good.' Aristotle *Protrepticus* Fr. 11, translation of Lennox (1985) p. 198.

48. One must be careful of anachronism here, as Plato never expresses his views in terms of laws.

49. See e.g. *Timaeus* 30a5 et passim, and cf. Aristotle *Metaphysics* 10 on *taxis* and armies.

50. See e.g. *Timaeus* 42d ff., where the demiurge is said to be blameless for anything that may happen after the initial ordering; cf. *Republic* 380c ff.

51. i.e. as in my example, if we throw a bucket of paint at a canvas.

52. See *Timaeus* 61d ff. on the shape and properties of fire.

53. See Cornford (1937) p. 165.

54. Sorabji (1980) pp. 61-2, cf. pp. 206-8

55. Sorabji (1980) p. 61, cf. pp. 207-8.

56. Sorabji (1980) pp. 61-2.

57. See Vlastos (1969), cf. Sedley (1998).

58. See *Timaeus* 67c ff.

59. See *Timaeus* 65c ff. on taste.

60. See e.g. *Timaeus* 56d ff.

61. Is the sorting of like to like an effect of the motion of particles (as with Descartes on gravity) or a cause in its own right (as with Newton on gravity)? I would argue that Plato is closer to the former and so would see like to like more as an effect than cause.

62. So A.E. Taylor (1939) p. 25 ff. The passage is open to several interpretations, though note that the standard middle period view that 'All things that come to be must also decay (*genomenoi panti phthora estin*)' (*Republic* 546a) is being pushed here and would be a suitable driving force for political and social corruption; cf. Adam (1902) vol. 2 p. 202, who asks 'What was the originating cause of degeneration? Plato finds the cause, not in anything peculiar to the Ideal city, but in a law which prevails through the whole of Nature – the law that everything created is doomed to decay'.

63. Taylor (1939) p. 25.

64. Popper though would disagree. He comments that 'Plato summed up his social experience, exactly as his historicist predecessor had done, by proffering a law of historical development. According to this law ... all social change is corruption or decay or degeneration' (1945 vol. 1) p. 19, cf. p. 37. This seems far too strong though. Both Plato's writing (e.g. *Republic* 473b, 502b ff., *Laws* 676b ff., 769d7-8) and his practice, if we are to believe *Epistle VII* (and it may be the case that only the digression is a forgery), show considerable optimism that actual societies can (at least temporarily) be improved rather than merely stopped from decaying.

65. As Plato once more invokes the (*paidia / spoude*) contrast at *Laws* 688bc, we must beware of his 'serious play' as with *Republic* 545d ff.

66. The Athenian Stranger hopes to root out ignorance and vices by an adequate education system (see books I, II and VII), and their are various strictures aimed against financial acquisitiveness. No citizen may possess gold or silver (*Laws* 741e ff.), one family may possess only one landholding, which is non-transferable (741b ff.), and there are strict limits on personal wealth.

67. It is perhaps worth noting that in the *Republic* the decline begins with incorrect breeding (*Republic* 546a ff.). The *Laws* too discusses this (775c ff.) but does not connect poor breeding with any inevitable decline.

68. See *Timaeus* 41d ff. These matters will be discussed in more detail in the next chapter.

69. The language of mastery/being mastered here is precisely that of the discussion of the relation of intellect, the revolutions and perception. Cf. *Timaeus* 44a where we find a similar use of *kratein*.

70. For an excellent to reply to Owen (1953) on the relative politics of the *Republic*, *Politicus* and the *Timaeus-Critias* which in his view indicate that the *Timaeus* is earlier, see C. Gill (1979).

71. One might note that the receptacle is entirely characterless in order not to distort what is in it (*Timaeus* 50d ff.), and that what is in the receptacle is marked out by forms and numbers (*Timaeus* 53b5).

72. See Hetherington (1996) pp. 275 ff. and p. 288.

73. Hetherington (1996) p. 275.

74. Some modern scientific realists, and indeed many seventeenth-century scientific realists, are realists about god, and might well say that god is most real, while still being scientific realists.

75. See e.g. *Timaeus* 40c and 47c2 on calculation and astronomy.

5. Plato and the development of Greek astronomy

1. See *Republic* 616b-617d.

2. See here Burkert (1972) pp. 330-1. See Anaxagoras Fr. 78, for Hippocrates, Aristotle *Meteorologica* 343a5 ff. and for Democritus, Aetius II, 16, 1.

3. Cf. *Republic* 617a, *Timaeus* 38e, *Laws* 822a, *Epinomis* 987b.

4. There is special relationship between sun, Venus and Mercury not shared by the other planets. See below Chapter 5.IV.

5. Cf. Knorr (1990) and van der Waerden (1951), (1974a), (1982).

6. On Philolaus and the Pythagoreans, see Aristotle *De Caelo* 293a18 ff., Aetius II, 7, 7 ff. See also Burkert (1972) p. 299 ff., Heath (1913) p. 99 ff. and Dicks (1970). The sirens on top of each circle making a note of constant pitch (*Republic* 617b) which together make a harmony would seem to be a Pythagorean influence.

7. Diagram following Adam (1902) and H.P.D. Lee (1955).

8. I agree with Dicks (1970) pp. 70-1 that the only evidence we have for an offset ecliptic in Philolaus is Aetius (DK 44 A21), and his evidence is suspect and should be rejected. The one suspect fragment appears to have sun, moon and earth all moving in an inclined orbit, and need special pleading to have the orbit of the earth offset to that of sun and moon. Against this one suspect fragment is the fact that longer accounts in Aetius, Aristotle and Simplicius give no indication of any offset orbits. Cf. Dreyer (1906) and Heath (1913).

9. See below Chapter 5.III on the retrograde motion of the planets.

10. See Vlastos (1975) Appendix H p. 106 on the etymology. But Heath (1913) p. 149 translates as 'third in speed of counter revolution the fourth appears to move'.

11. Burnet (1924) p. 304 note 1.

12. Dicks (1970) p. 112.

13. See Dicks (1970) pp. 112-13.

14. Knorr (1990) p. 326 note 11.

15. See *Timaeus* 36c and *Epinomis* 987b, cf. *Laws* 760de on right and left. Plato may not be entirely consistent in his usage.

16. See *Timaeus* 36b ff. on how the demiurge creates same and different and produces the angle between them.

17. The Greeks, probably drawing on Babylonian data, reckoned the maximum deviation in latitude to be that of the moon, 6^{0} either side of the ecliptic. See Neugebauer (1975) p. 626.

18. See Burkert (1972) p. 306, Aetius III, 14, 1.

19. I reject Burnet and Taylor's theory of a moving earth in the *Timaeus*. See Gregory (2000) and Chapter 6.

20. Deviation in latitude is exaggerated here for effect.

21. Modern explanation of retrogression for superior and inferior planets. If one follows the projection of the apparent position of the planets relative to the stars from earth, the planet will be seen to progress, stop, retrogress, stop and progress again due to the relative motions of the earth and the planet, moving in different orbits of the sun at different speeds.

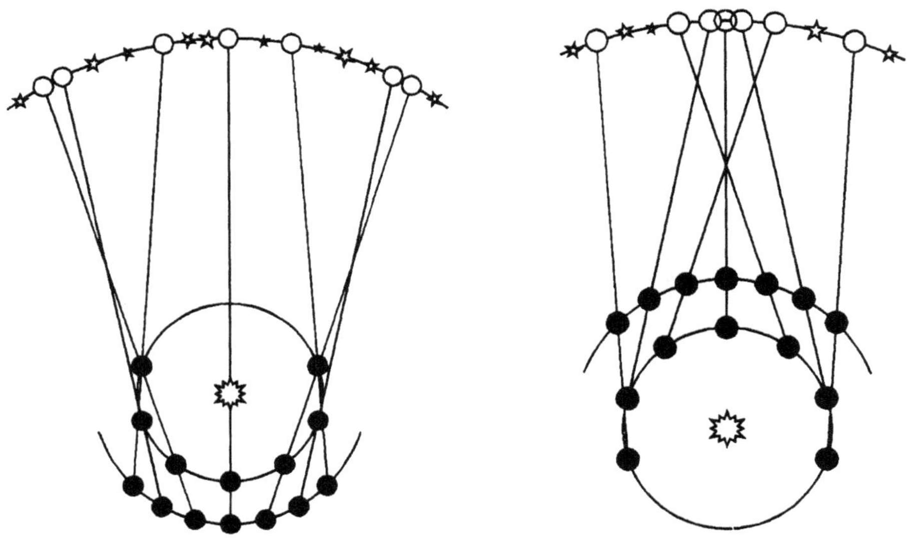

22. Inclinations to the ecliptic are earth 0°00′ (by definition), Mercury 7°00′, Venus 3°24′, Mars 1°51′, Jupiter 1°18′, Saturn 2°29′.

23. See Neugebauer (1975) p. 626.

24. There is debate as to which of the second and third spheres generates motion along the ecliptic and which generates deviation. See Dicks (1970) pp. 179-81.

25. It also of course generates some longitudinal motion.

26. We have only a partial account of the actual values Eudoxus used for the construction of his system, so reconstructions have to be conjectural.

27. See above on the myth of Er.

28. They are in fact on orbit around the sun.

29. Certainly he refers to the relation of Venus and Mercury to the sun at 38d-6 (and quite possibly he does so at *Timaeus* 40c8-9 as well), almost certainly *Timaeus* 40c5 refers to retrogression, and *Republic* 617b2 may well do so (see Heath (1913), Dicks (1970) and Knorr (1990) for debate here.

30. Vlastos (1975) Appendix J p. 108.

31. To give a full account of the orbits of Mercury and Venus one has of course to talk of retrogression, but that on its own will not explain why Mercury and Venus are always seen near the sun.

32. There are two interpretations of the contrary power which I do not discuss here as there are overwhelming objections to them. Martin argues that Mercury and Venus orbit at the same speed as the sun but in the opposite (contrary) direction. He tries to support this with similar uses of *enantia* at *Timaeus* 36d and *Republic* 617a7. However, the fact that Mercury and Venus do not orbit in a contrary direction from the sun is evident, and the contrary power is supposed to explain why Mercury, Venus and the sun overtake and are overtaken by each other, and this interpretation of the contrary power does no such thing. Alternatively, Mercury and Venus may orbit with the same direction and speed as the sun, but the contrary power may occasionally act in opposition to this movement. Depending on the strength of the contrary power, this might merely slow Mercury and Venus, or it might cause them to stop or even reverse their direction. However, such an interpretation could only have Mercury and Venus overtaken by the sun and not overtaking it, and they would appear at all angles relative to the sun.

33. See Knorr (1990).

34. Knorr (1990) p. 316.

35. Cornford (1937) p. 137.

36. See Empedocles Fr. 42.

37. The only modern discussion of the eclipse problem is in Taylor (1928) pp. 236-7.

38. There is debate as to which of the second and third spheres generates motion along the ecliptic and which generates deviation. See Dicks (1970) pp. 179-81.

39. It also of course generates some longitudinal motion.

40. That is our best conjecture on why Callippus added these spheres. See Neugebauer (1975) pp. 625-8, and cf. Mendell (1998) and Yavetz (1998).

41. The model of the *Timaeus*, with the moon and then the sun as the nearest celestial bodies, will not generate the transit of Mercury and Venus across the face of the sun, but such a phenomenon was unknown to the ancients.

42. See Cornford (1937) p. 135 on Proclus, Taylor (1928) p. 243.

43. See Dreyer (1906) p. 68 on this phenomenon.

44. The Babylonians were aware of the motion of the moon in latitude, plotting it against stars falling within the zodiac, see van der Waerden (1974b) pp. 98-9.

45. It should be noted that this is not a term that is later used for retrogression in the astronomical tradition.

46. See Simplicius *in De Caelo* 497, 18-22.

47. Neugebauer (1975) pp. 627-8.

48. However there is other evidence for 92, 89, 89, 95. See Neugebauer (1975) p. 627 note 9. The modern values are (days.hours) 93.15, 89.19, 92.19, 89.1 (Ricker (1967) p. 87).

49. See Simplicius *in De Caelo* 497, 18-22, cf. Simplicius' report of Eudemus 497, 17-22.

50. See Schiaperelli (1875), Neugebauer (1975) p. 625.

51. Simplicius *in De Caelo* 504, 17-20.

52. There is a problem with the text here, but the general meaning seems clear enough. See Taran (1975) ad loc. and Lamb (1927) ad loc.

53. The Loeb text omits these words, without comment, and for no good reason. They are in Burnet's text and are omitted in only one of the sources.

54. See Van der Waerden (1951), (1974a) and (1982), and see Gregory (2000).

55. Zeyl (2000) xliv-xlv.

56. The point of dividing the different is most likely to allow the seven planets to have different speeds, cf. Zeyl (2000) vlx note 89.

57. See e.g. Schiaperelli (1875), Heath (1913) pp. 210 ff., Dicks (1970), Mendell (1998), Yavetz (1998).

58. Taylor (1928) p. 202.
59. Cornford (1937) p. 107 note 2.
60. There is also a problem of the degree of sophistication we allow Plato with van der Waerden's theory. If we are to attribute epicycles to Plato, how far do we go with this? On what grounds do we accept or reject epicycles for Mercury and Venus, for retrograde motion generally, for deviation from the plane of the different, or for variations in planetary distance? Such considerations also apply if we attribute hippopedes to Plato.
61. See Taylor (1928) pp. 196-202.
62. Assuming we reject the attribution of either epicycles and hippopedes to Plato.
63. Philolaus probably accepted regular circular motion (see Geminus *Elementa Astronoma* 10.2-21), but did not have offset orbits (see above, section I). There is the matter of the central fire as well.
64. Vlastos (1975) p. 59.
65. See Gregory (2000).
66. See Neugebauer (1975) pp. 588 ff.
67. Burkert (1972) p. 324 considers Eudoxus to have influenced Plato, converting him to regular circular motion after the *Timaeus*, and Cavagnaro (1997) believes such an influence to have occurred before the *Timaeus*.
68. See Owen (1953) p. 325
69. See Skemp (1942) pp. 142-3.
70. Owen (1953), p. 325.
71. Apollodorus' practice of equating the high point of someone's achievement with the age of forty is of course highly dubious. Here he may have associated Eudoxus' acme with his temporary presidency of the academy in 367/366.
72. Gisinger (1921) pp. 5-6.
73. De Santillana (1949) pp. 248-62.
74. Lasserre (1966) pp. 137-42.
75. See Pliny, *Natural History* XXX, 3.
76. Owen (1953) p. 325.
77. For some further conjectures on Eudoxus and Plato, see Dicks (1970) p. 143 note 210.
78. See Owen (1953) pp. 325-6.
79. See Owen (1953) p. 326.
80. Dreyer (1906) p. 91.
81. Heath (1913) p. 196.
82. Wright (1974) p. 73.
83. There is no need for a mechanism here, if this is an instrumentalist solution.
84. Wright (1974) p. 167.
85. Wright (1974) p. 168.
86. *Epinomis* 991d is a good example here.
87. Cf. the comments on geometrical atomism at *Timaeus* 54ab.

6. Plato and the development of Greek cosmology

1. For debate here see Neugebauer (1957), (1975), Toulmin and Goodfield (1961), Vlastos (1975), Oppenheim (1978), Van der Waerden (1978), Lloyd (1979), (1987), Olsen (1982), Pingree (1992), Von Staden (1992).
2. Pedersen (1974) p. 26.
3. Burkert (1972) p. 335.

4. Aristotle *De Caelo* 294a28 ff. Cf. *Metaphysics* I.3.

5. See Popper (1963), cf. Kirk (1960), (1961) and Lloyd (1991) ch. 5.

6. Aristotle *De Caelo* 294b13 ff. While Furley (1987) p. 20 has called this the 'frisbee' theory of the stability of the earth, there are some important dissimilarities. A frisbee flies, but will not stay aloft when it is stationary. Moreover, when it flies the lift it requires is generated by the leading edge being slightly higher than the trailing edge, such that it is not parallel to the ground.

7. See Furley (1987) pp. 53 ff.

8. See Furley (1987) p. 24.

9. Though with a central earth and a like-to-like principle, Plato is practically centri-focal.

10. See Burkert (1972) pp. 299 ff.

11. Aristotle *De Caelo* 295b10 ff.

12. Simplicius *in De Caelo* 532, 13, Furley's translation and emphasis.

13. See Panchenko (1994), Schofield (1997), and Kahn (1999).

14. Hippolytus, *Refutatio* I, 6, 3, who says that for Anaximander 'the earth is aloft, not held in the power of anything but remaining where it is because of its like distance from all things'.

15. That the earth is a sphere would seem to be he likeliest reading of *peripheres* here, though Morrison (1959) raises doubts; cf. Calder (1958), Rosenmayer (1959), and Guthrie vol. 4 p. 336 note 1.

16. Plato uses *piptein* in this sense in other contexts, e.g. *Laches* 181b and *Philebus* 22e.

17. See Adam (1902) vol. 2 pp. 443-7 on the nature of *ta hupozomata*, and cf. *Laws* 945c.

18. See Morrison and Williams (1968) and Morrison and Coates (1986).

19. See e.g. *Timaeus* 36e.

20. For the main contributors to the debate see Heath (1913) pp. 174 ff., Cornford (1937) pp. 120 ff., Cherniss (1944) pp. 540 ff., Dicks (1970) pp. 132 ff., Burkert (1972) p. 326 note 16, Vlastos (1975) pp. 31 ff.

21. I side with Cornford (1937) p. 120 note 1 in removing the *ten* after *illomenen de* in Burnet's text.

22. See Burdach (1922) on the *il-* and *eil-* word groups (rolled around vs. packed around) which become confused in meaning and spelling. I do not see that either reading can support the up-and-down motion of Burnet and Taylor.

23. See Dicks (1970) p. 136.

24. See Burnet (1924) pp. 302-3 and Taylor (1928) pp. 226-40.

25. Taylor (1928) pp. 236-7.

26. See *Timaeus* 40c2-3.

27. All that is required here in fact is relative motion between the earth and the same. The simplest and most reasonable solution is that the same rotates once a day and the earth does not move at all, but the same could rotate twice and the earth once, etc. Certainly it is reasonable to assume that day is one revolution of the same, but note that month for Plato is not one revolution of the moon relative to itself, but relative to the sun, so perhaps day is one revolution of the same relative to the earth.

28. Aristotle *De Caelo* 293b30.

29. Plutarch, *Quaestiones Platonicae* VIII, 1, 1006c, *Numa* c. II.

30. See e.g. Heath (1913) pp. 183 ff. One would expect Aristotle to know of this and record it. The idea that the sun was the worthiest body and should be at the centre was also common in the Renaissance and the early part of the scientific revolution. Copernicus, Kepler and Harvey all gave the sun a central position on these grounds.

31. See Aristotle *De Caelo* 293a15 ff.

32. That is certainly the case prior to Galileo and Newton, when there were not the physics to deal with a moving earth.

33. See Cleanthes reported by Plutarch, *De Facie in Orbe Lunae* 922f-923a.

34. As indeed was the case up to the seventeenth century. In general, heliocentrism was not considered a serious possibility.

35. It is worth remembering that for both Plato and Aristotle the earth would not have been moving through a vacuum. The Greeks were also quite capable of realising that the earth would be moving at speeds unheard of for them.

36. The Copernican system was generally regarded as a useful computational tool but not a serious cosmological possibility. It took the work of Galileo both on the physics of motion and with the observational evidence of the telescope, along with the critical simplifying work of Kepler on ellipses to make the Copernican hypothesis a real challenge to geocentrism.

37. This despite the fact that the Greeks had very good observational astronomy. If they could discover the inequality of the seasons and the precession of the equinoxes, then being unable to detect parallax (especially in the relatively small Greek cosmos) was a serious matter.

38. This was also a problem for Galileo, as the telescopic evidence cut both ways. While there was evidence to support Copernicus, the failure to detect parallax meant that the cosmos had to be many times the previously supposed size.

39. Indeed the first step in establishing a distance ladder for beyond the solar system is being able to detect parallax for relatively near stars.

40. Even then a great deal of the telescopic evidence goes against Aristotle but does not specifically support Copernicus. All the telescopic evidence can be accommodated by the geoheliocentric thesis of Tycho Brahe.

41. Nor was it better observationally than the Ptolemaic system.

42. One can see Copernicus struggling to accommodate the idea of a moving earth within a revised Aristotelian physics.

43. Heavy objects would continue to move to the centre of the cosmos when in natural motion.

44. This is a further reason why Copernicanism was not accepted, and not even seen as a serious candidate until Galileo and Kepler.

45. *Apology* 26cd, *Laws* 886e, cf. Hippolytus *Refutatio* I, 8, 3-10.

46. Cf. Empedocles Fr. 31 (Hippolytus *Refutatio* VII, 29, 13) 'Of the shape of the cosmos, what is was like when ordered by Love, he speaks like this: "No twin branches spring from its back, there are no feet nor nimble knees, no parts of generation, but it was a Sphere and in all directions equal to itself." ' See too Xenophanes Fr. 15 & 16 for early criticism of anthropomorphism in religion.

47. Stobaeus 1.18.1, M.R. Wright's translation and brackets (1995) p. 62.

48. See e.g. Plutarch *On the Phenomenon of a Face in the Moon's Circle* 926d ff.

49. The six motions are up, down, left, right, backwards and forwards.

50. Cf. *Phaedrus* 245c ff. where self-motion is the essence and definition of soul and *Timaeus* 37b where it is that which is self moved.

51. Cf. *Epinomis* 982b on intelligence and regularity.

52. Both Plato and Aristotle would hold that these entities are fully alive, intelligent, etc. but would want to distinguish them from terrestrial living, intelligent entities.

53. Cf. Lloyd (1991) p. 149.

54. The *Laws* is also adamant that prayer or other offerings will be of no avail in trying to change the minds of the gods.

55. Cf. *Epinomis* 986e on observations having been proved true by the test of a great amount of time.

56. See e.g. Burtt (1932), Milton (1998). Leibniz comments that 'The general principles of physics and mechanics themselves depend upon the action of a sovereign intelligence and cannot be explained without taking it into consideration' (*Letter Explaining the Laws of Nature*, cf. *Discourse on Metaphysics* 19-20).

57. It was typical of the natural magic tradition that while they postulated unseen (and so occult, literally 'hidden') relations between things, these relations (forces, sympathies, correlations) behaved in a law-like manner, as opposed to the demonic and spiritual magic traditions.

58. For Aristotle too the heavens are divine, see e.g. *De Partibus Animalium* 644b22 ff.

59. See Lennox (1985).

60. While Plato's god is omnibenevolent he is not omnipotent and so Plato is not troubled by many of the paradoxes of omnipotence that trouble Christian theologians.

61. Shorey (1935) vol. 2 p. 127 note e calls this a 'Humian, Comtian, positivist, pragmatist view of causation'.

62. See below and *Laws* 720c, 938a.

63. Vlastos (1969) p. 106.

64. Vlastos (1969) p. 106.

65. I cannot see Plato subscribing to a great many of the assumptions which underpin Hume's analysis of causation.

66. Vlastos (1969) p. 106.

67. On the following issues, see Curd and Cover (1998) pp. 879 ff., Ayer (1956), Dretske (1977), Mellor (1980), Harré (2000a), (2000b).

68. If F-ness → G-ness, and a is F, why can we say that a is G as well?

69. See Dretske (1977) pp. 839 ff.

70. Vlastos (1969) p. 106, see Harré (2000b) and Newton-Smith (2000) on the current state of the debate.

71. As science continues to develop, one would not expect nor seek an exact match.

72. See Milton (1998). Plato may wish to challenge this line of thought though, as Socrates in the *Gorgias* believes Callicles to be incorrect in considering *nomos* and *phusis* to be opposed.

73. See *Phaedo* 103d.

74. See *Phaedo* 104a.

75. Both the relations between forms and the model of participation may well change during Plato's career. It is interesting to note in relation to the *Phaedo* that Dretske (1977) also refers to the relation between universals as 'brings with it'.

76. Note that fire (that is a collection of fire atoms) is not necessarily hot, a perceptual quality. It will be hot to humans, as it will divide up our bodies, but may not be so for the star gods whose bodies are mostly fire.

77. Vlastos (1969) p. 107.

78. See Cohen (1994) ch. 4 and Sambursky (1956).

79. Whether we consider these to be based on regularities or laws proper. See here on magic, scholasticism and the scientific revolution; Hutchison (1982), Webster (1982), Henry (1990), Nadler (1998), Copenhaver (1998).

80. Even if this process is not held to be fully reductive at each stage, it is held to be highly informative.

81. So for such influential thinkers as Descartes and Boyle. The later romantic reaction to the mechanical philosophy of the seventeenth century was enthusiastic about biological analogues, and the idea of the clash of opposites in nature, and opposed the view that nature was most beautiful in geometric and ordered forms with the idea that nature was most beautiful when primeval and untrammelled, storms and volcanoes being more beautiful than geometrically arranged gardens. See Hankins (1985) pp. 1-17.

82. See Hahm (1977).

83. See Furley (1987) p. 13.

84. It may be that it is not until one has metal machines that mechanism can become a good analogue for regularity and order.

85. Democritus Fr. 164, Sextus Empiricus *Against the Mathematicians* VII, 116-18, Barnes' translation and italics, Barnes (1987).

86. See Lloyd (1991) p. 148.

87. See Lloyd (1991) p. 149.

88. See Diogenes Laertius IX, 31, Aetius II, 7, 2, Simplicius *Physics* 327, 24.

89. Mechanical processes are in general reversible.

90. See e.g. *De Partibus Animalium* 644b20 ff.

91. On Anaxagoras see Hippolytus *Refutatio* I, 8, 3-10, Plato *Apology* 26de and *Laws* 886de, on the atomists see Diogenes Laertius IX, 31, Simplicius *Physics* 327, 24.

92. When Descartes revives the idea of vortices in the seventeenth century he has the advantage of doing so within heliocentrism. This mean that he needs only one vortex for the planets, created by the spin of the sun, and a vortex for each of the planets (again created by their spin, to explain terrestrial gravity and why planets such as earth and Jupiter carry moons with them). Note in respect to the discussion below that each vortex has a physical entity on which to centre itself.

93. I would agree in general with Taub (1993) that the Platonic influence on Ptolemy is significantly underestimated.

94. See Ptolemy *Planetary Hypotheses* I, 2, Goldstein p. 8.

95. See Ptolemy *Planetary Hypotheses* II, 7 et passim.

96. See Ptolemy *Planetary Hypotheses* II, 12.

97. See Ptolemy *Planetary Hypotheses* II, 7.

98. See Ptolemy *Planetary Hypotheses* II, 7.

99. I agree with Taub (1993) p. 123 and Pedersen (1974) p. 393 that Ptolemy's distinction between the terrestrial and the celestial is much less sharp than that of Aristotle.

100. See Vlastos (1975) p. 54 note 86.

101. See too the opening words of Copernicus *Commentariolus* (p. 57).

102. Copernicus *De Revolutionibus* I, 4.

103. Copernicus *Commentariolus* pp. 57-8.

104. Copernicus *De Revolutionibus*, Preface and Dedication to Pope Paul III.

105. Copernicus *De Revolutionibus*, Preface and Dedication to Pope Paul III.

106. Copernicus *De Revolutionibus* I, 1.

107. Copernicus' own system was on the whole no better than Ptolemy's, being in some respects better and in others worse. A comment from Galileo is interesting here: 'I cannot sufficiently admire the eminence of those men's intelligence who have received and held the Copernican system to be true, and with the sprightliness of their judgements have done such violence to their senses, that they have been able to prefer that which their reason dictated to them to that which sensible experiences represented most manifestly to the contrary.'

108. See e.g. Kuhn (1957), Koyre (1973), Ravetz (1990) and cf. Koestler (1959) for an alternative view.

109. Such a return to a strict interpretation of ancient thought is typical of Renaissance Neoplatonism.

110. See here, Kuhn (1957), Koyre (1973), Ravetz (1990), Field (1988).

7. Geometrical atomism – flux and language

1. Pyle (1997). Plato fares little better in other histories of atomism, see e.g. Gregory (1931).

2. Aristotle *Metaphysics* I.4 985b4 ff., KRS translation.

3. It is important to remember that such views are also expressed in other late works in non-mythological settings and without being framed with such strong TW assumptions.

4. Difficult, but note that whenever the world soul encounters anything sensible (*aistheton*, 37b6), opinions and beliefs which are stable and true occur (*doxai kai pisteis gignontai bebaioi kai alethes*, 37b8), something I say more about in Chapter 9. See Lee (1967) pp. 2-3 and Zeyl (1975) pp. 127-8 on whether Plato considers the task too difficult.

5. Interesting to compare Plato's use of an *ou mallon* argument here with that of the atomists. See Simplicius *Physics* 28, 8, and above Chapter 1.VII.

6. See e.g. Zeyl (1975) p. 127 ff.

7. I follow most commentators in removing *kai ten tode* at 49e3.

8. This formulation comes three times, at 49d4-5, 49d5-6 and 49e4. For Simplicius on the passage, see *in De Caelo* 637, 4-5.

9. So Gill (1987) p. 36, Silverman (1992) p. 89, Hunt (1998) p. 70.

10. Cf. Gill (1987) p. 36, Hunt (1998) p. 71.

11. Gill (1987) p. 36, and see Cherniss (1954) p. 128.

12. On this see Gill (1987) pp. 40 ff., cf. Silverman (1992) pp. 91 ff. and Algra (1994) 99-102. As will become clear, I too think there is something beyond the tripartite ontology of forms, receptacle and phenomena, but not until the imperceptible entities of geometrical atomism are introduced at 53c ff.

13. See Owen (1953) p. 322 ff., Reed (1972) p. 66 ff.

14. 'This argument defeats the lame plea of the *Tim.* that even if we cannot say what any mere *gignomenon* is we can describe it as *to toiouton*.' Owen (1953) p. 323 note 3.

15. See Cherniss (1954).

16. See e.g. Owen (1953) p. 322 ff., Bostock (1988) p. 147 ff.

17. There would seem to be change every instant for all the properties under consideration. A radical line here has been suggested to me by Verity Harte. Perhaps it is significant that Plato uses gold in this analogy, as opposed to say clay, for while one can easily imagine clay being instantly remodelled by hand and so being a reasonable analogue for radical flux, it is much harder to do so for gold, which would have to be worked, stamped or cast into a different shape.

18. See Guthrie (1962) vol. 5 pp. 263-4, Gill (1987) p. 50, Algra (1994) pp. 76 ff., Reed 1972) p. 78.

19. This may also be a useful way of approaching the comments on flux in the receptacle passages, which may not be consistent with each other and may present a variety of options up to a strong and radical flux. It may be that none of these are Plato's position, but that we are offered a range of alternatives for any proposed solution to the reference problem to be able to deal with.

20. Prior (1985) p. 113, cf. McCabe (1994) pp. 180 ff., Reed (1972) p. 75.

21. See Gill (1987) p. 46.

22. Contra Cornford (1937) pp. 184 and 200, I do not see the problem with translating *morphe* as shape here. The gold analogy begins specifically with triangles and other shapes being formed, and a verb in the following sentence, *diaschematizomenon* at 50c7, might well be though to echo *schemata* at 50b2. *Diaschematizomenon* does not have to mean shaped (Cornford (1937) p. 185 note 1) but in context that would seem likeliest with *morphe* as shape following on from the talk of shapes in the gold example. That *morphe* cannot mean shape in the *Phaedo*, if true, does not seem to me to be of any great relevance here.

23. Interesting to note here that the Greek piles up four negatives in the last part of the sentence and so is highly emphatic.

24. Cf. Aristotle, *Categories* V and *Metaphysics* V.8 and VII.3.

25. Cf. McCabe (1994) p. 187, Reed (1972) p. 75.

26. McCabe (1994) p. 187.

27. Gill (1987) p. 51.

28. Gill (1987) p. 51.

29. Note that *Theaetetus* 182c allows that we can speak of something if it is not undergoing both *phora* and *alloiosis*.

30. See Zeyl (1975) pp. 146-7.

31. See Gill (1987) p. 51.

32. *Kinesis* is often rendered as 'motion', but I prefer change, as *kinesis* is quite general, covering change in nature as well as change of position.

33. I take 'grows old' here to entail some sort of change and decay rather than mere temporal existence.

34. Reading *phoran* rather than *periphoran* at *Theaetetus* 181b6 with Burnet. Cf. *Parmenides* 138b on *phora* and *alloiosis*.

35. Let us for simplicity here exempt the account which describes the flux from change in itself, to avoid the problems that Plato ascribes to Heraclitus, and restrict our flux to the physical world. One can allow change into the account in a limited way, as long as it is ordered change, as then one can take as many derivatives with respect to time as are required to produce a temporally invariant description. The higher order account is then free from change as long as one is not bound by an 'everything, whatever it is, is in flux' clause.

36. See *Theaetetus* 182c ff.

37. See *Cratylus* 439b ff.

38. See Irwin (1977). On p. 4 he gives the following definition of a- and s-change: 'x s-changes iff at Time t_1 x is F and at Time t_2 x is not F, and x is not in the same condition at t_2 as it was at t_1 ... x a-changes iff x is F in one aspect, not F in another, and x is in the same condition when it is F and when it is not F.'

39. Cf. e.g. *Theaetetus* 202e ff., *Politicus* 277e ff., *Philebus* 18b ff., and cf. McCabe (1994) pp. 186-7.

40. See *Timaeus* 54d6, 55a8, 55b4, 57c9 and 61a7.

41. See *Timaeus* 54d ff. on how stoicheics form complex triangles.

42. See Friedlander (1958) pp. 246-60.

43. Cf. O'Brien (1984) pp. 341-58.

44. See Cornford (1937) pp. 230 ff. The problem with the Greek is whether *sustasin* at *Timaeus* 57c9 refers to the putting together of the stoicheic triangles in the sense of their internal construction or their external arrangement with other stoicheics into complexes. Cornford argues convincingly in favour of the latter.

45. See *De Caelo* 306a1-7 and *De Generatione et Corruptione* 332a27 ff. for Aristotle's criticisms of Plato's theory of transmutation.

46. See *Timaeus* 81b ff., and especially 89bc.

47. There is a second variable here in that bonds of the same durability may last different amounts of time depending on how much wear and tear they are subject to.

48. Earth, on this account, says Aristotle has the best right to be called an element because earth alone is indestructible (306a18-21).

49. See Cornford (1937) p. 329 note 2, Taylor (1928) pp. 585-6, cf. LSJ II.2 on *hriza*.

50. See Taylor (1928) p. 586, LSJ strongly supports Taylor on *chalao*.

51. *ek druochon* was a nautical metaphor for something newly made, like a ship 'fresh off the stocks'.

52. If someone asks what the timber and then the subsequent boat are, the safest answer I can give is wood; cf. the gold example, etc.

53. LSJ gives 'to place or set together, to combine, to organise, pass. to stand together', cf. Cornford (1937) pp. 230 ff who argues for this sense.

54. Cf. *Timaeus* 56e, 81b and 89c.

55. Archer-Hind (1888) notes to pp. 304-6 simply substitutes 'particles' for *trigona*.

56. This directly relates to *Timaeus* 81d; both passages discuss the marrow.

57. There is one difficulty with the interpretation that *trigona* at *Timaeus* 81b refers exclusively to the complex triangles which constitute atoms, which is that not all atoms are constituted out of complex triangles. This is the case for air, fire and water of course, but for the smallest 'isotope' of earth, the atomic faces are squares made of four stoicheic triangles. My interpretation requires that this is a slight oversight on Plato's part.

58. See for instance *Timaeus* 53d *duoin archetai trigonoin*, 54a *duoin trigonoin*, 55b and 57c *ton stoicheon*, 55e *ton kat' archas trigonon hupothenton*.

59. Cf. *Timaeus* 81c ff.

60. Zeyl (1975) p. 125.

61. See *Timaeus* 56bc.

62. Cf. McCabe (1993) p. 181.

63. See Silverman (1992) p. 87, cf. Schulz (1966) pp. 65-86 who considers that Plato reduces sensible qualities and ultimately has shapes and the receptacle.

64. As the *Theaetetus* is rather indeterminate on what it may be to give a *logos* at this point, I hedge between account and analysis for *logos* here.

65. Cf. e.g. *Theaetetus* 201e1, 202b7 and *Timaeus* 48bc, and *passim* references to the two basic triangles as stoicheia.

66. So contra McCabe (1994) pp. 186-7 Plato does not repudiate the letters and syllables analogy, nor does the *Timaeus* hold that analysis by decomposition is the exclusive approach to cosmology (p. 182).

67. I take it that the teleological account here would count, as G. Fine (1979b) p. 317 has put it, as logos$_K$, that is explanations that will help to produce knowledge rather than logos$_S$, mere sentences about elements which may or may not be true.

68. I would add the omission of teleology from the *Theaetetus* is as striking as the omission of forms cf. Cornford (1935) pp. 106-11, 158-63, Ross (1951) pp. 124-5.

69. Vlastos (1975) p. 68.

70. See Sayre (1983) pp. 113 and 154-5. So too one can characterise an instant of time as a Dedekind cut on a time line; cf. *Parmenides* 156d and Sayre p. 114.

71. Cf. Aristotle *De Anima* 404b23 and *Metaphysics* 1086a10 on forms and numbers.

72. See Gill (1987) pp. 51-2.

73. See Gill (1987) pp. 51-3.

74. The *Sophist* use of the letters and syllables analogy looks to the communion of forms, the *Theaetetus* and *Philebus* are (arguably!) more neutral and about world generally, the *Politicus* looks to the structure of politics and society.

75. See *Sophist* 253a5, *Philebus* 18c8

76. See especially *Timaeus* 81d4, 81d5.

77. See *desmos* at *Timaeus* 31c1, 36a9, 38e5, 41b5, 43a2, 73b3, 73d6, 77e3, 84a3 and cf. *Politicus* 309c.

78. Cf. Aristotle *Metaphysics* V.8 and VII.2-3.

79. Certainly a different ontology including non-perceptible physical entities, as I have argued.

80. See Chapter 1 for more on this passage.

8. Geometrical atomism – matter and space

1. See Algra (1994) p. 78, Sorabji (1988) pp. 32-5.

2. For problems with the matter theory, see Algra (1994) pp. 78-83.

3. Cf. Algra (1994) p. 86.

4. See the next section for occurrences of *chora* and *hedra* outside *Timaeus* 48e-53b.

5. For problems with the space theory, see Algra (1994) pp. 83-9. Algra does not discuss the *plokanon* metaphor, though it seems strongly spatial.

6. See Descartes *Principia Philosophia* II.4 ff.

7. On this see Algra (1992) pp. 81 ff. et passim, and cf. Guthrie (1962) vol. 5 p. 265.

8. See Algra (1994) p. 74 and Lee (1966) for an opposing view.

9. See O'Brien (1984) pp. 359-65.

10. See O'Brien (1984) p. 361.

11. See Cherniss (1944) pp. 154 ff.

12. See O'Brien (1984) p. 360.

13. Aristotle, *De Caelo* 306b3 ff.

14. Compare Descartes, who also has a packing problem, who allows all shapes and sizes of corpuscles, and so has a very fine matter in which the larger pieces are able to swim around.

15. So Taylor (1928) p. 399 and Cornford (1937) pp. 242-5.

16. See Taylor (1928) pp. 398-9, Cornford (1937) pp. 243-6, and see O'Brien (1984) pp. 361-2 for a critique of their positions. Cornford's translation of *Timaeus* 81b6 ff. plays down both the compressive effect of the body/cosmos and the possibility that chopping up blood particles will result in voids.

17. So O'Brien (1984) p. 362.

18. See here Aition (1972) on both ancients and moderns.

19. One might argue though that this cuts both ways, and the surrounding material metaphors are supposed to be explained by *en hoi*. There are possibly more spatial than material metaphors, but I doubt that a count will help here.

20. Cornford (1937) p. 187 note 1.

21. See Aristotle *Generation of Animals* 763b ff.

22. See Algra (1994) p. 94 note 49.

23. See *Theaetetus* 191c, 194de, 196ab, and cf. *Timaeus* 72c. *Keros* is also used in the *Theaetetus*, see e.g. 194c5 ff.

24. Cf. the use of *apotopousthai* 191d6 (like the impression from a seal ring).

25. Mohr (1985) p. 94.

26. Algra (1994) p. 85.

27. Algra (1994) p. 86.

28. See *Timaeus* 57c1, 58a7, 79d6, 82a3 and 83a4.
29. See *Timaeus* 59a3, 60c4, 62a, 67b, 72c, and 79b2.
30. See *Timaeus* 59d6 and 62b7.
31. See *Timaeus* 70c5, 74c1, 74d1 and 78d3.
32. It is perhaps important that geometrical atomism has its own analysis of these phenomena. If we want to explain ignification and liquefaction, geometrical atomism will tell us that we perceive fire and water when sufficient numbers of their respective atoms have accumulated together, and would also have a tale to tell about change in terms of the stoicheic triangles and their complexes which go to make up these atoms. So too gold, for geometrical atomism, gold can be analysed into elements, atoms and ultimately stoicheic triangles.
33. There is also a serious problem concerning the localisation of phenomena, as explained in the next section.
34. In some commentators this is explicit (e.g. Cornford (1935), (1937), Yolton (1949)), in many others it is implicit.
35. See Burnyeat (1991) and below.
36. See Taylor (1928) p. 430 ff.
37. See Cornford (1935) pp. 48-51 and (1937) p. 261 note 1, and cf. Nahnikian (1955) and O'Brien (1984) pp. 144-74. In my view it is significant that Cornford has an 'A' reading of the *Theaetetus* – see next chapter.
38. See McDowell (1973) pp. 138-40.
39. See G. Fine (1988a) pp. 20-1.
40. See e.g. *Theaetetus* 157ab.
41. See *Timaeus* 56bc on the perceptual status of the entities of geometrical atomism.
42. See *Theaetetus* 153e and cf. *Timaeus* 61c ff.
43. *Timaeus* 67c5-7, Fine's translation.
44. A. Fine (1988a) p. 21.
45. Cf. *Timaeus* 62b ff. on hard and soft.
46. Cf. O'Brien (1984), (1997).
47. See *Timaeus* 61e on the nature of fire.
48. Aristotle is critical of the role of geometrical atomism in Plato's conception of weight. At *De Caelo* 299b15 he gives us the principle that 'Weight cannot be constituted out of that which has no weight ... if it is weight that makes one weight heavier than another weight, then each indivisible part must have weight.' Aristotle argues that if the reason for a difference in weight is not this, but that earth is heavier than fire, then some surfaces will be light and some heavy. Aristotle's example would seem ill chosen, as Plato could accept that the surfaces that make up earth are heavier than those of fire (or water or air) as earth does not transmute and has different stoicheic triangles. Aristotle also needs to be more discriminating about what he is referring to by *epipedos*. Does he mean complex triangles, which might well differ from each other under the 'isotope' theory, or does he have the stoicheic triangles in mind? The basic principle he proceeds from seems suspect, as in general we cannot infer from a whole having a property that all the parts do as well, so some more argument in the case of weight is required.
49. Cf. Burnyeat (1991) p. 8, cf. *Theaetetus* 160de.
50. Assuming a 'B' reading here.
51. See *Theaetetus* 152a.
52. On this, and whether the *Theaetetus* theory is Plato's, see Day (1997).
53. Cf. *Theaetetus* 182c ff.
54. See *Timaeus* 43e ff., and next chapter.

55. We are likely to mix attributions of sameness and difference, or judge left to be right.

56. Aristotle, *Physics* 209b13 ff.

57. See Aristotle, *Physics* 209b23 ff., 211a36 ff., and cf. Sorabji (1988) p. 36.

58. Aristotle, *De Caelo* 305b29-31.

59. Aristotle, *De Caelo* 306b15 ff.

60. Aristotle, *De Generatione et Corruptione* 329a13-17.

61. Cf. E.N. Lee (1966) pp. 349 ff.

62. Aristotle *De Generatione et Corruptione* 329a14-15, 329a21-24.

63. It is worth noting that the evidence here too is from *De Caelo* and *De Generatione et Corruptione*, rather than the *Physics*.

64. See e.g. Silverman (1992) p. 97 note 26.

65. Cf. T.M. Robinson (1970) p. 93, Mohr (1985) pp. 92-3.

66. See *Timaeus* 45a ff., and 71a.

67. See Kung (1988), cf. Guthrie (1962) vol. 6 p. 265 note 3.

68. See Algra (1994) p. 92 note 43.

69. See Kung (1988) p. 167. Indeed, Plato also describes the relation as *aporotata*.

70. Cornford (1937) p. 209, Cornford's capital for space.

71. Cf. O'Brien (1984), (1997).

72. See e.g. Descartes *Principia Philosophia* II.10, 11, 16.

73. See e.g. Descartes *Principia Philosophia* II.20.

74. See Burtt (1932) pp. 96 ff.

75. See e.g. Descartes *Principia Philosophia* IV.20 ff. (gravity).

76. Generated by rubbing amber, here *electron* 81c2.

77. The lodestone, here *ton Heracleion lithon*, 81c2.

78. Crombie (1963) vol. 2 p. 222.

79. Though Descartes is careful to say that his particles are impenetrable.

80. See Archer-Hind (1888) pp. 244-5, Taylor (1928) pp. 470-2, Cornford (1937) pp. 271-5.

81. Vlastos (1967).

82. Taylor (1928) p. 370.

83. Cornford (1937) pp. 231-9.

84. Artman and Schafer (1993) p. 258.

85. See Artman and Schafer (1993) pp. 256-61.

86. See *Meno* 82b ff.

87. Diagrams here following Artman and Schafer (1993).

88. See Artman and Schafer (1993) p. 260.

89. Vlastos (1975) pp. 68-9.

90. Vlastos (1975) pp. 71-2. Vlastos does not suggest this. Cf. M.R. Wright (1995) pp. 102-3.

91. Vlastos (1975) pp. 85 ff.

92. Though Aristotle is critical that Plato gives no proper account here, see *De Caelo* III.

93. Boyle, *A Disquisition about the Final Causes of Natural Things*, *Works* (1688) vol. 5 p. 409.

94. Newton, *Optics* IV.260.

95. Descartes, *Principia Philosophia* IV.133 ff.

96. See Popper (1952) pp. 75 ff., and e.g. Aristotle *Metaphysics* 985b31 ff.

97. See e.g. Aristotle *Metaphysics* 987b11-13.

98. Plato is well aware of the irrationality of several numbers; see *Theaetetus*

147c ff., where the square roots of three, five, and seventeen are mentioned, along with a general term ('oblong numbers') for all numbers with irrational roots.

99. Cf. also *Epinomis* 990d ff. on numbers and modelling.

100. See Popper (1945) p. 248 note 9 and (1952) p. 87ff., also M.R. Wright (1995) p. 54.

101. See Popper (1952) p. 88.

102. Proclus *Euclid Commentary* p. 71, Popper's translation.

103. Popper (1952) p. 88.

9. Epistemology in the *Timaeus* and *Philebus*

1. Cf. *Theaetetus* 189e ff., *Sophist* 263e ff. and *Philebus* 38a ff. on the relation of judging as silent statement and statement as spoken judgement.

2. There is a question here as to whether this 'pair' refers to intelligence and knowledge or to (opinions and beliefs) and (intelligence and knowledge); see Cornford (1937) p. 96 note 1.

3. Cf. *Timaeus* 35a ff.

4. See *Timaeus* 42a ff. As well as desire, fear and anger, mortals are subject to 'all such emotions as follow with them, and all that are of a contrary nature'.

5. Cf. *Timaeus* 34a, 47c, 90cd.

6. See *Timaeus* 43b.

7. See *Timaeus* 36e.

8. Cf. *Sophist* 266c ff.

9. But woe betide the indolent! *Timaeus* continues at 44d 'Should he be neglectful, he lives a lame existence, he returns to Hades unperfected and lacking in intelligence.' This translation of *holokleros hugies te pantelos* as 'perfect and entirely sound in mind' is justified by comparison with the state of the negligent who have to return to Hades who are left *ateles kai anoetes*, 'unperfected and lacking in intelligence'.

10. Cf. *Theaetetus* 176bc and Sedley (1997) here.

11. It might be argued that *Sophist* 255c ff. explicitly distinguishes between one and two place predicates, although I tend to the view that the distinction there is between the 'is' of identity and the 'is' of predication; cf. Owen (1971) and Reeve (1985) pp. 54-5, Cornford (1935) pp. 282-5, Malcolm (1967) p. 143 note 24. Cf. Bostock (1986) pp. 49-51 and 181-3 on properties and relations in the *Phaedo*.

12. For complete uses, see *Timaeus* 35a and passim, for incomplete see e.g. *Timaeus* 37a and 44a.

13. Cf. Frede (1996).

14. See *Timaeus* 37a ff., 44a ff.

15. Plato treats thought and judgement as if they have the logical characteristics of speech and assertion; cf. *Theaetetus* 189a ff., *Sophist* 263e ff., Burnyeat (1991) pp. 84-5.

16. See e.g. *Philebus* 38c5 ff., 42a ff. and *Sophist* 234b and the discussion of what we can see at a distance. Note though that Plato's concern there is to develop accounts of what we can visualise.

17. I side with Cooper (1970) p. 123 ff. against Cornford (1935) pp. 102-9 that the main contrast at *Theaetetus* 184-6 is between perception and judgement rather than objects of opinion (sensibles) and objects of knowledge (forms).

18. See *Theaetetus* 185a, 185c.

19. See *Theaetetus* 185b, 185c.

20. See *Theaetetus* 186a.

21. Cf. *Epinomis* 978b ff., which if anything is even clearer and more forthright on this matter.

22. See too Aristotle *De Anima* 404b16, cf. 406b25 ff., and cf. *Sophist* 254a ff.

23. The theme of *bebaiotes*, 'fixity' or 'security' plays an important part in later Plato. Cf. *Philebus* 11e, 14c, 15b and 62b, *Sophist* 250c, the discussion of the *Cratylus* concerning the fixity of names (and the digression of *Epistle VII*), and MacKenzie (1986a) pp. 141-2.

24. Cf. *Theaetetus* 186c, where Socrates states that: 'Knowledge is not then in the sensations, but in the process of reasoning (*sullogismoi*) concerning them; for it is possible to apprehend being and truth in this manner, but impossible otherwise.'

25. There is of course an important parallel and contrast here with *Republic* VII; in both Plato first treats eyesight and then moves to draw conclusions about hearing. In the *Timaeus* he continues 'Again, the same account applies to both sound and hearing, as they were given by the gods with the same intent and purpose' (*Timaeus* 47c). Cf. especially *Republic* 530d, 'As our eyes are framed for astronomy, so are our ears for harmony, and these two are in some way kindred disciplines'.

26. At 37bc where the world soul is producing stable and true opinion, and knowledge and reason, the same's motion is described as *eutrochos*, and that of the different is *orthos*.

27. See Frede (1988), cf. Zeyl (2000) xxx ff.

28. Frede's translation (1988) p. 41, cf. *Republic* 507b.

29. See *Phaedo* 96d, and the solution at 100e ff. It is sometimes suggested that the myth of the *Phaedo* is a 'likely story' in the manner of the *Timaeus*. However, while Socrates in the *Phaedo* does not claim certainty for his tale, there is no suggestion that all that can be achieved is a likely story, merely that this is an hypothesis that would take an inordinate amount of time to prove; see *Phaedo* 108d, 114d.

30. See Aristotle *De Caelo* 306a9 ff.

31. Cf. especially Parmenides Fr. 1/28 ff. and *Timaeus* 29b.

32. Parmenides Fr. 8/60, 'I tell you the entire arrangement as seems likely, so that no opinion of mortal men will ever surpass you'.

33. See Parmenides Fr. 8/50 and cf. *Timaeus* 29c ff.

34. Cf. Parmenides Fr. 8/60 and *Timaeus* 29c, and passim comments in the *Timaeus* that this is the most likely account.

35. See *Theaetetus* 181d ff.

36. See Aristotle *Metaphysics* 1078b16 ff. on Plato's early Heraclitean influences.

37. It is also worth noting that Socrates allows a long speech rather than question and answer, and is well dressed, again contrary to his usual custom, cf. Schoos (1999) p. 103.

38. Cf. McCabe (1994) p. 185.

39. Sayre (1983) ix.

40. This is not to say that Plato does not have views, either positive or negative, that he wishes to get across. See Burnyeat (1980) pp. 186 ff., cf. Annas (1981), Burnyeat (1981) pp. 97-139, MacKenzie (1986b) p. 22 ff., McCabe (1993).

41. To use Burnyeat's example, it is one thing to know that root 3 is irrational, but quite another to understand why that should be so.

42. Cf. *Phaedrus* 275e.

43. If we are to believe Dionysius of Halicarnassus in his *On Literary Composition*, Plato painstakingly wrote and re-wrote each of his works.

44. Paradigm cases here are the *Laches*, with physical and intellectual conceptions of courage, and the *Cratylus* with its conventionalist and naturalist accounts of naming. Cf. *Sophist* 251a, where we must fend off both arguments at once, and force a passage between them.

45. See here MacKenzie (1982a).

46. See my comments in Chapter 3 on complex irony.

47. One implication of this is that we cannot attribute the book paradox in its fully-fledged form to Plato. It is employed as a tool to provoke the reader to investigate the text at a different level.

48. Note also the resonances between the fertility metaphors between *Theaetetus* at 149b ff. and 210b, and those at *Phaedrus* 276b ff.

49. So I would disagree with Arieti (1991) who sees the *Timaeus* as a playful parody of scientific thinkers.

50. See *Phaedrus* 229d ff. Note also that there are no criteria here for judging the matter one way or the other.

51. Typho was a monster with a hundred heads, so I translate *haplousteron* as single-minded.

52. See *Phaedrus* 229bc; the questions are whether the myth is to be believed, and the exact spot where the abduction is supposed to have happened.

53. See *Phaedrus* 229c. Was Boreas in fact the north wind which blew Orethuia to her doom?

54. Cf. Stewart (1960) pp. 98 ff. and T.M. Robinson (1970) p. 60 on myths 'which *engage* a man'.

55. Cf. Osborne (1988), (1996).

56. Cf. Lloyd (1991) pp. 342-3.

57. Cf. *Timaeus* 21d and 26d.

58. See especially *Phaedrus* 275bc, cf. Vlastos (1939) p. 380.

59. Cf. *Phaedrus* 274a ff.

60. See *Timaeus* 27a.

61. Here I would concur with Graham (1990) p. 21, who comments that 'If we think of the *Timaeus* as setting out the general conditions for a teleological programme of explanation in nature, there is no particular reason why Plato should have to interpret further the model for us. Here, he might say, are the tools for a teleological natural science. I present them in mythical form as a kind of archetype for you. You must decide on how to interpret the model and the categories of the model.'

62. Cf. Burnyeat (1982) pp. 4-8 and especially the notes on pp. 6-7.

63. Burnyeat (1991) p. 8.

64. Burnyeat (1991) p. 8.

65. Burnyeat (1991) p. 10.

66. See Cornford (1937).

67. Note that while I believe we can have similar 'A' readings of the *Theaetetus* and *Timaeus*, the alternative readings need not be similar.

68. See *Theaetetus* 210bc.

69. Though the *Timaeus* does not offer a definition of knowledge.

70. Both *Theaetetus* and *Timaeus* would appear to allude to the *Phaedo* at many points and provide revised views.

71. Cf. Gosling (1975) pp. 153-4, 170-80, 196-205 for a discussion of *techne* in the *Philebus*.

72. Cf. my earlier discussions of *techne*, *empeiria* and *tribe*.

73. See e.g. *Gorgias* 465a, 500e.

74. Xenophon *Oeconomicus* IV, 2.

75. Cf. *Philebus* 56bc.

76. According to Aristotle, Plato thought that 'Concrete things (*ta pragmata*) are apprehended in some cases by mind, in others by knowledge, by opinion, or by sensation, and the numbers are the forms (*eide*) of these things (*pragmata*)' (*De Anima* 404b26), though cf. *Metaphysics* 987a32, 1078b11.

Bibliography

Adam, J. (1902) *The Republic of Plato*, 2 vols., Cambridge U.P.
Aition, E.J. (1972) *The Vortex Theory of Planetary Motions*. Baird.
Alexander, H.G. (1956) *The Leibniz-Clarke Correspondence*. Manchester U.P.
Algra, K. (1994) *Concepts of Space in Greek Thought*. Brill, Leiden.
Annas, J. (1981) *An Introduction to Plato's Republic*. Oxford U.P.
——— (1982a) Aristotle on Inefficient Causes. *Philosophical Quarterly* 32, pp. 311-26.
——— (1982b) Knowledge and Language: The Theaetetus and the Cratylus. In M. Schofield and M.C. Nussbaum (eds.), *Language and Logos: Studies in Ancient Greek Philosophy Presented to G.E.L. Owen*, Cambridge U.P., pp. 95-114.
Anton, J.P. (1981) *Science and the Sciences in Plato*. Buffalo N.Y.
Archer-Hind, R.D. (1888) *The Timaeus of Plato*. Macmillan and Co, London.
Arieti, J.A. (1991) *Interpreting Plato*. Rowman and Littlefield.
Ariew, R. (1984) The Duhem Thesis, *British Journal for the Philosophy of Science* 35, pp. 313-25.
Artman, B. and Schafer, L. (1993) On Plato's 'Fairest Triangles' (Timaeus 54a). *Historia Mathematica* 20, pp. 255-64.
Ayer, A.J. (1956) What is a Law of Nature? *Revue Internationale de Philosophie* 67, pp. 144-65
——— (1963) *What is a Law of Nature?* In Curd and Cover (1998).
Ayers, M. and Garber, D. (1998) *The Cambridge History of Seventeenth-Century Philosophy*. Cambridge U.P.
Baeumker, C. (1890) *Das Problem der Materie in der Griechieschen Philosophie*. Munster.
Ballew, L. (1974) Straight and Circular in Parmenides and the 'Timaeus'. *Phronesis* 19, pp. 189-209.
Baltussen, H. (1993) *Theophrastus on Theories of Perception: Argument and Purpose in De Sensibus*. Utrecht U.P.
Barker, A. (1978) Sumphonoi Arithmoi: A Note on Republic 531c1-4. *Classical Philology* 73, pp. 337-42.
——— (1994) Ptolemy's Pythagoreans, Archytas, and Plato's Conception of Mathematics. *Phronesis* 39, 1994, pp. 113-35.
Barnes, J. (1979) *The Presocratic Philosophers*. 2nd ed., 1982, Routledge and Kegan Paul, London and N.Y.
——— (1987) *Early Greek Philosophy*. Penguin.
Barrow, J.D. (1991) *Theories of Everything*. Vintage.
Barrow, J.D. and Tipler F.J. (1986) *The Anthropic Cosmological Principle*. Oxford U.P.
Bergstrom, L. (2000) *Quine*. In Newton-Smith (2000).

Bedu-Addo, J.T. (1979) The Role of the Hypothetical Method in the Phaedo. *Phronesis* 24, pp. 111-32.
Bobrovnikov, N.T. (1984) *Astronomy Before the Telescope I*. Pachart, Tucson.
——— (1990) *Astronomy Before the Telescope II*. Pachart, Tucson.
Bodemar, C.W. (1964), Regeneration and the Decline of Preformationism in Eighteenth-Century Embryology. *Bulletin of the History of Medicine* 38, pp. 20-31.
Bodnar, I.M. (1997) Alexander of Aphrodisias on Celestial Motions. *Phronesis* 42, pp. 190-205.
Bolton, R. (1975) Plato's Distinction between Being and Becoming. *Review of Metaphysics* 29, 1975/6, pp. 66-95.
Bonjour, L. (1986) *The Structure of Empirical Knowledge*. Harvard U.P.
Borelli, G.A. (1680) *De Motu Animalium*.
Bostock, D. (1978) Plato on Time and Change in the Parmenides. *Phronesis* 23, 1978, pp. 229-41.
——— (1986) *Plato's Phaedo*. Oxford U.P.
——— (1988) *Plato's Theaetetus*. Oxford U.P.
Bowen, A.C. (1991) *Science and Philosophy in Classical Greece*. Garland, N.Y.
Boyd, R. (1973) Realism, Underdetermination, and a Causal Theory of Evidence. *Nous* 7, pp. 1-12.
Boyle, R. (1688) *The Works of the Honourable Robert Boyle*, ed. T. Birch, London.
Brandwood, L. (1990) *The Chronology of Plato's Dialogues*. Cambridge U.P.
Brisson, L. (1987) *Platon, Lettres*. Paris.
——— (1992) *Timée: Critias*. Flammarion, Paris.
——— (1994) *Le Même et l'autre dans la structure ontologique du Timée du Platon*. Academia Verlag, Sankt Augustin.
——— (1995) Interpretation du mythe du Politique. In Rowe (1995a), pp. 349-63.
——— (1997) Perception sensible et raison dans la Timée. In Calvo and Brisson (1997).
——— (1999) *Plato the Myth Maker*. Chicago U.P.
Brisson, L. and Meyerstein, F.W. (1995) *Inventing the Universe: Plato's Timaeus, the Big Bang, and the Problem of Scientific Knowledge*. SUNY.
Brocker, W. (1963) Der Philosophische Exkurs in Platons Siebtem Brief. *Hermes* 91, pp. 416-25.
——— (1965) Nachtrag zum Philosophischen Exkurs in Platons Siebentum Brief. *Hermes* 93, pp. 132.
Brown T.M. (1974) From Mechanism to Vitalism in Eighteenth-Century English Physiology. *Journal of the History of Medicine* 7, pp. 179-216.
Bruins, E.M. (1951) La Chimée du Timée. *Revue de Métaphysique et Morale* 56, pp. 269-82.
Brumbaugh, R.S. (1961) Plato and the History of Science. *Studiem Generale* 14/9, pp. 520-7.
Bulmer-Thomas, I. (1984) Plato's Astronomy. *Classical Quarterly* 34, pp. 107-12.
Burdach, K. (1922) Die Lehre des Platonischen Timaios (40b) von der kosmischen Stellung der Erde. *Neue Jahrbuch für der klassischen Altertum* 49, pp. 254-78.
Burge, E.L. (1971) The Ideas as Aitiai in the Phaedo. *Phronesis* 16, pp. 1-13.
Burkert, W. (1972) *Lore and Science in Ancient Pythagoreanism*. Harvard U.P.
Burnet, J. (1911) *Plato's Phaedo*. Oxford U.P.
——— (1924) *Greek Philosophy: Thales to Plato*. London.
Burnyeat, M.F. (1976) Plato on the Grammar of Perceiving. *Classical Quarterly*, NS 26, pp. 29-51.
——— (1980) Socrates and the Jury: Paradoxes in Plato's Distinction between

Knowledge and True Belief. *Proceedings of the Aristotelian Society*, Supp. Vol. 54, pp. 173-91.
——— (1981) Aristotle on Understanding Knowledge. In E. Berti (ed.), *Aristotle on Science: The Posterior Analytics*, Padua U.P.
——— (1982) Idealism and Greek Philosophy: What Descartes Saw and Berkeley Missed. *Philosophical Review* 91, pp. 3-40.
——— (1987) Platonism and Mathematics: A Prelude to Discussion. In A. Graeser (ed.), *Mathematics and Metaphysics in Aristotle*, Bern, pp. 213-40.
——— (1991) *The Theaetetus of Plato*. Hackett, Indianapolis/ Cambridge.
Burtt, E.A. (1932) *The Metaphysical Foundations of Modern Physical Science*. Kegan Paul, Trench and Trubner, London.
Bury, R.G. (1897) *The Philebus of Plato*. Cambridge U.P.
Bynum, W.F., Browne E.J. and Porter, R. (1981) *The Dictionary of the History of Science*. Macmillan, Hong Kong.
Calder, W.M. (1958) The Spherical Earth in Plato's Phaedo. *Phronesis* 3, pp. 121-5.
Callahan, J.F. (1948) *Four Views of Time in Ancient Philosophy*, Harvard U.P.
Calvert, B. (1970) Forms and Flux in Plato's Cratylus. *Phronesis* 15, pp. 26-47.
Calvo, C. and Brisson, L. (1997) Interpreting the Timaeus-Critias. *Proceedings of the IV Symposium Platonicum*. Academia Verlag, Sankt Augustin.
Carone, G.R. (1997) *The Ethical Function of Astronomy in Plato's Timaeus*. In Calvo and Brisson (1997).
Cavagnaro, E. (1997) *The Timaeus of Plato and the Erratic Motion of the Planets*. In Calvo and Brisson (1997), pp. 351-62.
Cherniss, H.F. (1936) The Philosophical Economy of the Theory of Ideas, in R.E. Allen (ed.), *Studies in Plato's Metaphysics*, Routledge and Kegan Paul, London and N.Y. 1965, pp. 1-12.
——— (1944) *Aristotle's Criticism of Plato and the Academy*, Baltimore U.P.
——— (1945) *The Riddle of the Early Academy*. California U.P.
——— (1954) A Much Misread Passage in the Timaeus (Timaeus 49c7-50b5). *American Journal of Philology* 75, pp. 113-30.
——— (1957a) The Relation of the Timaeus to Plato's Later Dialogues. In R.E. Allen (ed.), *Studies in Plato's Metaphysics*, Routledge and Kegan Paul, London and N.Y., 1965, pp. 339-78.
——— (1957b) Timaeus 38a8-b5. *Journal of Hellenic Studies* 77, pp. 18-23.
——— (1975) The Sources of Evil According to Plato. *American Philosophical Quarterly* 12, pp. 244-58.
Cherry, R.S. (1967) Timaeus 49c-50b. *Apeiron* 2, pp. 1-11.
Claghorn, G.S. (1954) *Aristotle's Criticism of Plato's Timaeus*. Nijhoff.
Cleary, J.J. (1987) Science, Universals and Reality. *Ancient Philosophy* 7, 1987, pp. 95-130.
——— (1997) *Plato's Teleological Atomism*. In Calvo and Brisson (1997).
Clegg, J.S. (1976) Plato's Vision of Chaos. *Classical Quarterly* NS 26, pp. 52-61.
Code, A. (1988) Reply to Michael Frede's 'Being and Becoming in Plato'. *Oxford Studies in Ancient Philosophy* 6, pp. 53-60.
Cohen, H.F. (1994) *The Scientific Revolution*. Chicago U.P.
Cohen, S.M. and Keyt, D. (1992) Analysing Plato's Arguments: Plato and Platonism. *Oxford Studies in Ancient Philosophy*, Supp. Vol. 1992, *Methods of Interpreting Plato and his Dialogues*, pp. 173-200.
Cooper, J.M. (1970) Plato on Sense Perception and Knowledge: Theaetetus 184 to 186. *Phronesis* 15, pp. 123-46.
——— (1977) The Psychology of Justice in Plato. *American Philosophical Quarterly* 14, pp. 151-7.

Copenhaver, B. (1998) *The Occultist Tradition and its Critics*. In Ayers and Garber (1998).
Cornford, F.M. (1932) Mathematics and Dialectic in the Republic VI-VII. *Mind* 41, (I) pp. 37-52, (II) pp. 173-90.
——— (1935) *Plato's Theory of Knowledge*. Routledge and Kegan Paul, London.
——— (1937) *Plato's Cosmology*. Routledge and Kegan Paul, London.
——— (1939) *Plato and Parmenides*. Routledge and Kegan Paul, London and N.Y.
——— (1941) *The Republic of Plato*. Clarendon, Oxford.
——— (1950) *Before and After Socrates*. Cambridge U.P.
——— (1952) *Principium Sapientiae*. Cambridge U.P.
Cottingham, J., Stoothoff, R. and Murdoch, D. (1985) *The Philosophical Writings of Descartes*, 3 vols., Cambridge U.P.
Cresswell, M.J. (1971) Plato's Theory of Causality: Phaedo 95-106. *Australian Journal of Philosophy* 49, pp. 244-9.
Crombie, I.M. (1963) *An Examination of Plato's Doctrines*, 2 vols., Routledge and Kegan Paul, London.
Curd, M. and Cover, J.A. (1998) *Philosophy of Science: The Central Issues*, Norton, N.Y.
Cushing, J.T. (1998) *Philosophical Concepts in Physics*. Cambridge U.P.
Dampier-Whetham, W.C. (1948) *A History of Science*. Cambridge U.P.
Davies, P.C.W. (1982) *The Accidental Universe*, Cambridge U.P.
——— (1992) *The Mind of God*. Penguin.
Davis, M. (1967) The Statesman as a Political Dialogue. *American Journal of Philology* 88, pp. 319-31.
Davis, P.J. (1979) The Fourfold Classification in Plato's Philebus. *Apeiron* 13, pp. 124-34.
Day, J.M. (1997) The Theory of Perception in Plato's Theaetetus 152-183. *Oxford Studies in Ancient Philosophy* 15, pp. 51-80.
Day, T. and Kincaid, H. (1994) Putting Inference to the Best Explanation in its Place. *Synthèse* 98, pp. 271-95.
De Santillana, G. (1949) Eudoxus and Plato, *Isis* 32, pp. 248-62.
——— (1961) *The Origins of Scientific Thought*, vol. I, *From Anaximander to Proclus – 600 BC to 300 AD*. Weidenfeld and Nicolson, London.
De Vogel, C.J. (1955) The Present State of the Socratic Problem. *Phronesis* 1, pp. 26-35.
Dicks, D.R. (1970) *Early Greek Astronomy to Aristotle*. Thames and Hudson, Bristol.
Diels, H. (1922) *Die Fragmente der Vorsokratiker*. Weidman, Berlin.
Dillon, J. (1995) The Neoplatonic Exegesis of the Statesman Myth. In Rowe (1995a), pp. 364-74.
Dombrowski, D.A. (1981) Atlantis and Plato's Philosophy. *Apeiron* 15, pp. 47-60.
Dretske, F.I. (1977) *Laws of Nature*. In Curd and Cover (1998).
Dreyer, J.L.E. (1906) *A History of Astronomy*. 2nd ed., Dover 1953.
Driscoll, J. (1979) The Platonic Ancestry of Primary Substance. *Phronesis* 24, pp. 253-69.
Duhem, P. (1906a) *Le Système du monde: histoire des doctrines cosmologiques de Platon à Copernic*, vol. I. Herman, Paris 1954.
——— (1954) *The Aim and Structure of Physical Theory*. Originally *La Théorie physique, son objet et sa structure*, translated by P. Wiener, from the second edition (Rivière, Paris 1914), Princeton.
——— (1969) *To Save the Phenomena*. Originally 'Sozein ta Phainomena; Essai sur

la Notion de Théorie Physique de Platon à Galilée' (*Annales de Philosophie Chretienne*, 1908), trans. E. Doland and C. Maschler, Chicago.
Edwards, P. (1972) *The Encyclopaedia of Philosophy*, 8 vols., Macmillan, N.Y. 1972.
Einstein, A. (1973) *Ideas and Opinions*. Souvenir Press.
Erler, M. (1995) *Kommentar zu Brisson und Dillon*. In Rowe (1995a), pp. 375-80.
Evans, J. (1998) *The History and Practice of Ancient Astronomy*. Oxford U.P.
Everson, S. (1990) *Epistemology (Companions to Ancient Thought 1)*. Cambridge U.P.
——— (1991) *Psychology (Companions to Ancient Thought 2)*. Cambridge U.P.
——— (1994) *Language (Companions to Ancient Thought 3)*. Cambridge U.P.
——— (1998) *Ethics (Companions to Ancient Thought 4)*. Cambridge U.P.
Fahrnkopf, R. (1977) Forms in the Philebus. *Journal of the History of Philosophy* 15, pp. 202-7.
Farrington, B. (1936) *Science in Antiquity*. Oxford U.P.
Ferguson, J. (1963) Sun, Line and Cave Again. *Classical Quarterly* NS 13, pp. 188-93.
Ferrari, G.R.F. (1995) Myth and Conservatism in Plato's *Statesman*. In Rowe (1995a), pp. 389-98.
Field, G.C. (1933) Plato and Natural Science. *Philosophy* 8, pp. 131-41.
Field, J.V. (1988) *Kepler's Geometrical Cosmology*. Athlone Press, London.
Findlay, J.N. (1974) *Plato: The Written and Unwritten Doctrines*. Humanities Press, N.Y.
Fine, A. (1984) The Natural Ontological Attitude. In Curd and Cover (1998).
——— (1996) *The Shaky Game: Einstein, Realism and the Quantum Theory*. Chicago U.P.
Fine, G. (1977) Plato on Naming. *Philosophical Quarterly* 27, 1977, pp. 289-301.
——— (1978) Knowledge and Belief in *Republic* V. *Archiv für Geschichte der Philosophie* 60, pp. 121-39.
——— (1979a) False Belief in the Theaetetus. *Phronesis* 24, pp. 70-80.
——— (1979b) Knowledge and Logos in the Theaetetus, *Philosophical Review* 88, pp. 366-97.
——— (1983) Plato and Aristotle on Form and Substance. *Proceedings of the Cambridge Philological Society* 209, pp. 23-47.
——— (1984) Separation. *Oxford Studies in Ancient Philosophy* 2, pp. 31-87.
——— (1986) Immanence. *Oxford Studies in Ancient Philosophy 4*, pp. 71-98.
——— (1987) Forms as Causes: Plato and Aristotle. In A. Graeser (ed.), *Mathematics and Metaphysics in Aristotle*, Bern pp. 69-112.
——— (1988a) Plato on Perception: A Reply to Professor Turnbull, 'Becoming and Intelligibility in Plato'. *Oxford Studies in Ancient Philosophy* 6, pp. 15-28.
——— (1988b) Owen's Progress. *Philosophical Review* 97, pp. 373-99.
——— (1990) Knowledge and Belief in Republic V-VII. *Companions To Ancient Thought 1, Epistemology*, ed. S. Everson, Cambridge U.P. pp. 85-115.
——— (1992) *Inquiry in the Meno*. In Kraut, R. (1992), pp. 200-26.
Frank, E. (1923) *Plato und die Sogennaten Pythagoreer*. Halle an der Salle.
Frankel, H. (1942) Zeno of Elea's Attacks on Plurality. *American Journal of Philology* 63, pp. 1-25 and 193-206.
Frede, D. (1989) The Soul's Silent Dialogue: A Non-Aporetic Reading of the Theaetetus. *Proceedings of the Cambridge Philological Society* 215, pp. 20-49.
——— (1996) The Philosophical Economy of Plato's Psychology: Rationality and Common Concepts in the *Timaeus*, pp. 29-58 in Frede and Striker (1996).
Frede, M. (1967) *Prädikation und Existenzaussage*. Vandenhoeck and Ruprecht, Gottingen.

——— (1980) The Original Notion of Cause. In J. Barnes et al. (ed.) *Doubt and Dogmatism*. Oxford U.P.
——— (1988) Being and Becoming in Plato. *Oxford Studies in Ancient Philosophy* 6, pp. 37-52.
Frede, M. and Striker, G. (1996) *Rationality in Greek Thought*. Oxford U.P.
Friedlander, P. (1958) *An Introduction to Plato*. Routledge and Kegan Paul, London and N.Y., ch. 14 pp. 246-60.
——— (1969) *Plato, The Dialogues*, 2 vols, Princeton U.P.
Furley, D.J. (1967) *Two Studies in the Greek Atomists*. Princeton U.P.
——— (1981) The Greek Theory of the Infinite Universe. *Journal of the History of Ideas* 42, pp. 571-86.
——— (1986) The Cosmological Crisis in Classical Antiquity. *Proceedings of the Boston Area Colloquium on Ancient Philosophy* 2, pp. 1-19.
——— (1989) *Cosmic Problems*. Cambridge U.P.
——— (1987) *The Greek Cosmologists*. Cambridge U.P.
Gaiser, K. (1963) *Platons Ungeschriben Lehre*, 2nd ed., Stuttgart U.P, 1968.
——— (1980) Plato's Enigmatic Lecture 'On the Good'. *Phronesis* 25, pp. 5-37.
——— (1986) Platons Zusammenschau der Mathematischen Wissenschaften, *A & A* 32, pp. 89-124.
Gallop, D. (1963) Plato and the Alphabet, *Philosophical Review* 72, 364-76.
——— (1975) *Plato's Phaedo*. Oxford U.P.
Gentzler, J. (1991) Sumphonein in Plato's Phaedo. *Phronesis* 36, pp. 265-76.
Gill, C. (1979) Plato and Politics: The Critias and the Politicus. *Phronesis* 24, pp. 148-67.
Gill, C. and McCabe, M.M. (1996) *Form and Argument in Later Plato*. Oxford U.P.
Gill, M.L. (1987) Matter and Flux in Plato's Timaeus. *Phronesis* 32, pp. 34-53.
Gillies, D. (1998) *The Duhem Thesis and the Quine Thesis*. In Curd and Cover (1998), pp. 302-19.
Gillispie, C.C. (1970-1980) *The Dictionary of Scientific Biography* (ed.), 16 vols, Scribner, N.Y.
Gisinger, F. (1921) Die Erdbeschreibung des Eudoxus von Knidos. *Stoicheia*, Heft 6.
Goldstein, B.R. and Bowen, A.C. (1983) A New View of Early Greek Astronomy. *Isis* 74, 1983, pp. 330-40.
Goodrich, W.J. (1903) On Phaedo 96a-102a and on the Deuteros Plous 1. *Classical Review* 17, pp. 5-11.
——— (1904) On Phaedo 96a-102a and on the Deuteros Plous 2. *Classical Review* 18, pp. 381-5.
Gosling, J.C.B. (1960) Republic V: *ta polla kala*. *Phronesis* 5, pp. 116-28.
——— (1968) *Doxa* and *dunamis* in Plato's Republic. *Phronesis* 13, pp. 119-30.
——— (1975) *Plato: Philebus*. Oxford U.P.
Graham, D.W. (1990) Socrates, the Craft Analogy and Science. *Apeiron* 23, pp. 1-24.
Gregory, A.D. (1996) Astronomy and Observation in Plato's Republic. *Studies in History and Philosophy of Science* 27, pp. 451-71.
——— (2000) Plato and Aristotle on Eclipses. *Journal for the History of Astronomy* xxxi, pp. 245-59.
Gregory, J.C. (1931) *A Short History of Atomism: From Democritus to Bohr*. Black.
Gribbin, J. (1986) *In Search of the Big Bang*. Heinemann, London.
——— (1993) *In the Beginning*. Penguin, London.
Gribbin, J and Rees, M. (1990) *Cosmic Coincidences*. Heinemann, London, 1990.
Grote, G. (1875) *Plato and the Other Companions of Socrates*. John Murray, London.

Gulley, N. (1954) Plato's Theory of Recollection. *Classical Quarterly* NS 4, pp. 194-213.
——— (1960) The Interpretation of Plato, Timaeus 49d-e. *American Journal of Philology* 81, pp. 53-64.
——— (1962) *Plato's Theory of Knowledge*. Butler and Tanner, Frome and London.
——— (1968) *The Philosophy of Socrates*. London.
Guthrie, W.K.C. (1962-81) *A History of Greek Philosophy*, 6 vols., Cambridge U.P.
——— (1976) Rhetoric and Philosophy: The Unity of the Phaedrus. *Paideia*, Special Plato Issue, pp. 117-24.
Hackforth, R. (1952) *Plato's Phaedrus*. Cambridge U.P.
——— (1955) *Plato's Phaedo*. Cambridge U.P.
——— (1959) Plato's Cosmogony. *Classical Quarterly* NS 9, pp. 17-22.
——— (1972) *Plato's Philebus*. Cambridge U.P.
Hahm, D.E. (1977) *The Origins of Stoic Cosmology*. Ohio.
Hahn, R. (1978) On Plato's Philebus 15b1-8. *Phronesis* 23, pp. 158-72.
——— (1980) Material Causality, Non-Being, and Plato's Hypodoche: A Re-View of the Timaeus in Terms of the Divided Line. *Apeiron* 14, pp. 59-67.
Hall, T.S. (1969) *Ideas of Life and Matter: Studies in the History of General Physiology, 600 BC – 1900 AD*, 2 vols., Chicago U.P.
Hankins, T.L. (1985) *Science and the Enlightenment*. Cambridge U.P.
Harman, G. (1965) Inference to the Best Explanation. *Philosophical Review* 74, pp. 88-95.
Harré, R. (1972) *The Philosophies of Science*. Oxford U.P.
——— (1981) *Great Scientific Experiments*. Oxford U.P.
——— (1993) *Laws of Nature*. Duckworth.
——— (1995) *Realism Rescued: How Scientific Progress is Possible*. Duckworth and Open Court.
——— (2000a) *Dispositional Powers*, pp. 97-101 in Newton-Smith (2000).
——— (2000b) *Laws of Nature*, pp. 213-23 in Newton-Smith (2000).
Hawking, S. (1988) *A Brief History of Time*. Guild, London.
——— (1993) *Black Holes and Baby Universes and Other Essays*. Bantam, London.
Heath, T. (1913) *Aristarchos of Samos*. Dover edition, New York 1981.
——— (1921) *A History of Greek Mathematics*. Oxford U.P.
——— (1947) *Euclid: The Elements*. St. Johns College Press, Annapolis.
Henry, J. (1990) Magic and Science in the Sixteenth and Seventeenth Centuries. In Olby, Cantor, Christie and Hodge (1990).
Hetherington, N.S. (1993a) *Cosmology: Historical, Literary, Religious and Scientific Perspectives*. Garland, N.Y.
——— (1993b) *Encyclopedia of Cosmology: Historical, Philosophical and Scientific Foundations of Modern Cosmology*. Garland, N.Y.
——— (1996) Plato and Eudoxus: Instrumentalists, Realists or Prisoners of Themata? *Studies in History and Philosophy of Science* 27, pp. 271-89.
Hicken, W.F. (1958) The Character and Provenance of Socrates' 'Dream' in the Theaetetus, *Phronesis* 3, pp. 126-45.
Hirsch, U. (1997) *Sinnesqualitäten und ihre Namen* (zu *Tim*. 61-69). In Calvo and Brisson (1997).
Hocutt, M. (1974) Aristotle's Four Becauses. *Philosophy* 49, pp. 385-9.
Holland, A.J. (1973) An Argument in Plato's Theaetetus: 184-6. *Philosophical Quarterly* 23, pp. 10-116.
Hoskin, M. (1997) *The Cambridge Illustrated History of Astronomy*. Cambridge U.P.
——— (1999) *The Cambridge Concise History of Astronomy*. Cambridge U.P.

Huby, P.M. (1959) Phaedo 99d-102a. *Phronesis* 4, pp. 12-14.
Hume, D. (1738) *A Treatise on Human Nature*. Everyman, London 1955.
Hunt, D.P. (1998) The 'Problem of Fire': Referring to Phenomena in Plato's Timaeus. *Ancient Philosophy* 18, pp. 70-80.
Hutchison, K. (1982) What Happened to Occult Qualities in the Scientific Revolution. *Isis* 73, pp. 233-53.
Irwin, T.H. (1974) *Plato's Moral Theory*. Oxford U.P.
——— (1977) Plato's Heracleiteanism. *Philosophical Quarterly* 27, pp. 1-13.
——— (1988) Reply to David. L. Roochnik. *Platonic Writings, Platonic Readings*, ed. C.L Griswold, Routledge, London and N.Y., pp. 194-9.
Joad, C.E.M. (1933) Plato's Theory of Forms and Modern Physics. *Philosophy* 8, pp. 142-54.
Jones, W.H.S. (1948) *Hippocrates*, vol. I. Loeb, Harvard U.P.
Jordan, R.W. (1984) Plato's Arguments for Forms. *Cambridge Philological Society* Supp. Vol. 9.
Jowett, B. (1953) *The Dialogues of Plato*, 4 vols., 4th ed., Oxford U.P.
Kahn, C.H. (1970) On Early Greek Astronomy. *Journal of Hellenic Studies* 90, pp. 99-116.
——— (1974) Pythagorean Philosophy Before Plato. In *The Pre-Socratics*, ed. A.P.D. Mourelatos, Anchor Books, Garden City.
——— (1981) Some Philosophical Uses of 'to be' in Plato. *Phronesis* 26, pp. 105-34.
——— (1991) *Some Remarks on the Origins of Greek Science and Philosophy*, pp. 1-10 in Bowen (1991).
——— (1999) Review of Greek Philosophy from the Beginning to Plato. *Oxford Studies in Ancient Philosophy*, pp. 325-41.
Keyt, D. (1961) Aristotle on Plato's Receptacle. *American Journal of Philology*, 82, pp. 291-300.
——— (1963) The Fallacies of the Phaedo. *Phronesis* 8, pp. 167-71.
——— (1969) Plato's Paradox that the Immutable is Unknowable. *Philosophical Quarterly* 19, pp. 1-14.
——— (1971) The Mad Craftsman of the Timaeus. *Philosophical Review* 80, pp. 230-5.
Kirk, G.S. (1960) Popper on Science and the Presocratics. *Mind* NS 69, pp. 318-19.
——— (1961) Sense and Common-sense in the Development of Greek Philosophy. *Journal of Hellenic Studies* 81, pp. 105-17.
Kirk, G.S., Raven, J.E. and Schofield, M. (1983) *The Presocratic Philosophers*. 2nd edition, Cambridge U.P.
Klagge, J.C. (1992) Editor's Prologue, *Oxford Studies in Ancient Philosophy*, Supp. Vol., *Methods of Interpreting Plato and his Dialogues*, pp. 1-12.
Kline, M. (1972) *Mathematical Thought from Ancient to Modern Times*. Oxford U.P.
Knorr, W. (1982) The Interaction of Mathematics and Philosophy in Antiquity. In N. Kretzman, (ed.), *Infinity and Continuity in Ancient and Medieval Thought*. Ithaca 1982.
——— (1989) The Practical Element in Ancient Exact Sciences. *Synthèse* 81, pp. 313-28.
——— (1990) Plato and Eudoxus on the Planetary Motions. *Journal for the History of Astronomy* 21, pp. 313-29.
——— (1993) *Plato's Cosmology*. In Hetherington (1993b).
Koestler, A. (1959) *The Sleepwalkers: A History of Man's Changing Vision of the Universe*. Penguin.
Koyre, A. (1943) Galileo and Plato. *Journal of the History of Ideas* 4, pp. 400-28.
——— (1973) *The Astronomical Revolution*. Methuen, London.
Krafft, F. (1965) *Der Mathematikos und der Physikos*. Bemerkungen zu der Ange-

blichen Platonischen Aufgabe, die Phänomene zu Retten. *Beiträge zur Geschichte der Wissenschaft und Technik* 5, pp. 5-24.
Kramer, H.J. (1959) *Arete bei Platon und Aristoteles*. Heidelberg U.P.
——— (1964) Retraktionen zum Problem des Esoterischen Platon. *Museum Helveticum* 21, pp. 137-67.
——— (1966) Aristoxenus über Platons peri tagathou. *Hermes* 94, pp. 111-26.
Kraut, R. (1992) *The Cambridge Companion to Plato*. Cambridge U.P.
Kung, J. (1981) Aristotle on Thises, Suches and the Third Man Argument. *Phronesis* 26, pp. 207-47.
——— (1985) Tetrahedra, Motion and Virtue. *Nous* 19, pp. 17-27.
——— (1988) Why the Receptacle is Not a Mirror. *Archiv für Geschichte der Philosophie* 70, pp. 167-78.
Kuhn, T.S. (1957) *The Copernican Revolution*. Random House, N.Y.
Lamb, W.R.M. (1927) *Epinomis*. Loeb Plato vol. XII.
Lane, M. (1995) A New Angle on Utopia: The Political Theory of the Statesman. In Rowe (1995a) pp. 276-91.
Lang, H.S. (1992) *Aristotle's Physics and its Medieval Varieties*. SUNY.
Lasserre, F. (1964) *The Birth of Mathematics in the Age of Plato*. London.
——— (1966) *Die Fragmente des Eudoxus von Knidos*. De Gruyter, Berlin.
Laudan, L. (1998) Demystifying Underdetermination. In Curd and Cover (1998) pp. 320-53.
Ledger, G.R. (1989) *A Computer Analysis of Plato's Style*. Oxford U.P.
Lee, E.N. (1966) On the Metaphysics of the Image in Plato's Timaeus. *Monist* 50, pp. 341-68.
——— (1967) On Plato's Timaeus, 49d4-e7. *American Journal of Philology* 88, pp. 1-28.
——— (1971) On the Gold Example in Plato's Timaeus. In *Essays in Ancient Philosophy*, ed. J.P. Anton and G.L. Kustas, Albany, N.Y. pp. 219-35.
——— (1976) Reason and Rotation. Circular Movement as the Model of Mind (Nous) in Later Plato. In *Facets of Plato's Philosophy*, *Phronesis* Supp. Vol. II, pp. 70-102.
Lee, H.P.D. (1955) *The Republic*. Penguin.
——— (1965) *Plato: Timaeus and Critias*. Penguin.
Leibniz, G.W.F. (1686) *Discourse on Metaphysics*. In G.H.R. Parkinson (ed.) *Leibniz: Philosophical Writings*. Dent, London, 1973.
——— (1687) *Letter Explaining the Laws of Nature*. In G.H.R. Parkinson (ed.) *Leibniz: Philosophical Writings*. Dent, London, 1973.
Lennox, J.G. (1985) Plato's Unnatural Teleology. In D.J. O'Meara (ed.) *Platonic Investigations*. Washington D.C. U.P., pp. 195-218.
Leplin, J. (2000) Realism and Instrumentalism. In Newton-Smith (2000).
Leslie, J. (1989) *Universes*. Routledge, London.
Letwin, O. (1981) Interpreting the Philebus. *Phronesis* 26, pp. 187-206.
Liddle, A.R. (1999) *An Introduction to Modern Cosmology*. Wiley, John and Sons.
Lipton, P. (2000) *Inference to Best Explanation*. In Newton-Smith (2000).
Lloyd, G.E.R. (1966) *Polarity and Analogy*. Cambridge U.P.
——— (1968) Plato as a Natural Scientist, *Journal of Hellenic Studies* 28, pp. 78-92.
——— (1970) *Early Greek Science to Aristotle*. Chatto and Windus, London.
——— (1978) Saving the Appearances. *Classical Quarterly* 88, pp. 202-22. See also Lloyd (1991) pp. 248-53.
——— (1979) *Magic, Reason and Experience*. Cambridge U.P.
——— (1987) *The Revolutions of Wisdom*. California U.P.

——— (1990) Plato and Archytas in the Seventh Letter. *Phronesis* 35, pp. 159-74.
——— (1991) *Methods and Problems in Greek Science*, Cambridge U.P., pp. 333-51.
——— (1996) *Aristotelian Explorations*. Cambridge U.P.
Lombard, L.B. (1978) Relational Change and Relational Changes. *Philosophical Studies* 34, pp. 63-79.
Long, A.A. and Sedley, D. (1987) *The Hellenistic Philosophers*, 2 vols., Cambridge U.P.
Long, A.A. (1977) Chance and Law in Epicureanism. *Phronesis* 22, pp. 63-88.
Lovibond, S. (1991) *Plato's Theory of Mind*. In Everson (1991).
McCabe/ MacKenzie, M.M. (1982a) Paradox in Plato's Phaedrus. *Proceedings of the Cambridge Philological Society*, NS 28, pp. 64-76.
——— (1982b) Parmenides' Dilemma. *Phronesis* 28, pp. 1-13.
——— (1986a) Putting the Cratylus in its Place. *Classical Quarterly* 36, pp. 124-50.
——— (1986b) Impasse and Explanation: from the Lysis to the Phaedo. *Archiv für Geschichte der Philosophie* 88, pp. 15-45.
——— (1988) The Virtues of Socratic Ignorance. *Classical Quarterly* 38, pp. 331-50.
——— (1993) Myth, Allegory and Argument. *Apeiron* 25, *The Language of the Cave*, pp. 47-67.
——— (1994) *Plato's Individuals*. Princeton U.P.
——— (1997) Chaos and Control: Reading Plato's Politicus. *Phronesis* 42, pp. 94-117.
McDowell, J.H. (1973) *Plato: Theaetetus*. Oxford U.P.
MacKenzie, *see* McCabe.
Malcolm, J. (1962) The Line and the Cave. *Phronesis* 7, pp. 38-45.
——— (1967) Plato's Analysis of 'to on' and 'to me on' in the Sophist. *Phronesis* 12, pp. 130-46.
Martin, A. and Primaversi, O. (1999) *L' Empedocle de Strasbourg*. Bibliothèque Nationale et Universitaire de Strasbourg/Berlin. De Gruyter.
Mellor, D.H. (1980) Necessities and Universals in Natural Laws. In Mellor (ed.), *Science, Belief and Behaviour: Essays in Honour of R.B. Braithwaite*. Cambridge U.P., pp. 105-25.
Mendell, H. (1998) Reflections on Eudoxus, Callippus and their Curves: Hippopedes and Callippopedes. *Centaurus* 40, pp. 177-275.
Milton, J.R. (1998) *Laws of Nature*. In Ayers and Garber.
Mittlestrass, J. (1962) *Die Rettung der Phänomene*. De Gruyter, Berlin.
——— (1988) On Socratic Dialogue. In *Platonic Writings, Platonic Readings*, ed. C.L Griswold, Routledge, London and N.Y., pp. 126-42.
Modrak, D.K. (1981) Perception and Judgement in the Theaetetus. *Phronesis* 26, pp. 35-54.
Mohr, R.D. (1977) Plato, Statesman 284c-d: An Argument from the Sciences. *Phronesis* 22, pp. 232-4.
——— (1978a) The Formation of the Cosmos in the Statesman Myth. *Phoenix* 27, pp. 250-2.
——— (1978b) The Gold Analogy in Plato's Timaeus (50a4-b5). *Phronesis* 23, pp. 243-52.
——— (1980a) Image, Flux and Space in Plato's Timaeus. *Phoenix* 34, pp. 138-52.
——— (1980b) The Sources of Evil Problem and the *arche kineseos* Doctrine in Plato. *Apeiron* 14, pp. 41-56.
——— (1980c) The Sources of Flux in Plato's Timaeus. *Apeiron* 14, pp. 96-114.
——— (1981) Disorderly Motion in Plato's Statesman. *Phoenix* 35, pp. 199-215.

——— (1983) Philebus 55c-62a and Revisionism. *Canadian Journal of Philosophy* 9, Supp. Vol., pp. 165-70.

——— (1985) *The Platonic Cosmology*. Brill, Leiden.

——— (1985a) Plato's Theology Reconsidered: What the Demiurge Does. *History of Philosophy Quarterly* 2, pp. 131-44.

——— (1986) Plato on Time and Eternity. *Ancient Philosophy* 6, pp. 39-46.

Moravscik, J. (1974) On Explanation. *Synthèse* 28, pp. 22-5.

Morgan, M.L. (1984) Sense-Perception and Recollection in the Phaedo. *Phronesis* 29, pp. 237-51.

——— (1989) How Does Socrates Solve the Paradox of Inquiry in the Meno? *Essays in Ancient Greek Philosophy III, Plato*, ed. J.P. Anton and A. Preus, New York U.P., pp. 169-82.

Morrison, J.S. (1959) The Shape of the Earth in Plato's Phaedo. *Phronesis* 4, pp. 101-19.

——— (1977) Two Unresolved Difficulties in the Line and Cave. *Phronesis* 22, pp. 212-31.

Morrison, J.S. and Coates, J.F. (1986) *The Athenian Trireme: The History and Reconstruction of an Ancient Greek Warship*. Cambridge U.P.

Morrison, J.S. and Williams, R.T. (1968) *Greek Oared Ships, 900-322 BC.*, Cambridge U.P.

Morrow, G.R. (1950) Necessity and Persuasion in Plato's Timaeus. In *Studies in Plato's Metaphysics*, ed. R.E. Allen, Routledge and Kegan Paul, London 1965, pp. 421-38.

——— (1962) *Plato's Epistles*. Bobbs-Merrill, Indianapolis and N.Y.

——— (1970) Plato and the Mathematicians: An Interpretation of Socrates' Dream in the Theaetetus. *Philosophical Review* 79, pp. 309-33.

——— (1968) Plato's Theory of the Primary Bodies in the *Timaeus* and the Later Doctrine of Forms. *Archiv für Geschichte der Philosophie* 60, pp. 12-28.

Mourelatos, A.P.D. (1981a) Plato's 'Real Astronomy': Republic 527d-531d. In *Science and the Sciences in Plato*, ed. P. Anton, Buffalo N.Y, pp. 33-73.

——— (1981b) Astronomy and Kinematics in Plato's Project of Rationalist Explanation. *Studies in the History and Philosophy of Science* 12, pp. 1-32.

——— (1991) Plato's Science – His View and Ours of His, pp. 11-42, in Bowen (1991).

Mueller, I. (1981) Ascending to Problems: Astronomy and Harmonics in Republic VII. *Science and the Sciences in Plato*, ed. P. Anton, Buffalo N.Y, pp. 103-22.

——— (1989) Joan Kung's Reading of the Timaeus. *Apeiron* 22, pp. 1-25.

——— (1992) Mathematical Method and Philosophical Truth. In Kraut (1992).

Murphy, N.R. (1936) The Deuteros Plous in the Phaedo. *Classical Quarterly* 30, pp. 40-7.

Musgrave, A. (1985) Realism versus Constructive Empiricism. In Curd and Cover (1998).

——— (1989) NOA's Ark – Fine for Realism. In Curd and Cover (1998).

——— (1991) The Myth of Astronomical Instrumentalism. In G. Munevar (ed.) *Beyond Reason: Essays on the Philosophy of Paul Feyerabend*. Klusener Academic, Dordrecht.

Naddaf, G. (1997) Plato and the *Peri Phuseos* Tradition. In Calvo and Brisson (1997).

Nadler, S. (1998) Doctrines of Explanation in Late Scholasticism and in the Mechanical Philosophy. In Ayers and Garber (1998).

Nahnikian, G. (1955) Plato's Theory of Sensation. *Review of Metaphysics* 9, pp. 129-48.

Nehamas, A. (1975) Plato on the Imperfection of the Sensible World. *American Philosophical Quarterly* 12, pp. 105-17.
——— (1984) Episteme and Logos in Plato's Later Thought. *Archiv für Geschichte der Philosophie* 66, pp. 11-36.
——— (1985) Meno's Paradox and Socrates as a Teacher. *Oxford Studies in Ancient Philosophy* 3, pp. 1-30.
Neugebauer, O. (1957) *The Exact Sciences in Antiquity*. 2nd edition, Providence.
——— (1975) *A History of Ancient Mathematical Astronomy*. Springer-Verlag, Berlin.
Newton, I. (1687) *Principia Mathematica Philosophia Naturalis*. Third edition (1726) with variant readings assembled and edited by A. Koyre and I.B. Cohen, Cambridge U.P., 1972.
——— (1704) *Opticks*. London, 1730.
Newton-Smith, W. (1980) The Underdetermination of Theories by Data. In N.R. Hilpinen (ed.), *Rationality in Science*. Dordrecht, Reidel.
——— (1981) *The Rationality of Science*. Routledge, London.
——— (2000) *A Companion to the Philosophy of Science*. Blackwell, Massachusetts.
Nikolic, M. (1974) Eudoxus and Dedekind. *Boston Studies in the History and Philosophy of Science* 15.
North, J. (1994) *The History of Astronomy and Cosmology*. Fontana.
O'Brien, D. (1968) Derived Light and Eclipses in the Fifth Century. *Journal of Hellenic Studies* 88, pp. 114-327.
——— (1984) *Theories of Weight in the Ancient World*, vol. II, *Plato: Weight and Sensation*. Brill.
——— (1997) Perception et intelligence dans la *Timée* de Platon. In Calvo and Brisson (1997).
Olby, R.C., Cantor, G.N., Christie, J.R.R. and Hodge, M.J.S. (1990) *Companion to the History of Modern Science*. Routledge, London.
Olsen, R. (1982) *Science Deified and Science Defied*, 2 vols., California U.P.
Oppenheim, A.L. (1978) Man and Nature in Mesopotamian Civilisation. In Gillispie vol. 15, 1978.
Osborne, C. (1988) Topography in the Timaeus: Plato and Augustine on Mankind's Place in the Natural World. *Proceedings of the Cambridge Philological Society* 214, pp. 104-14.
——— (1996) Space, Time, Shape and Direction: Creative Discourse in the *Timaeus*. In Gill & McCabe (1996), pp. 179-211.
Owen, G.E.L. (1953) The Place of the Timaeus in Plato's Dialogues. Reprinted in R.E. Allen (ed.), *Studies in Plato's Metaphysics*, Routledge and Kegan Paul, 1965, pp. 313-39.
——— (1957a) A Proof in the Peri Ideon. Reprinted in *Logic, Science and Dialectic*, ed. M. Nussbaum, Duckworth, London, 1986, pp. 165-79.
——— (1957b) Logic and Metaphysics in Some Earlier Works of Aristotle. Reprinted in *Logic, Science and Dialectic*, ed. M. Nussbaum, Duckworth, London, 1986, pp. 180-99.
——— (1966a) The Platonism of Aristotle. Reprinted in *Logic, Science and Dialectic*, ed. M. Nussbaum, Duckworth, London, 1986, pp. 200-20.
——— (1966b) Plato and Parmenides on the Timeless Present. *The Monist* 50, pp. 317-40.
——— (1970) Notes on Ryle's Plato. Reprinted in *Logic, Science and Dialectic*, ed. M. Nussbaum, Duckworth, London, 1986, pp. 85-103.

——— (1971) Plato on Not-Being. In *Logic, Science and Dialectic*, ed. M. Nussbaum, Duckworth, London, 1986, pp. 104-37.
Pais, A. (1982) *Subtle is the Lord*. Oxford U.P.
Palter, R. (1970) An Approach to the Early History of Astronomy. *Studies in History and Philosophy of Science* 1, pp. 93-133.
Panchenko, D. (1994) Homoios and Homoiotes in Thales and Anaximander. *Hyperorboreus* 1, pp. 28-55.
Pangle, T.L. (1980) *The Laws of Plato*. Chicago U.P.
Pannekoek, A. (1961) *A History of Astronomy*. George Allen and Unwin.
Parry, R.D. (1979) The Unique Worlds of the Timaeus. *Journal of the History of Philosophy* 17, pp. 1-10.
——— (1991) The Intelligible World Animal in Plato's Timaeus, *Journal of the History of Philosophy* 29, pp. 13-32.
Parry, R.D. (1979). The Unique World of the Timaeus. *Journal for the History of Philosophy* 17, pp. 1-10.
——— (1983) The Craft of Justice. *Canadian Journal of Philosophy* 9, pp. 19-38.
——— (1991) The Intelligible World-Animal in Plato's Timaeus. *Journal for the History of Philosophy* 29, pp. 13-32.
——— (1996) *Plato's Craft of Justice*. SUNY, Albany.
Patterson, R. (1981) The Unique Worlds of the Timaeus. *Phoenix* 35, pp. 105-19.
——— (1985) On the Eternality of Platonic Forms. *Archiv für Geschichte der Philosophie* 67, pp. 27-45.
Pedersen, O. (1974) *Early Physics and Astronomy*, , 2nd edition 1993, Cambridge U.P.
Penner, T. (1987) *The Ascent from Nominalism*, Dordrecht.
Philip, A. (1981) Recurrences thematiques et topologie dans le 'Phèdre'. *Revue de Métaphysique et Morale* 86, pp. 452-76.
Pingree, D. (1992) Hellenophilia vs. the History of Science. *Isis* 83, pp. 554-63.
Plass, P. (1960) Socrates' Method of Hypothesis in the Phaedo. *Phronesis* 5, pp. 103-15.
Pohle, W. (1971) The Mathematical Foundations of Plato's Atomic Physics. *Isis* 62, pp. 36-46.
——— (1973) Dimensional Concepts in Plato's Physics. In *Exegesis and Argument*, ed. Lee, Mourelatos and Rorty, Assen 1973, pp. 306-23.
Popper, K.R. (1945) *The Open Society and its Enemies*, vol. I, *The Spell of Plato*, 5th edition, Routledge and Kegan Paul, London 1966.
——— (1963) *Conjectures and Refutations*, Routledge and Kegan Paul, London.
Post, L.A. (1963) Letter VII. In E. Hamilton and H. Cairns, *Plato: The Collected Dialogues*. Princeton U.P.
Prior, W.J. (1985) *Unity and Development in Plato's Metaphysics*. Croom Helm, Beckenham.
Proclus. (1966) *Commentaire sur le Timée*, 5 vols., ed. A.J. Festugière, Libraire Philosophique, J. Vrin, Paris.
Ptolemy. (1898) *Suntaxis Mathematica*, 2 vols., ed. J.L. Heiberg, Teubner Series.
——— (1980) *Tetrabiblos*, ed. & trans. F.E. Robbins, Loeb Series, Harvard U.P.
——— (1907) *Planetary Hypotheses – Opera Astronomica Minora*, ed. J.L. Heiberg, Teubner Series.
——— The Arabic Version of Ptolemy's Planetary Hypotheses, ed. B.R. Goldstein, *Transactions of the American Philosophical Society*, NS 57, 1967, pp. 3-55.
Pyle, A. (1997) *Atomism and its Critics*. Thoemmes, Bristol.
Quine, W.V.O. (1953) Two Dogmas of Empiricism. In *From a Logical Point of View*. Harper.

——— (1960) *Word and Object*. Cambridge, Massachusetts, M.I.T. Press.
——— (1969) *Ontological Relativity and Other Essays*, New York, Columbia U.P.
——— (1975) On Empirically Equivalent Systems of the World, *Erkenntnis* 9, pp. 313-28.
——— (1986) *The Philosophy of W.V.O. Quine*. Open Court.
Raven, J.E. (1953) Sun, Divided Line and Cave. *Classical Quarterly* NS 3, pp. 131-87.
Ravetz, J. (1990) *The Copernican Revolution*. In Olby, Cantor, Christie and Hodge (1990).
Reale, G. (1997) Plato's Doctrine of the Origin of the World, with Special Reference to the Timaeus. In Calvo and Brisson (1997).
Reed, N.H. (1972) Plato on Flux, Perception and Language. *Proceedings of the Cambridge Philological Society* 18, pp. 65-77.
Reeve, C.D.C. Motion, Rest and Dialectic in the Sophist. *Archiv für Geschishte der Philosophie* 67, pp. 47-64.
Reydam-Schils, G. (1997) Grasping Sensibles Without Sense-Perception. In Calvo and Brisson (1997).
Richardson, H. (1926) The Myth of Er. *Classical Quarterly* 20, pp. 113-33.
Ricker, R.A.R. (1967) *The Paths of the Planets*. London.
Ritter, C. (1910) *Platon, seine Leben, seine Schriften, seine Lehre*. Munich U.P.
Robin, L. (1908) *La Théorie Platonicienne des Idées et des Nombres d'après Aristote*. Presses Universitaires de France, Paris 1908.
Robins, I. (1995) Mathematics and the Conversion of the Mind, Republic VII 522c1-531e3, *Ancient Philosophy* 15.
Robinson, R. (1953) *Plato's Earlier Dialectic*. 2nd edition, Oxford U.P., 1984, Introduction, pp. 1-6.
Robinson, T.M. (1970) *Plato's Psychology*. Toronto U.P.
——— (1979) The Argument of Timaeus 27d ff. *Phronesis* 24, pp. 105-9.
Roochnik, D. (1986) Socrates' Use of the Techne Analogy. *Journal of the History of Philosophy* 24, pp. 295-310.
——— (1987) Plato's Use of Atechnos. *Phoenix* 41, pp. 255-63.
——— (1994) Counting on Number: Plato on the Goodness of Arithmos. *American Journal of Philology* 115, pp. 543-63.
——— (1996) *Of Art and Wisdom: Plato's Understanding of Techne*. Pennsylvania State U.P.
Rose, L.E. (1966) The Deuteros Plous in Plato's Phaedo. *Monist* 50, pp. 464-73.
Rosen, S. (1979) Plato's Myth of the Reversed Cosmos. *Review of Metaphysics* 33, pp. 60-85.
——— (1983) *Plato's Sophist*. Yale U.P.
Rosenmayer, T.G. (1959) The Shape of the Earth in the Phaedo: A Rejoinder. *Phronesis*, pp. 71-2.
——— (1960) Plato's Hypothesis and the Upward Path. *American Journal of Philology* 81, pp. 355-66.
Ross, W.D. (1951) *Plato's Theory of Ideas*. Oxford U.P.
Rowan-Robinson, M. (1996) *Cosmology*, 3rd edition, Oxford U.P.
Rowe, C.J. (1993a) Explanation in Phaedo 99c6-102a8. *Oxford Studies in Ancient Philosophy* 11, pp. 49-69.
——— (1993b) *Plato: Phaedo*. Cambridge U.P.
——— (1995a) Reading the Statesman. *Proceedings of the III Symposium Platonicum*. Academia Verlag, Sankt Augustin.
——— (1995b) *Plato: Statesman*. Aris and Phillips, Warminster.

Runia, D. (1997) The Literary and Philosophical Status of Timaeus' Proemium. In Calvo and Brisson (1997).
Runciman, W.G. (1962) *Plato's Later Epistemology*. Cambridge U.P.
Ryle, G. (1960) Letters and Syllables in Plato. *Philosophical Review* 69, pp. 431-51.
——— (1966) *Plato's Progress*. Cambridge U.P.
——— (1990) Logical Atomism in Plato's Theaetetus. *Phronesis* 35, pp. 21-46.
Sambursky, S. (1956) *The Physical World of the Greeks*. Routledge and Kegan Paul, London.
——— (1965) Plato, Proclus, and the Limits of Science. *Journal of the History of Philosophy* 03, pp. 1-11.
Santa Cruz, M.I. (1997) Le Discours de la Physique: *eikos logos*. In Calvo and Brisson (1997).
Saunders, T.J. (1970) *Plato: The Laws*. Penguin, London.
Sayre, K.M. (1969) *Plato's Analytic Method*. Chicago U.P.
——— (1983) *Plato's Late Ontology: A Riddle Resolved*. Princeton U.P.
——— (1992) A Maieutic View of Five Late Dialogues. *Oxford Studies in Ancient Philosophy*, Supp. Vol., *Methods of Interpreting Plato and His Dialogues*, pp. 221-44.
——— (1998) The Role of the Timaeus in the Development of Plato's Late Ontology. *Ancient Philosophy* 18, pp. 93-124.
Schiaperelli (1875) Le Sfere Omocentriche di Eudosso, di Callippo et di Aristotele. *Publicazioni del R. Observatorio di Brera in Milano* IX.
Schicker, R. (1995) Aspekte der Rezeption des *Politikos* im Mittel- und Neuplatonismus. In Rowe (1995a), pp. 381-8.
Schleiermacher, F. (1804) *Introductions to Plato's Dialogues*, trans. W. Dobson, New York 1973.
Schofield, M. (1997) *The Ionians*. In Taylor (1997).
Schoos, D.J. (1999) Timaeus' Banquet. *Ancient Philosophy* 19, pp. 97-107.
Schulz, D.J. (1966) Das Problem der Materie in Platons Timaios. *Abhandlungen zur Philosophie, Psychologie und Pädogogik* 31, Bonn.
Schulz, D.J. (1966) *Das Problem der Materie in Platons Timaios*. Bonn.
Scott, D. (1987) Platonic Anamnesis Revisited. *Classical Quarterly* 37, pp. 346-66.
Scolnikov, S. (1997) Freedom and Education in Plato's Timaeus. In Calvo and Brisson (1997).
Sedley, D. (1982) Two Conceptions of Vacuum. *Phronesis* 27 pp. 175-93.
——— (1989) Teleology and Myth in the Phaedo. *Proceedings of the Boston Colloquium on Ancient Philosophy* 5, pp. 359-83.
——— (1991) Is Aristotle's Teleology Anthropocentric? *Phronesis* 36, pp. 179-96.
——— (1997) Becoming Like God in the *Timaeus* and Aristotle. In Calvo and Brisson (1997), pp. 327-40.
——— (1998) Platonic Causes. *Phronesis* 43, pp. 114-32.
Shiner, R. (1974) *Knowledge and Reality in Plato's Philebus*. Assen.
Shipton, K.M.W. (1979) A Good Second-best: Phaedo 99b ff. *Phronesis* 24, pp. 33-53.
Shoemaker, S. (1984) *Identity, Cause and Mind*, Cambridge U.P.
Shorey, P. (1927) Platonism and the History of Science. *Proceedings of the American Philosophical Association* 46, pp. 159-82.
——— (1935) *The Republic*, 2 vols., Loeb Classical Library, Harvard U.P.
Silk, J. (1989) *The Big Bang*, 2nd edition, Freeman.
Silverman, A. (1990) Plato on Perception and 'Commons'. *Classical Quarterly* 40, pp. 148-50.
——— (1992) Timaean Particulars. *Classical Quarterly* 42, pp. 87-113.
Simplicius. *Aristotelis in De Caelo Commentaria*, ed J.L. Heiberg, Berlin 1894.

Skemp, J.B. (1942) *The Theory of Motion in Plato's Later Dialogues*, 2nd edition, Adolf M. Hakkert, Amsterdam, 1967.

——— (1952) *Plato: The Statesman*, 2nd edition, Bristol Classical Press, Bristol 1987.

Smoot, G. (1993) *Wrinkles in Time*. Little, Brown and Company.

Sober, E. (2000) *Simplicity*. In Newton-Smith (2000), pp. 433-41.

Solmsen, F. (1940) Plato and the Unity of Science. *Philosophical Review* 44, pp. 566-71.

——— (1960) *Aristotle's System of the Physical World*. Cornell U.P.

——— (1963) Nature as Craftsman in Greek Thought. *Journal of the History of Ideas* 24, pp. 473-96.

——— (1971) The Tradition about Zeno of Elea Re-examined. *Phronesis* 16, pp. 116-41.

Sorabji, R. (1980) *Necessity, Cause and Blame. Perspectives on Aristotle's Theory*. Duckworth, London.

——— (1983) *Time, Creation and the Continuum: Theories in Antiquity and the Early Middle Ages*. Duckworth, London.

——— (1988) *Matter, Space and Motion: Theories in Antiquity and their Sequel*. Duckworth, London.

Stenzel, J. (1959) *Zahl und Gestalt bei Plato und Aristoteles*. Herman Gentner Verlag, Bad Homburg vor der Hohe.

Stewart, J.A. (1960) *The Myths of Plato*. London.

Stocks, J.L. (1911) The Divided Line of Plato Rep. VI. *Classical Quarterly* 5, pp. 73-88.

Stokes, M.C. (1992) Plato and the Sightlovers of the Republic. In *The Language of the Cave*, ed. A. Barker and M. Warner, *Apeiron* 25, pp. 103-33.

Stough, C.L. (1976) Forms and Explanation in the Phaedo. *Phronesis* 21, pp. 1-30.

Strang, C. (1956) Plato and the Third Man. *Proceedings of the Aristotelian Society*, Supp. Vol. 37, 1963, pp. 147-63.

——— (1986) Plato's Analogy of the Cave. *Oxford Studies in Ancient Philosophy* 4, pp. 19-34.

Striker, G. (1970) *Peras und Apeiron: Das Problem der Formen in Platons Philebus*. Vandenhoeck and Ruprecht, Gottingen.

Szlezak, T.A. (1979) The Acquiring of Philosophical Knowledge According to Plato's Seventh Letter. In *Arktouros, Hellenic Studies Presented to B.M.W. Knox*, Berlin U.P., pp. 354-63.

Taran, L. (1971) The Creation Myth in Plato's Timaeus. In *Essays in Ancient Greek Philosophy*, ed. J.P Anton and G.L Kustas, pp. 372-406.

——— (1975) *Academica: Plato, Philip of Opus, and the Pseudo-Platonic Epinomis*. American Philosophical Society, Philadelphia.

Taub, L. (1993) *Ptolemy's Universe*. Open Court.

Taylor, A.E. (1926) *Plato: The Man and his Work*. Meridian Books, Cleveland 1956.

——— (1928) *A Commentary on Plato's Timaeus*. Oxford U.P.

——— (1939) The Decline and Fall of the State in Republic, VIII. *Mind*, NS 48, pp. 23-38.

——— (1956) *Plato: Philebus and Epinomis*. Nelson and Sons, London.

Taylor, C.C.W. (1969) Forms as Causes in the Phaedo. *Mind* 78, pp. 45-59.

——— (1997) *From the Beginning to Plato. Routledge History of Philosophy*, vol. 1. Routledge, London.

——— (1998) Plato's Ethics. In Everson (1998).

Teloh, H. (1981) *The Development of Plato's Metaphysics*. University Park.

Theiler, W. (1925) *Zur Geschichte der Teleologischen Naturbetrachtung bis auf Aristote*. Zurich.
Thesleff, H. (1982) *Studies in Platonic Chronology*. Societas Humanarum Litterarum, Helsinki.
Toulmin, S. and Goodfield, J. (1961) *The Fabric of the Heavens*. Hutchison, London.
Turbayne, C.M. (1976) Plato's Fantastic Appendix: The Procreation Model of the *Timaeus*. *Paideia* 50.
Turnbull, R.G. (1955) Aristotle's Debt to the Natural Philosophy of the Phaedo. *Philosophical Quarterly* 8, pp. 131-43.
——— (1978) Knowledge and the Forms in the Later Platonic Dialogues. *Proceedings and Addresses of the American Philosophical Association* 51, pp. 735-58.
——— (1988) Becoming and Intellegibility. *Oxford Studies in Ancient Philosophy* 6, pp. 1-14; Response to Professor Fine's Critique of 'Becoming and Intelligibility'. *Oxford Studies in Ancient Philosophy* 6, pp. 29-36.
Vallejo, A. (1997) *No, It's Not a Fiction*. In Calvo and Brisson (1997).
Van der Waerden, B.L. (1951) *Die Pythagoreer*. Amsterdam.
——— (1974a) The Earliest Form of the Epicycle Theory, *Journal for the History of Astronomy* 5, pp. 175-85.
——— (1974b) *Science Awakening* II. Oxford U.P.
——— (1978) *Basic Ideas and Methods of Babylonian and Greek Astronomy*. In Gillispie vol. 15, 1978.
——— (1982) The Motion of Venus, Mercury and the Sun in Early Greek Astronomy. *Archive for the History of the Exact Sciences* 26, pp. 99-113.
Van Eck, J. (1994) Skopein en Logois. *Ancient Philosophy* 20, pp. 19-40.
Van Fraassen, B. (1980) *The Scientific Image*. Oxford U.P.
——— (1989) *Laws and Symmetry*. Oxford U.P.
Vartanian, A. (1957) Trembly's Polyp, La Mettrie and Eighteenth-Century French Materialism. In P.P. Wiener and A. Noland (eds.), *Roots of Scientific Thought*, New York U.P.
Vidal-Naquet, P. (1978) Plato's Myth of the Statesman, the Ambiguities of the Golden Age and History. *Journal of Hellenic Studies* 98, pp. 132-41.
Vlastos, G.L. (1939) The Disorderly Motion in the Timaeus. In *Studies in Plato's Metaphysics*, ed. R.E. Allen, Routledge and Kegan Paul, London, 1965, pp. 379-420.
——— (1950) The Physical Theory of Anaxagoras. *Philosophical Review* 59.
——— (1964) Creation in the Timaeus: Is it a Fiction? In *Studies in Plato's Metaphysics*, ed. R.E. Allen, Routledge and Kegan Paul, London 1965, pp. 401-19.
——— (1965a) Anamnesis in the Meno. *Dialogue* 4, pp. 143-67.
——— (1965b) A Metaphysical Paradox. Reprinted in *Platonic Studies*, 2nd edition, Princeton U.P., 1981, pp. 43-57.
——— (1967) Plato's Supposed Theory of Irregular Atomic Figures. *Isis* 57, pp 204-9.
——— (1969) Reasons and Causes in the Phaedo. Reprinted with revisions in *Platonic Studies*, 2nd edition, Princeton U.P. 1981, pp. 76-110.
——— (1972) The Unity of the Virtues in the Protagoras. Reprinted with revisions in *Platonic Studies*, 2nd edition, Princeton U.P., 1981, pp. 415-58.
——— (1975) *Plato's Universe*. Clarendon, Oxford.
——— (1981) The Role of Observation in Plato's Conception of Astronomy. In *Science and the Sciences in Plato*, ed. P. Anton, Buffalo N.Y, pp. 1-31.
——— (1985) 'Socrates" Disavowal of Knowledge. *Philosophical Quarterly* 35, pp. 1-31.
——— (1991) *Socrates*. Cambridge U.P.

von Fritz, K. (1971) *Grundproblem der Geschichte der antiken Wissenschaft*. De Gruyter, Berlin.
Von Leyden, W. (1964) Time, Number and Eternity in Plato and Aristotle. *Philosophical Quarterly* 14, pp. 35-52.
Von Staden, H. (1992) Affinities and Elisions. *Isis* 83, pp. 578-95.
Wasserstein, A. (1962) Greek Scientific Thought. *Proceedings of the Cambridge Philological Society*, NS 8, pp. 51-63.
Waterfield, R.A.H. (1980) The Place of the Philebus in Plato's Dialogues. *Phronesis* 25, pp. 270-305.
Webster, C. (1982) *From Paracelsus to Newton*. Cambridge U.P.
Wedberg, A. (1955) *Plato's Philosophy of Mathematics*. Stockholm.
Weinberg, S. (1983) *The First Three Minutes*, 2nd edition. Fontana, London.
Westerink, L.G. (1976) *The Greek Commentaries on Plato's Phaedo*, 2 vols, Amsterdam.
White, N. (1974) Inquiry. *Review of Metaphysics* 28, 1974/75, pp. 289-310.
——— (1976) *Plato on Knowledge and Reality*. Hackett, Indianapolis.
Whitehead, A.N. (1933) *Adventures of Ideas*. The Free Press, N.Y.
——— (1941) Mathematics and the Good. In P. Schilpp (ed.), *The Philosophy of Alfred North Whitehead*, Open Court.
Wicksteed, P.H. and Cornford, F.M. (1929) *Aristotle: Physics*, 2 vols, Loeb Classical Library, Harvard U.P.
Wieland, W. (1982) *Platon und die Formen des Wissens*, Gottingen, pp. 13-94.
Wiggins, D. (1986) Teleology and the Good in Plato's Phaedo. *Oxford Studies in Ancient Philosophy* 4, pp. 1-18.
Wightman, W.P.D. (1950) *The Growth of Scientific Ideas*. Edinburgh.
Williamson, H. (1904) *The Phaedo of Plato*. Macmillan and Co, London.
Wittaker, J. (1968) The 'Eternity' of the Platonic Forms. *Phronesis* 13, pp. 131-44.
Wolfsdorf, D. (1999) Plato and the Mouth-Piece Theory. *Ancient Philosophy* 19, pp. 13-24.
Worrall, J. (1982) Scientific Realism and Scientific Change. *Philosophical Quarterly* 32, pp. 210-31.
Wright, L. (1974) The Astronomy of Eudoxus: Geometry or Physics? *Studies in History and Philosophy of Science* 4, pp. 165-72.
——— (1976) *Teleological Explanations*. Berkeley U.P.
Wright, M.R. (1995) *Cosmology in Antiquity*. Routledge, London.
Yavetz, I. (1998) On the Homocentric Spheres of Eudoxus. *Archive for the History of the Exact Sciences* 52, pp. 221-78.
Yoh, M. (1975) On the Third Attempted Definition of Knowledge, Theaetetus 201c-210b, *Dialogue* 14, 420-42.
Yolton, J. (1949) The Ontological Status of Sense Data in Plato's Theory of Perception. *Review of Metaphysics* 3, pp. 21-58.
Zembaty, J.S. (1983) Plato's Timaeus. Mass Terms, Sortal Terms and Identity Through Time in the Phenomenal World. *Canadian Journal of Philosophy*, Supp. Vol. 9.
Zeyl, D. (1975) Plato and Talk of a World of Flux: Timaeus 49a6-50b5. *Harvard Studies in Classical Philology* 79, pp. 125-48.
——— (2000) *Timaeus*. Hackett.
Zhmud, L. (1994) Die Beziehungen zwischen Philosophie und Wissenschaft in der Antike, *Sudhoffs Archiv* 78, pp. 1-13.
——— (1997) *Wissenschaft, Philosophie und Religion im Frühen Pythagoreismus*, Berlin.
——— (1999) Plato as 'Architect of Science'. *Phronesis* 43, pp. 211-44.

Index Locorum

References in bold are to the works cited; references in italics are to the pages of the Notes section.

Index of Names

References in italics are to the pages of the Notes section.

General Index

References in italics are to the pages of the Notes section.

www.ingramcontent.com/pod-product-compliance
Ingram Content Group UK Ltd.
Pitfield, Milton Keynes, MK11 3LW, UK
UKHW050943290926
15470UKWH00005B/1615

9 780715 62987